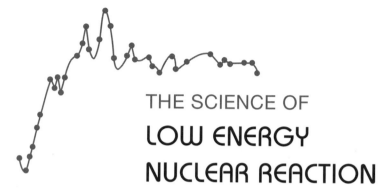

THE SCIENCE OF
LOW ENERGY
NUCLEAR REACTION

A Comprehensive Compilation of Evidence
and Explanations about Cold Fusion

THE SCIENCE OF
LOW ENERGY
NUCLEAR REACTION

**A Comprehensive Compilation of Evidence
and Explanations about Cold Fusion**

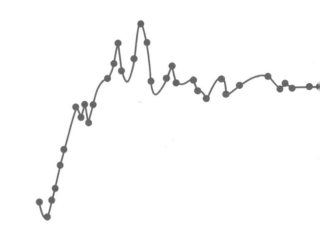

Edmund Storms

World Scientific

NEW JERSEY · LONDON · SINGAPORE · BEIJING · SHANGHAI · HONG KONG · TAIPEI · CHENNAI

Published by

World Scientific Publishing Co. Pte. Ltd.

5 Toh Tuck Link, Singapore 596224

USA office: 27 Warren Street, Suite 401-402, Hackensack, NJ 07601

UK office: 57 Shelton Street, Covent Garden, London WC2H 9HE

British Library Cataloguing-in-Publication Data
A catalogue record for this book is available from the British Library.

First published 2007
Reprinted 2008

THE SCIENCE OF LOW ENERGY NUCLEAR REACTION
A Comprehensive Compilation of Evidence and Explanations about Cold Fusion

ISBN-13 978-981-270-620-1
ISBN-10 981-270-620-8

Printed by FuIsland Offset Printing (S) Pte Ltd, Singapore

FOREWORD 1

There is no topic more relevant or important today than energy. With stable, cheap and non-polluting energy sources we have the means to refine and even manufacture the air, water and food we need to live and travel in comfort. Ours is the first generation with a globally uncertain energy future, with no assurance that energy resources will continue to expand to meet our needs. With this background it is no surprise that the results reported on March 23, 1989, by my good friends Martin Fleischmann and Stanley Pons in Salt Lake City, Utah, were greeted with such a universal outpouring of interest, hope and expectation. When all other resources based on chemical and conventional nuclear fission are drained, our only prospect for a sustained energy future and thus for the continued growth or even survival of civilization is the use of fusion power. Such power is provided by the fusion energy of the sun or it can be obtained from either hot fusion or cold fusion here on earth.

The first of these options, tapping the fusion energy of the sun, will certainly help to satisfy our future energy demand although solar power is never likely to meet the needs of dense urban and industrial regions. Lack of progress in addressing materials issues make the prospects for practical energy production *via* hot fusion less likely now than when first proposed over 50 years ago. This is a course on which we cannot pin our hope for the future of mankind. One is left with the third option, cold fusion, using whichever acronym one prefers. There is no man better positioned to describe the issues, history and results, and assess realistically the opportunities and future of cold fusion research than Ed Storms. This is more than a science book or a book about a scientific phenomenon, in the broader context this book addresses our future.

Searching for a fitting way to complement Ed Storms masterful work I thought to introduce it as the first textbook in the emerging field of cold fusion and condensed matter nuclear science. I have no doubt that this book "The Science of Low Energy Nuclear Reaction" will serve as a valuable reference and trusted guide for many years to come. There is a wealth of experimental detail and it was a pleasure for me to read this manuscript and be reminded of the excellent and difficult work undertaken and completed by so many old friends over the years. Storms also includes a very valuable discussion in Chapter 7 covering what is measured and how, providing a useful primer for experimentalists newly entering the field.

One might, however, reasonably argue that the first textbook in any field signals the end of the discovery phase and the onset of the new mature phase of consolidation. Textbooks also tend to be difficult to read. For both reasons I am personally happy, as I am sure the reader will be, that this book is not a textbook.

It is a well-written, easily readable account of the birth and early childhood of a field whose limits and applications have not yet been revealed. To have been part of this development, to have shared this experience with Ed Storms and so many other talented and committed people, has been the greatest experience of my scientific career.

Michael C. H. McKubre, Ph.D.
Director, Energy Research Center,
SRI International (formerly Stanford Research Institute),
Menlo Park, California.

FOREWORD 2

This book has several interesting facets, including the subject, the author, the content and the potential readers. Each of them deserves some attention.

The topic of cold fusion has been as controversial and exciting as it is complex and promising. It burst on the scene over 18 years ago. The subject got great and immediate worldwide attention both because of its promise as a new source of energy and because it was announced at a press conference. Within a few months, it was regarded by the mainstream scientific community as a sorry mistake. Nevertheless, a few hundred scientists in a score of countries continued investigating the subject both experimentally and theoretically. About three dozen conferences specifically on cold fusion, and many sessions at other conferences, have been held to present and discuss the results and ideas. Now, well over one thousand papers discuss the subject. The experimental evidence for nuclear reactions at ordinary temperatures is now great and robust. The subject has come to be called Low Energy Nuclear Reactions (LENR), and is viewed as a part of the discipline of Condensed Matter Nuclear Science. Solid lattices are necessary for production of LENR, hence the descriptor "condensed matter". The International Society for Condensed Matter Nuclear Science (www.iscnms.org) was founded in 2003. Still, reproducibility and control of LENR are both far less than desirable. Engineering optimization and commercial exploitation remain for the future. Understanding the cause of LENR still eludes those who have researched the subject. People working in the field are intrigued by both the scientific challenge and the potential applications. It is possible that nuclear energy from a unit about the size of a modern furnace will supply the electricity and hot water needs of a home, including recharging of cars. Large and costly central power plants and distribution systems may still be needed for offices and factories, but LENR energy sources could prove scalable even to levels needed for those applications.

A small group of scientists has spear headed the international research effort on LENR. Ed Storms has been a key member of that group. He has worked on LENR since the 1989 announcement by Fleischmann and Pons, first at the Los Alamos National Laboratory and then in a remarkable home laboratory in Santa Fe, New Mexico. Ed is a virtuoso experimenter, with capabilities that range from expert glass blowing to the setup and use of complex electrochemical experiments. His scientific contributions are numerous and important. He is probably best known for studies of materials and their surfaces, which are relevant to LENR. Ed has been an active and respected contributor to many of the conferences on cold fusion and LENR. He is known for his clear and crisp presentations of worthwhile data, not only on the results of LENR experiments, but also on instrumentation for them. Ed has also performed a great service to the community by writing well organized and comprehensive reviews of the subject. His contributions have reached the broader scientific community and general public, as well. Ed's overview "A Student's Guide to Cold Fusion" is very popular, as indicated by the large number of times it is downloaded from the web site www.lenr.org. It is a happy circumstance that Ed has now taken the time to organize and write a long and detailed book on LENR.

This book lives up to its subtitle "A Comprehensive Compilation of Evidence and Explanations about Cold Fusion". The descriptor "comprehensive" applies both to the thorough coverage of the field and to the extensive references. The ten chapters and six appendices range from a personal history of Ed's activities in the field to important technical details on electrochemical experiments to produce LENR. The fourth chapter on "A Look at What is Known or Believed" is an up-to-date summary of the results of most of the noteworthy experiments and theoretical developments in the field. The next chapter makes the important case for LENR reactions occurring on or near the surfaces of critical materials. The following two chapters are summaries of the "goes into" and "goes out of" aspects of the field. The first of these deals with the conditions that have been found conducive to the production of LENR. The wide variety of approaches to the production of effects, which do not agree with accepted current theories, is detailed. The second surveys the types of measurements that have been made during and after LENR experiments. Measurements of heat, nuclear ash and radiations are all important and receive attention. Then an entire chapter is devoted to the theories, more than two dozen, that have been advanced to explain some of the effects observed in LENR experiments. The appendices provide abundant detail on particular aspects of electrochemical LENR experiments. The references in the book are remarkable in both their breadth and number. The 1070 citations take up 78 pages, or about one-quarter of the entire book. It is unlikely that a better referenced book on LENR will appear in the near future.

Given the variety and amount of information in this book, it will enlighten a wide variety of readers. Scientists now active in the field will find much of the book to be a useful reference, both for techniques and results, some of which have not been published previously. Other scientists, who are considering work on LENR, will find the compact summaries of what has been done and found, and the details on techniques, immensely useful as they decide what to do and begin new studies. The large number of citations will be appreciated both by researchers already in the field and by others who are attracted to the central puzzle of LENR and its exceptional promise. Many of the current workers on LENR believe that the topic will become of much wider interest in the next few years. When that happens, this book will be a primary source of information for managers of both government and industrial programs.

While this is not formally a textbook, it can serve as one until the field attracts authors skilled in the particular art of teaching everything from scientific concepts and engineering applications. Students will find this book to be a rich source of details on the wide range of topics and the many levels of detail for many years in the future. Historians should appreciate both Ed's personal account of his work on LENR, and the broad and well-organized review of the field at this time.

In summary, the field of LENR now ranks as one of the best current scientific puzzles. The expected understanding and utilization could be among the more important advances of this century. Ed's experimental research and useful reviews have already made him a recognized leader in the field. He has written a book that is remarkable in both its breadth and depth. The numerous references are significant and useful. Many types of people should find the book to be an interesting and valuable resource. History may judge this book to be among best early books on a subject that turns
.out to be historic.

David J. Nagel, Research Professor
The George Washington University
School of Engineering and Applied Science
Washington, DC

ACKNOWLEDGMENTS

I am grateful to all of the very stubborn and creative people who continue to study cold fusion after many leaders in the scientific profession said the claims were utter nonsense. Your efforts to prove this judgment wrong made this book possible and will eventually make a better world for us all. Prof. John Bockris, who contributed much important information to the field, while being rejected by his fellow professors, kept me focused on what is important to put in such a book. Any deficiencies you find are there because I failed to take his advice. Robin van Spaandonk read the book through the eyes of an outsider to the field and, by asking the right questions, helped make the description much easier to understand. He helped me appreciate how the work of Dr. Randell Mills can be applied and how hydrino formation might play a role in initiating the observed nuclear reactions. Jed Rothwell, Mike Carrell, Steve Krivit, and Tavo Holloway helped make the text more readable and found their share of typos. In addition, Jed generously provided some of the equipment needed for my studies. His work in making the www.LENR.org website function so well has provided me and many other people with valuable information found nowhere else. Dieter Britz was very kind in providing copies of many papers that were otherwise difficult to obtain. Barbara Storms applied her skill as an editor in an effort to keep me from making too many grammatical mistakes. With her help and a good spell checker, perhaps the battle has not been completely lost. Many people helped me get started and keep working, including Dave Nagel who purchased one of my calorimeters for the Navy when I needed this encouragement; Bruce Mathews, my group leader at LANL, who gave me the freedom to continue my efforts there until I retired; Charles Becker, who invited me to join ENECO and encouraged my work in Santa Fe; Lewis Larsen, who invited me to join Lattice Energy, LLC, for awhile and provided a very useful SEM; and Charles Entenmann whose extraordinary and farsighted generosity kept me and many other people actively contributing to the field. Finally, I thank Carol, my wife, for making this effort possible by her encouragement and loving support.

PREFACE

**Figure 1. Stanley Pons and Martin Fleischmann with examples of their cold fusion cells.
(From Special Collections Dept., J. Willard Marriott Library, University of Utah)**

March 23, 1989 can now be acknowledged as a major event in the long history of scientific discovery, on par with the discovery of fission, which gave us the atomic bomb and electrical power from nuclear reactors. On this date, Profs. Martin Fleischmann and Stanley Pons announced to the world a new nuclear process they claimed was a method to fuse two deuterons together. For many reasons, their work was rejected or used as an example of bad science. Only now,

18 years later, and after a lot of hard work by hundreds of open-minded scientists can the importance of the discovery be fully understood and appreciated. It is the challenge of this book to assemble the evidence provided by numerous studies done in laboratories located all over the world and to show that a new and important discovery did in fact take place, contrary to what many people were led to believe.

Who are these men who were threatened and mocked after making such an historical discovery? Very few people knew Stanley Pons as the chairman of the Chemistry Department at the University of Utah. However, many people in science recognized Martin Fleischmann's name and reputation. He is a major contributor to and teacher in the field known as electrochemistry. Born in Czechoslovakia and narrowly escaping the Nazi plague, he settled in England where he taught at the University Southampton from 1967 until he retired. He was awarded just about every scientific honor England has to offer a scientist. Hearing his name, many people trained in chemistry took notice, at least at first. However, as often happens when important discoveries are made, a vocal group of influential people rejected the new idea. Fortunately, a few stubborn people continued to work in obscurity and have now proved the claims are real. In so doing, they risked their reputation and, in a few cases, their livelihood. Even Pons had to emigrate to France to avoid the harsh treatment provided by his own countrymen.

This book is mainly about the science of what was discovered in laboratories located in many countries by hundreds of researchers. My own research has provided me with a useful vantage-point for evaluating this work. Several goals have been attempted. Many people have contacted me wanting to learn about the subject and how they might see the cold fusion effect for themselves. Hopefully, this book can answer their many questions and show them where to look for more information. My second goal is to summarize the large accumulation of information. Such a summary is necessary because many observations are not accessible in easily searched journals and conventional databases. As a result, evidence for the effect is scattered and its full meaning is difficult to appreciate. Much work is only described in obscure conference proceedings that can be obtained from sources listed in Appendix D. My third goal is to describe what took place at the Los Alamos National Laboratory (LANL), where I worked when the announcement was made by Fleischmann and Pons. Everyone who was involved in trying to replicate the claims at LANL remembers the intellectual excitement at the Laboratory as being in the highest tradition of science. Such unique events are worth remembering and sharing as rare examples of what can and should happen. Finally, I hope when the considerable collection of observations are viewed in their totality, rational evaluation will replace blind skepticism and unfounded ignorance. My opinions alone need not be accepted because more than 1060 publications have been cited in which the primary information can be found. Indeed, only by viewing a wide assortment of observations can an understanding be achieved. This situation is rather like trying to visualize a complex jigsaw puzzle that makes sense only after a large number of pieces have been assembled. In this case, some of the pieces are so strange it is hard to believe they belong to the same puzzle. In addition, many critical pieces are still missing. As a result, little agreement can be found among scientists about what the puzzle actually looks like. My personal view is offered in the hope that it will make the puzzle a little easier to understand.

Unlike many scientific fields these days, this one is driven by observation rather than by theory. No theory explains all of what is known to be true, even though many explanations have been proposed—some plausible and some not. At this stage, theories are expected to be incomplete and very limited in their application, rather like the maps provided by early explorers or like biology before genes were understood. To make matters worse, many times people do not make clear which part of their theory is based on accepted knowledge and which part is based on imagination—again very much like early maps. Nevertheless, it is important to realize that acceptance of data is not dependent on an explanation being provided, any more than a river can be ignored just because it is not on the map. Data stands on its accuracy, consistency, and eventually on universal experience. Therefore, my main effort will be to show what is known empirically and separate this clearly from what is not known without trying to fill the gap with excessive imagination. This is a treasure hunt using clues to shrink the large area of ignorance to a smaller area where we can start digging. As we dig, small nuggets of understanding will emerge, which should be carefully examined. These nuggets should not be tossed aside just because the entire ore-body has not been uncovered.

After reading all that has been published about the subject and enjoying many successful replications, I'm absolutely certain the basic claims are correct and are caused by a previously unobserved nuclear mechanism operating in complex solid structures. Consequently, this book is not an unbiased description of the controversy. This is not to say that all studies are correct. In fact, many studies contain significant errors and a few are completely wrong. Indeed, I have great sympathy for those who reject the claims. These problems would have been reduced if papers had experienced competent peer review. Instead, papers were either rejected out of hand by most peer-reviewed journals or published with only minor changes in conference proceedings. As a result, other scientists, even the open-minded ones, have reason to ignore the work. Nevertheless, enough good work has been published to clearly show the reality of the phenomenon. This good work needs to be acknowledged and supported without the distraction poor measurements provide. My task here is to make this process easier by showing the agreement between well-documented studies.

I would like to apologize to those who consider themselves "skeptics," which is an honorable title I sometimes assume for myself. In the future, perhaps by the time you read this book, cold fusion will be an accepted phenomenon and the idea of someone doubting its reality will be as improbable as someone doubting that the Earth goes around the Sun. At such time, a reader might find my harping on the reality of cold fusion to be silly and unnecessary. Unfortunately, at the present time, many people still think the idea is nonsense and approach the subject the way the Church approached Heliocentric astronomy 500 years ago. I hope this book will be accepted as a better telescope. If this considerable body of work is dismissed as error, what does this say about the competence of modern science? Is it rational to believe that many modern tools only give the wrong answer when they are applied to cold fusion?

While at LANL, as described in Chapter 2, I had a unique view of how events unfolded, at least within LANL. During this time, I investigated the science of cold fusion and, after "retiring" in 1991, continued the work in my own laboratory. This experience taught me to accept the reality of these "impossible" claims. Chapter 3 summarizes some of these lessons. Evidence provided by hundreds of others is discussed in later chapters, where a huge collection of experience is evaluated and put into perspective. Even people working in the field are not fully aware of what has been discovered. As the reader will soon learn, the novel effects occur only in unique and very small locations. These locations are discussed in Chapter 5. Methods used to initiate the anomalous effects are described in Chapter 6 and detection of the resulting behavior is discussed in Chapter 7. Development of a proper theory has been one of the great challenges of the field. Consequently, some explanations are offered and evaluated in Chapter 8. As will become clear, cold fusion is not cold, except in comparison to hot fusion, and it is not normal fusion. Unlike hot-fusion, which is used to "explain" cold fusion by applying high-energy physics, an explanation should be based on solid-state physics and chemistry. Consequently, the observations need to be viewed through a different lens than is applied to hot fusion. Chapter 9 tries to show what all this information means and what should be done next. A very brief summary of the phenomenon is provided in Chapter 10, which might be worth reading first. The implications of this discovery are so profound that people need to accept its reality and be prepared to enjoy the consequences of its eventual application. The only uncertainty remaining is which country will first gain the benefits and how soon.

As for my background, I came to the Los Alamos National Laboratory (LANL), Los Alamos, New Mexico, first in 1956 and again in 1957 as a summer student and returned as a staff member in 1958 after getting a Ph.D. in radiochemistry from Washington University, St. Louis. Prof. Joseph Kennedy, my research professor, had been the director of the Chemical and Metallurgical Division at the secret laboratory in Los Alamos during the war and co-discovered plutonium. Thanks to his encouragement, I joined a steady stream of graduates from the University he was recruiting for the peace-time Laboratory. This was a time when the Laboratory was changing from the primitive conditions existing during the war to what was to become a major national laboratory located in a place of unusual beauty. It was an ideal place to do creative work because competent people who knew science and scientists were administering the laboratory at that time. I was hired to study the thermodynamic and phase relationship properties of very high melting point materials[1] used in reactors designed to provide power or propulsion in space, a useful and exciting subject even though the intended machines were never built. Nevertheless, my work was productive and satisfying, resulting in more than 100 publications, a book[2], and teaching sabbaticals at several universities including the University of Vienna, Austria. I did not need another project and I was content to believe the theories everyone else accepts in nuclear physics. Besides, Carol, my wife, could have done without the scientific mistress cold fusion later became for me. To some extent, this book describes a personal awakening to the realization that what is taught and thought to be true in nuclear physics is only partly correct. A totally unexplored environment in which nuclear interaction can take place apparently exists within solid materials.

Edmund Storms
Santa Fe, NM
January 2007

TABLE OF CONTENTS

LIST OF FIGURES

LIST OF TABLES

CHAPTER 1

Introduction

Manmade nuclear fusion was produced well before Fleischmann and Pons came along, but only with great difficulty and expense. Over the last 60 years, the machines have grown in complexity and have now culminated in the proposed ITER[a] hot fusion reactor, shown in Figure 2, which is being constructed in France as well as a similar independent project in China.[1] Even though this research has cost over 16.5 billion dollars,[2,b] efforts up to now have not produced more power than it takes to run the huge machine.[3]

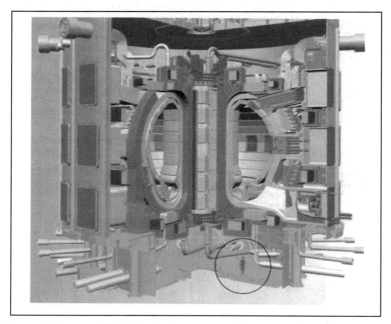

Figure 2. Proposed ITER hot fusion reactor. (From www.iter.org)

In contrast, the new cold fusion method appeared to be so simple that any household could afford the device and hope to use it to supply all domestic energy. Instead of a huge central generator of unimaginable complexity and cost, the Fleischmann-Pons device is something a bright high-school student might understand and afford to construct. Essentially it is a glass container in

---

[a] ITER is an attempt to harness thermonuclear power and is supported by the European Union, Japan, China, India, South Korea, Russia, and the United States. The latest version is being built at an estimated cost of 12 billion dollars.
[b] Source: Department of Energy as of year 2000 in FY2000 dollars adjusted for inflation.

which two metal rods are immersed in an electrolyte made by mixing heavy-water (D_2O) with a lithium salt (LiOD). A conventional constant-current power supply forces current to flow between the rods. This current releases deuterium from the heavy-water, allowing nuclear reactions to occur at the negative electrode if the conditions are right. From then on, the process gets complicated, but still looks simple. Of course, a practical source of energy would be bigger and more complex, perhaps the size of a refrigerator located in every home. Unfortunately, despite this outward simplicity, the actual mechanism producing the "cold fusion" reaction has, so far, defied complete understanding and has resisted easy replication. As a result, the claims have been very hard for many people to believe. It is this failure to understand and accept a discovery of potentially immense importance that makes this unusual saga worth our time and interest.

Just what is cold fusion and how does it differ from hot fusion? Cold fusion and hot fusion both involve a nuclear process, but the mechanisms and end products are different. Hot fusion requires a very hot plasma (ionized gas), which is used to force the nuclei close enough to fuse. This process takes considerable energy, supplied by the very high temperature, because all nuclei repel one another as a consequence of the Coulomb barrier. In contrast, cold fusion occurs in a relatively cool solid in which a process operates much like a catalyst, *i.e.* by neutralizing or lowering the barrier without the need for high energy. Once this barrier is overcome, hot fusion makes energetic neutrons and tritium in equal amounts, while cold fusion makes mainly helium (^{4}He), with few emissions of any kind being detected outside of the apparatus. In short, brute force is used to cause hot fusion, while cold fusion requires a complex solid environment in which a process similar to seduction can operate. Hot fusion has been studied for more than 60 years by many and, arguably, is well understood. Cold fusion has been studied for only 18 years by a few and is hardly understood at all. Cold fusion is a clean energy source resulting in essentially no radioactivity. Hot fusion results in the generation of considerable amounts of radioactive elements. Cold fusion has been difficult to replicate. Hot fusion has been difficult to be make useful.

Initially, cold fusion caught popular imagination because two researchers working in their own laboratory and using their own money discovered what seems to be an ideal energy source. For a brief time, Fleischmann and Pons became known to people all over the world. Even now, almost 18 years later, many people remember them with interest. As energy prices soar, this memory becomes bittersweet because over the years the press and a few outspoken scientists have given the impression that this beautiful promise had failed and Fleischmann and Pons had just made a stupid mistake. Imagine the outrage people feel when they realize the claims are real after all, not a mistake, and this ideal energy source would now be available to mankind were it not for a false myth being spread by people who insist the claims are false. Despite this opposition, active studies continue in at least eight countries[c], and evidence that Fleischmann and Pons are right continues to mount.

Books describing the history and politics of cold fusion were written by David Peat,[4] Eugene Mallove,[5] Hal Fox,[6] Frank Close,[7] John Huizenga,[8] Gary Taubes,[9] Nate Hoffman,[10] Charles Beaudette,[11] Bart Simon,[12] Steve Krivit/N. Winocur,[13] Roberto Germano,[14] and Thomas Stolper[15]. Books discussing certain aspects of the science are provided by Hideo Kozima,[16,17] Tadahiko Mizuno,[18] and Joe Champion.[19] One by Jed Rothwell[20] even predicts the future based on cold fusion being successfully applied.[d] These sources of information can be added to the efforts of Hal Fox (Fusion Information Center, Inc.), who published a monthly update of the field called "Fusion Facts" from 1990 to 1994. Later he became the founder and editor of *New Energy News* and *Journal of New Energy*, which published many important papers. Bruce Lewenstein[21-27] (Cornell University) has addressed the sociology of the saga, which historians of science will be examining for many years in an effort to understand what went wrong with the scientific process when it was applied to this subject. The archive he created at Cornell University is a valuable resource of public information. Similar information can now be found at www.newenergytimes.com where Steve Krivit publishes his insightful editorials about the field. Dieter Britz provided a very valuable collection of published papers on the web, which has now been incorporated into the www.LENR-CANR.org website, administered by Jed Rothwell. This website gives access to all of the published information and links to other sources. The www.iscmns.org website, maintained by William Collis, provides information about the International Society of Condensed Matter Nuclear Science and gives access to the new journal, *Journal of Condensed Matter Nuclear Science*. Current information can also be obtained at http://world.std.com/~mica/cft.html, which is maintained by Mitchell Swartz. *Infinite Energy*, founded by the late Eugene Mallove, continues to publish articles about cold fusion under the competent editorship of Christy Frazier. Scott Chubb has been effective in having sessions devoted to cold fusion made part of various American Physical Society meetings. As a result of the efforts made by all of these people, combined with the extensive written literature, facts about the subject are slowly seeping into the collective consciousness.

---

[c] China, France, Israel, Italy, Japan, Russia, US, and Ukraine.
[d] Additional information and sources can be found at http://worldcat.org/search?q=su%3ACold+fusion.&qt=hot_subject.

CHAPTER 2

History as Seen from the Los Alamos National Laboratory—and Beyond

March 27th, 1989 at Los Alamos, was a normal spring day without the anxiety and uncertainty we all have learned to endure in recent years. On that date my life took a sharp turn onto a path seldom traveled. I was invited to attend the first Electrochemical Fusion Meeting being held at the Laboratory. From what I learned, my regular job quickly lost its appeal. If the claims being made by Profs. Fleischmann and Pons were correct my past efforts would become completely obsolete. In addition, the yet to be successful hot fusion program, after spending nearly ten billion dollars by that time, would no longer be needed. Even more important, all other energy sources, on which the infrastructure and economy of the world are based, could be abandoned and replaced by this much cleaner, cheaper, and longer lasting producer of primary power. Everyone recognized the very high stakes this situation created. Consequently, it was very important to discover whether the claims were correct or not. President George Bush, the elder, ordered this to be done immediately! Thus, the Energy Research Advisory Board (ERAB) was organized[a] to make this evaluation.

A day after the public announcement, work was under way at LANL by Shimshon Gottesfeld (MEE-11)[b] and people in his group in an attempt to duplicate the electrolytic cell used by the Utah researchers. Like many other attempts, this effort was eventually unsuccessful.[1] People were quickly organized and administered by Rulon Linford (ET/MFE) with a speed that is no longer possible at LANL. Everyone scurried off to find palladium and heavy-water before the limited supplies were snatched up by someone else. Most people had no idea what these materials even looked like much less how they would behave. Excitement was building as more people heard about the

---

[a] According to Glen Seaborg, who advised President Bush, ERAB was merely a formal method to be used to reject the whole idea of cold fusion.
(http://newenergytimes.com/news/2006/NET15.htm#seaborg) The members of the ERAB panel were: John Huizenga, co-chairman (University of Rochester), Norman Ramsey, co-chairman (Harvard University), Allen Bard (University of Texas), Jacob Bigeleisen (SUNY), Howard Birnbaum (University of Illinois), Michel Boudart (Stanford University), Clayton Callis (ACS), Mildred Dresselhaus (MIT), Larry Faulkner (University of Illinois), Richard Garwin (IBM), Joseph Gavin, Jr. (Grumman Corp.), William Happer, Jr. (Mitre Corp.), Steven Koonin (Caltech), Peter Lipman (USGS), Barry Miller (AT&T), David Nelson (Harvard University), John Schiffer (ANL), Dale Stein (Michigan Technology University), and Mark Wrighton (MIT).
[b] The group designations are those in use at the time.

"discovery" and wanted to get in on the action. If real, such an important discovery hardly ever happens during a scientist's career, especially one involving a behavior seemingly so simple it could be easily explored. During most of April, large and animated meetings were held every week as people tried to understand what Fleischmann and Pons had done and how the claimed effects might be duplicated. A barely readable version of their paper had been sent by FAX all over the world,[2] sometimes with missing pages. Although this provided the bedrock description of the method, it generated more questions than it answered. The basic assumptions of the paper were that deuterium[c] could be reacted with palladium[d] to achieve a very high internal pressure of deuterium within the metal by using electrolysis; that the D^+ ions contained in the palladium were lightly bound, thereby allowing a significant number of close collisions; that tritium[e] and neutrons would result from fusion caused by these collisions; and that they had detected the expected energy. Adding to their credibility, Fleischmann and Pons said they had detected gamma rays[f] produced by the expected neutron emission, they alluded to expected tritium production in an amount consistent with the neutron emission, and they gave examples of excess energy in amounts claimed to be too large to result from a chemical process. Because the reactions were thought to take place within the bulk metal, anomalous power was reported as watt/cm^3, which made the results look even more impressive because very small samples were used. Frustration quickly grew because the method used to measure energy was not described in detail, the exact procedure for treating the palladium was not given, and the neutron and tritium measurements looked very suspect. Everyone agreed, for a paper claiming such an extraordinary effect, their description was very inadequate. Consequently, an effort was made to contact the authors directly. After much negotiation, Stanley Pons visited LANL on April 18, 1989 and gave a brief lecture without revealing enough to make replication successful. He gave the impression that the effect was easy to produce and only required loading the palladium to a very high, but unstated, deuterium content. After his talk, debate

---

[c] Deuterium is a non-radioactive isotope that exists in all water at about 6000 parts per million concentration. It is presently separated from water on a very large scale for other uses.

[d] Palladium is a rare and relatively inert metal used for jewelry and as a chemical catalyst.

[e] Tritium is an isotope of hydrogen that is radioactive with a half-life of 12.3 years. A neutron is a basic nuclear particle having no charge and a half-life of 10.25 minutes when not in the nucleus. Both tritium and the neutron emit electrons (beta) when they decay and both are produced in nature mainly by natural processes, but only in very tiny amounts.

[f] Gamma rays are electromagnetic radiation, similar to but more energetic than light. These photons are emitted by a nucleus when it needs to lose energy.

raged as to how much information Stan was holding back and how much he simply did not know. Ten months later, Martin Fleischmann also paid a visit and tried to answer the growing list of questions. Martin is one of those geniuses who seems very clear during a lecture, but after subsequent thought a listener realizes more questions were created than answered. In any case, we were forced to learn from our own efforts. The hoped for collaboration was eventually dashed by lawyers who were writing patents for the pair. As a result, one more door to acceptance was slammed shut.

After a few days of contemplation, people arrived at many of the basic insights that later work would confirm. We concluded that the effect, if it existed at all, would occur within the surface. Dual occupancy of lattice sites by deuterium was probably important and this would require a very high D/Pd ratio. Neutron flux would be very low because no one at LANL expected the hot fusion mechanism to operate under such "normal" conditions. Production of ^{4}He would be expected as tritium was made. and then consumed by fusion with the few protons present in the heavy-water. This assumption has yet to be demonstrated. By this time, many ideas and methods published much later by other people were already being debated. For example, Melvin Prueitt suggested ultrasonic cavitation might be used to initiate the effect. Lawrence Cranberg was quick to suggest the branching ratio between tritium and neutron production (see Table 5) might be anomalous at low energy and this quantity needed to be measured below the limit at that time of 13 keV. The possibility of fractofusion[g] [3,4] being the initiator of a nuclear process was first suggested by R. Ryan as an explanation and later explored with success at LANL[5] and at ENEA, Frascati in Italy.[6] Fractofusion, as discussed in detail by Preparata[7] (University di Milano, Italy), describes the fusion process when it occurs in cracks as high energy is generated for a brief time while the crack forms. Many other mechanisms needed to explain the fusion process were debated with a split forming between those who found the claims to be impossible based on conventional theory and those who proposed new mechanisms. This same split continues today.

Pressure was being placed on all government laboratories to resolve this issue as quickly as possible. By April 19, multiple programs were underway at Argonne National Laboratory (ANL), Pacific Northwest Laboratory (PNL), Lawrence Livermore National Laboratory (LLNL), Sandia National Laboratory, Albuquerque (SNLA), Oak Ridge National Laboratory (ORNL), Lawrence Berkeley Laboratory (LBL), Brookhaven National Laboratory (BNL), Naval Weapons Center at China Lake, Naval Research Laboratory (NRL), Westinghouse Savannah River Company, and Ames Laboratory. In addition, 56 people, involving 8 teams, were working on the problem at LANL. Of course,

---

[g] Fractofusion is the mechanism used to describe observed neutron emission when materials are pulverized in the presence of deuterium.

non-government laboratories as well as groups in other countries were also working hard. Loading of titanium by D_2 gas was being explored at ENEA, Frascati in Italy, and eventually by Howard Menlove at LANL. Electrolytic cells were being studied at Chalk River Laboratories in Canada, at Max-Planck Institute for Plasma Physics in Germany, at the University of Sao Paulo and at the Institute of Space Research in Brazil, in China at Beijing University, in Denmark at University of Aarhus, at Universitá di Roma and Instituto Superiore di Sanita in Italy, and at the BARC and IGARC Atomic Research Laboratories in India. In the Netherlands, studies were undertaken at the Institute for Plasma Physics, Rijnhuizen, Nieuwegein, at the University of Groningen, at the University of Delft, and at the University of Utrecht. Studies were also underway in the US at AT&T Bell Laboratory, IBM, Westinghouse Electric Co., MIT, Stanford University, SRI International, Yale, Washington State, University of Michigan, Princeton Plasma Laboratory, Georgia Institute of Technology, California State Polytechnic University, Purdue, Cal. Tech., and the Universities of Arizona, Florida, Illinois, Texas, and Portland State. Prof. Steve Jones[8], who claimed to have initiated the same nuclear process, continued his approach at Brigham Young University and new programs, other than those of Fleischmann and Pons, were started at the University of Utah. Texas A & M had a particularly strong program because Prof. John Bockris had known Martin Fleischmann since his student days and had quickly obtained detailed information. People at many other institutions were exploring the theoretical implications. As this list of laboratories demonstrates, interest was widespread and spontaneous, with studies started in at least 50 major laboratories worldwide involving at least 600 scientists. In addition, many articles in the press and on TV spread interest to the general public. All of the major news magazines featured Fleischmann and Pons on their front covers. A congressional hearing on April 26, 1989 focused attention within the government. Even today, many people, who were old enough at the time to understand the issues, remember Pons and Fleischmann and still hope the method will be applied to solve our pressing needs.

By April 1989, excitement was building in the scientific community. The first hasty replication efforts were scheduled to be described during several meetings including a special section on Cold Fusion at the American Physical Society (APS) meeting in Baltimore on May 1, at the Electrochemical Society meeting in Los Angeles on May 7, and at a LANL-DOE workshop in Santa Fe on May 23. In addition, the Energy Research Advisory Board (ERAB), officially created on April 24 by Secretary of Energy James Watkins, was anxious to reach an official conclusion, with a first draft due by July 31. This rush to judgment was to be a major flaw in the effort to evaluate the claims and to achieve eventual acceptance. As was only appreciated later, electrolysis for hundreds of hours is frequently required to cause any observable effect. Deadlines for reports and meetings seldom allowed enough time for nature to create the right conditions.

People at laboratories all over the US and, indeed, all over the world had dropped what they had been doing and frantically put together unfamiliar devices using unfamiliar materials. Making these devices work as expected in the time available was a hopeless challenge, even when nature co-operated. Unfortunately, most people were looking for neutrons and finding very few. It turns out neutrons are the least abundant of the anomalous nuclear products. To make matters worse, measurement of heat production is unfamiliar to most scientists including many who chose to make this measurement. Consequently, most of the devices suffered from many unrecognized errors. To make matters worse, many experimental procedures were based largely on rumor and speculation. Looking back, success seems to have been mostly the result of luck.

People were starting to realize that the claimed neutron flux reported by Fleischmann and Pons in their paper was much too small to be consistent with the amount of energy reported, based on the so called "dead graduate student effect"—an unpleasant event that would occur if the expected flux were present even if the neutron detector missed most of the radiation. By late April, the reported neutron flux was known to be wrong and other serious challenges were being raised. The battle was shifting in favor of the skeptics. As a result, the evidence reported by Fleischmann and Pons was being given less weight by people compared to what they were finding using their own devices, which was generally negative. Nevertheless, work continued with growing intensity. At one point, the director of LANL, Dr. Siegfried Hecker, confided to me that he had not seen so much enthusiasm at the Laboratory since World War II. "Physicists are actually talking to chemists," he observed with amazement. This attitude was being duplicated all over the world. To be sustained, this huge bubble of enthusiasm needed some very significant confirming results, especially in face of growing doubt about the evidence from Utah.

As expected, people at LANL had different ideas about how the process could be initiated, a few of which did not involve electrolysis. For example, titanium, a metal known to crack upon reacting with deuterium, was examined to see if neutrons might result from fractofusion when cracks formed,[9] as suggested by work done in Italy.[10] Neutrons were sought when PdD was explosively compressed. Even the electrolytic method took many different forms. Cathodes were made from palladium or titanium, obtained from many different sources and in many different forms, some of which were made "in house." Anomalous behavior was based on heat measurement as well as on tritium or neutron production, but seldom at the same time using the same cell. Even the very unique "Jones" cell design[8] was explored, but without seeing anything unusual. Excess tritium was even added to some cells to see if the T-D fusion reaction could be initiated—without success.

A few experts at LANL doubted the effects were real because they had considerable past experience, mainly in weapons research, with reacting deuterium and tritium gases with various metals, yet they never saw anomalous behavior. This negative attitude did not stop people from taking advantage of the in-depth understanding of the Pd-H and Pd-D chemical systems unique to LANL and applying these insights to their studies. Unfortunately, most experiments at LANL failed even though people explored a wide range of conditions using world-class neutron and tritium detection equipment and had as much help as Fleischmann and Pons were permitted to give. By late May, only two teams at LANL were bragging about their results. These could be added to confirmations coming from Texas A & M for heat and the first replication of tritium production, Stanford University for heat, Oak Ridge National Laboratory for heat, Italy for neutrons, and India for tritium. However, the many failures and the serious errors found in the Fleischmann and Pons paper fueled a growing doubt about the original claims. Too many people had spent too much time to get so little. They were beginning to feel they had been had. Nevertheless, work continued at most laboratories with undiminished intensity. The implications of the claims and the few successes were too important to ignore.

In early April 1989, Fleischmann and Pons addressed the American Chemical Society meeting in Dallas where they were met with an enthusiastic response. Later, on April 26, a few people described successful efforts to replicate the effect at a meeting arranged by the Materials Research Society in San Diego. However, the first major opportunity for most people to make their work known to the scientific public occurred at the American Physical Society (APS) meeting on May 1 in Baltimore, at which Martin Fleischmann and Steve Jones were invited to present their work. Prof. Fleischmann chose not to attend. In his absence, criticism of the Utah work got out of hand as experimental results of Prof. Lewis and co-workers[11,12] (Cal. Tech.) were accepted as showing the heat measurement of Fleischmann and Pons to be fatally flawed and the theory described by Prof. Koonin[13] (University of California) was accepted as proving the claimed results were impossible. In short, the audience came to the conclusion that the claims made by Fleischmann and Pons, the considerable effort to replicate the claims, and the hopes and dreams for a clean source of energy were all based on a stupid mistake and on an impossible process. This alleged insult to good science and apparent waste of time made many people very angry, which they showed in uncharacteristic ways. Only much later was the work of Lewis, itself, found to contain important errors[14,15] and the theory being used by Koonin was shown to be much too simple. Nevertheless, major damage was done to the field and to Fleischmann and Pons's reputations by the hubris and misdirected self-confidence of a few people. Thus, a myth was created that even today continues to have a negative impact. As Prof.

Goodstein[16,17] then vice provost of California Institute of Technology observed, "these people executed a perfect slam-dunk that cast cold fusion right out of the arena of mainstream science."

The next public presentation was on May 7, at the Electrochemical Society meeting, where Fleischmann and Pons were able to present their side of the raging debate to scientists who better understood the experimental issues and who were willing to listen. Nevertheless, attacks by their fellow scientists were beginning to extract a personal toll. After all, a scientist does not and should not expect to receive vicious personal attacks including death threats just for revealing what they had observed. Then came the LANL-DOE Workshop in Santa Fe, NM, on May 23. This two-day meeting gave everyone working independently at various laboratories in the US a chance to meet, compare results, and debate issues with civility in a very beautiful and serene environment Even so, strong skepticism hung around like an unwashed guest requiring an occasional angry remark or insult to be endured. Unfortunately, neither Pons nor Fleischmann attended, which again increased their isolation. Another workshop was held in Washington in October, where Pons gave a talk and where supporting results obtained by other people, including our own, were presented. Edward Teller provided some balance to the heated arguments by promising he could explain the Fleischmann and Pons results by making minor modifications to accepted theory. Unfortunately, such attempts made by numerous people were not believed. Many conferences have been held on the subject over the years, some sponsored by the APS, with no repetition of the outrageous behavior demonstrated at the APS Baltimore meeting. Nevertheless, the damage was done and examples of uninformed comments by individuals still remain.[18]

Gradually support for the work changed from using money taken from other programs to money provided solely for investigating cold fusion. The DOE gave $350,000 to LANL, which was distributed among those few studies showing some hope of success, including mine. Because the group in which I worked (MST-3) had experts in tritium detection, we set about trying to make tritium using the Fleischmann and Pons approach with the help of a technician and Carol Talcott, who was an expert in palladium hydride chemistry and who would eventually become my wife. During the latter part of 1989 and early 1990, we electrolyzed every imaginable source of palladium and surface deposit in an effort to achieve a high D/Pd ratio, which was thought to be a requirement. We measured the gas and liquid for anomalous tritium, we added tritium to cells to see how it would behave, and we exposed cells to a contaminated environment to see how much tritium would be picked up. More than 250 cells were explored, with abnormal tritium found in only thirteen. In spite of such poor success, the accumulated experience showed this extra tritium to result from a nuclear reaction within the cell. A paper was submitted for publication[19] after extensive review by a dozen scientists at LANL. Another 6 scientists took a shot during the review process by *Fusion Technology*. Normal papers are seldom reviewed by more than 3 people at a journal and perhaps by one person at LANL. The paper was eventually published in the midst of complaints from skeptics about a lack of peer review being applied to cold fusion papers.

At the same time, Tom Claytor (ESA-MT, LANL)[20,21] was making tritium by a different process. For his initial method, he passed a pulsed current through alternate layers of palladium powder and silicon powder, all pressed into a solid sandwich, thereby hopefully making super-heavy electrons. Such electrons, if they could be made in the silicon, might enter palladium and enhance fusion between the dissolved deuterium. Later, he produced an electric discharge in deuterium gas between palladium electrodes. Both methods produced unusual amounts of tritium if the proper palladium were used. As later work would also show, the material characteristics of the palladium are critical to all methods. The Laboratory brought in experts from universities and other national laboratories on several occasions in an attempt to discover how such a novel process was possible. The conclusion was always the same—tritium was being made by a nuclear reaction even though this was impossible according to conventional theory. His work was eventually shut down because, after the ERAB released its report in November 1989, the DoE officially concluded that evidence for the novel claims was "not convincing", hence not worth special funding—or funding of any kind as it later turned out. Fortunately, his work continued for a while because it could be supported by internal funding at LANL, which was granted by knowledgeable scientists who were not controlled by the DoE. Sadly, the DoE eventually had its way. Nevertheless, administrators at the Laboratory who knew the effect to be real allowed some research to continue for a while. For example, I worked on the subject periodically until I left the Laboratory for good in 1993.

The effect of the ERAB report was much greater than a reader would conclude from a literal reading of the text. From the printed words, a person might believe a proposal would be given serious consideration for funding if submitted through the normal peer review system. A more accurate attitude is expressed in the book, "Cold Fusion, The Scientific Fiasco of the Century", by John Huizenga[22] (University of Rochester). Huizenga was the co-chairman of the ERAB panel. He wrote the ERAB document and ran the show. His book recommends how the government should treat cold fusion, shows how the myth got started, and gives an insight into why early evaluation was so flawed. Nevertheless, Steven Jones (BYU) and I attempted to test the offer made in the ERAB report by submitting a proposal.[23] We proposed to replicate heat production using the calorimeters of Lee Hansen at BYU and the knowledge about how to initiate heat production I had acquired. This proposal was rejected just as later submissions by other people were rejected. In general, rejection is based on the belief that the claims for anomalous energy and nuclear products are impossible and are based on bad science, hence not worth funding.

This "official" document has also affected the attitude of editors of many conventional scientific journals. These journals play an essential role in science, because they allow ordinary researchers to learn about and to understand what is being discovered. For some strange reason, ordinary scientists do not consider any information to be believable unless it has survived the peer review process provided by journals. Apparently, they do not consider themselves competent to make this evaluation for themselves. Consequently, when papers are rejected, most scientists ignore the information even though it might be easily available from non-reviewed sources.

Most people attempting to publish anything about the subject continue to have a similar experience, and editors sympathetic to the field have even been encouraged to quit. Even Julian Schwinger[24,25], a Nobel laureate, was so outraged by the way the APS treated his papers, he resigned in protest. An editor pays no price for rejecting a good paper, but can be severely chastised for publishing a paper considered poor by a few outspoken critics. The entire system of publication is skewed in favor of the passionate skeptic who opposes a new idea. Very few editors have the courage to fight this system, an exception being George Miley (University of Illinois) while he was the editor of *Fusion Technology*. Even though Miley occasionally had his doubts, he provided a means for good papers to be made available to the scientific profession, in the best tradition of scientific publication. In the process, he came close to losing his job as editor because of pressure brought by less open-minded people. While the peer review system has its flaws, it plays an important role by encouraging authors to correct mistakes and make descriptions clear and concise. When used properly without self-interest, arrogance, or a closed mind, the system is very helpful. When used to stifle new ideas, we all pay a dear price.

In addition to the limits imposed on information by journal publication, universities and other organizations applied pressure. For example, John Bockris almost lost his distinguished professor position at Texas A & M because certain professors there did not believe his results.[26,27] His treatment was nasty and bitter, not what you would expect to occur at a major university. Such threats to academic freedom have the effect of reducing available information about any and all "impossible" ideas, something professors, of all people, should understand and resist. Many who continued their interest in cold fusion at most institutions suffered similar treatment. Fortunately, people and ideas were still treated with respect at LANL while I was there, even though skepticism was rampant.

During 1990 and early 1991, people were busy refining their work and submitting papers. The National Cold Fusion Institute, directed by Fritz Will, was established in Salt Lake City and was producing a steady stream of supporting results.[28-31] The Electric Power Research Institute (EPRI) began funding a large program to measure heat at SRI International under the direction of Michael McKubre. They also granted research funds to Texas A & M. Most laboratories found ways to replicate the Fleischmann and Pons claims if they worked long enough. Optimism was growing when The First Annual Conference on Cold Fusion[h] was held in Salt Lake City on March 28, 1990. Later, in July of 1990, the University of Hawaii held the World Hydrogen Energy Conference, where Douglas Morrison once again used the concept of "Pathological Science"[32] and continued to emphasize this concept in widely-read reports he circulated, which helped greatly to bolster the myth. Morrison attended all of the ICCF conferences until his death and was sincerely unable to accept any observation, no matter how well done, if it violated accepted theory. In October, Jones organized a meeting at BYU called "Anomalous Nuclear Effects in Deuterium/Solid Systems" where a growing number of supporting studies were presented. Huizenga attended the conference without experiencing any change in attitude. Nevertheless, those of us who were investigating this unique phenomenon were sure the general scientific profession would wake up to the reality very soon. Dueling books were published in 1991. Eugene Mallove's book "Fire from Ice" supported cold fusion while Frank Close's book "Too Hot to Handle" was skeptical. The battle lines were being drawn based on a basic difference in approach within the scientific profession. Some people are simply unable to evaluate observations with an open-mind because, to them, truth is only defined by theory. If theory and observation are in conflict, theory wins. In this case, the absence of neutrons proved that the effect does not occur even when tritium and extra heat are measured, because theory requires neutrons be produced. In their minds, the extra heat must be a measurement error and the tritium must be contamination. Evidence to the contrary was simply ignored. This is how faith-based science operates, but not the kind of science we are taught to respect. On the other hand, reality-based science acknowledges what nature reveals and then attempts to find an explanation. Rejection occurs only if a satisfactory explanation cannot be demonstrated. This demonstration is still in progress for cold fusion.

After I wrote a comprehensive scientific review of what was known about the subject up to May 1991[33] and after other independent reviews became available[34-38], I asked for and received funding from my division leader, Delbert Harbur (MST-DO), to build a calorimeter. Use of internal funding for this effort took courage on the part of Harbur, for which he was later criticized. By that time, positive results were not sufficient to protect a person from criticism for

---

[h] This conference series later became known as the International Conference on Cold Fusion (ICCF) and recently as the Conference on Condensed Matter Nuclear Science. Twelve of these conferences have been held in seven countries as of 2006 (see Appendix D).

even making an effort to study the effect. A calorimeter was eventually built in which a palladium sample sent from Japan by Akito Takahashi of Osaka University (Japan) was studied. This batch had shown production of anomalous energy when measured in Japan[39-42], in Italy[43], and later in the US.[44] As promised, the sample produced impressive energy in my calorimeter. A second sample sent from Japan, made from a different batch, was found to be completely normal—no excess heat was observed. Inquiry revealed that the second batch had been made by nearly the same method as the first batch by Tanaka Kikinzoku Co., Japan, but not in exactly the same way. A third batch was made by a process identical to the first, which again produced anomalous energy, but in a lesser amount. This work was published in 1993.[45] By that time, the source of palladium and its treatment was known to be very important for success.

On May 5, 1993, I was asked to testify before the Committee on Science, Space, and Technology, U. S. House of Representatives[46]. This hearing was held to address the requested budget of the hot fusion program, about which Congress had serious doubts. This program had promised too much and delivered too little while costing a fortune, a condition that has not changed. Rep. Dick Swett of New Hampshire, a friend of Eugene Mallove, wanted to acquaint the government with energy sources other than hot fusion. So he invited Dr. Randell Mills, Dr. Bogdan Maglich, and me to describe, respectively, the hydrino energy source, a better design for a hot fusion reactor, and cold fusion. This also provided an opportunity for letters from various laboratories describing successful replication, as well as many important papers, to be placed in the published record. For me, as well as for the Laboratory, this was a unique experience. Normally, when Congress requests testimony about what the Laboratory is doing, people who are hired and trained for this purpose are sent. Seldom has an ordinary staff member been invited to give testimony. Even though I had retired as a staff member in 1991, I was working as a consultant, hence still under Laboratory jurisdiction. The Laboratory administration did not want to take a stand on the subject, yet they could not refuse the request. People at high levels were very worried about what I might say. A statement was carefully crafted with help from people who normally testify. It made clear that my statement was my own opinion and, in spite of my heated objections, I was to make clear I was not speaking for the Laboratory in my support for cold fusion. With this document in hand and anxiety growing at the Laboratory, Carol and I flew to Washington. Only a few Congressmen showed up for the hearing, but they, nevertheless, gave the hot fusion people a very hard time indeed. Our testimony about other possibilities was an anticlimax. Except for the personal fun and excitement, the event was a waste of time with no change resulting in the government's approach.

By 1993, Gary Taubes[47] had published his very negative and occasionally inaccurate book "Bad Science, the Short Life and Weird Times of Cold Fusion" and John Huizenga[48] published the first edition of "Cold Fusion, The Scientific Fiasco of the Century." The debate was shifting strongly in favor of a skeptical attitude in the press, with most articles following the lead provided by these two authors. A myth was now firmly in place and rejection of the field was assured. Researchers in the field who were claiming positive results experienced growing hostility and active censorship by their superiors. As Taubes made clear to me, he did not know and did not care if cold fusion were real or not. He only wanted an interesting story that would make him enough money to do what he really wanted to do, which was to write plays for Broadway. As a result, the book made no effort to be fair or even accurate in many important ways—it was designed to sell. On the other hand, Huizenga, as a competent scientist and teacher, is certain that the view of nature he had mastered and taught to his students is correct. He was not about to change his mind, especially while so many questions remain unanswered. The basic characteristics of these two authors provide a good example of why facts about cold fusion were and continue to be distorted. Self-interest and arrogant ego continue to influence how the field is being presented to the public. Such treatment is not a surprise because new discoveries often suffer at the hands of these character defects as described in two excellent books by Milton[49] and by Cohen[50]

In 1994, I was asked by Charles Becker to join the board of directors of ENECO, where I remained until 1998. Charles is a visionary businessman who put his money where he saw the best return for both himself and for mankind. Unfortunately, the vision was a little premature. The company is now developing efficient thermal-electric energy conversion devices, which may some day be applied to cold fusion. ENECO was derived from Future Energy Applied Technology, Inc., founded by Hal Fox. Hal hoped to get a US patent for the work of Fleischmann and Pons, learn how to scale up the effect, and obtain income by leasing the patent to other companies. A small laboratory was set up in Salt Lake City where various ideas were tested in an effort to replicate and amplify the effect. The company also supported research in my Santa Fe laboratory. In addition, Paul Evans, a lawyer, was hired to move various patent applications, including the one describing the Fleischmann and Pons work, through the Patent Office. Little did we realize that the Patent Office, in the person of Harvey Behrend (since retired), would refuse to issue patents on the subject—of course because he followed the wishes of his boss and various political appointees. Reams of deposition, detailed published evidence, and numerous official forms were sent to the Patent Office only to be rejected because the *New York Times*, in one article or another, said the claims were not real. Scientific evidence we presented had no effect. More than a million dollars was spent satisfying the legal requirements and demands of Mr. Behrend, to no avail. Finally, the company ran out of money to go further and the patent application was given back to the University of Utah where it was allowed to lapse. As a result, the basic patent for one of the great discoveries of the 20th century was not accepted in the US even though patents in Japan and in other countries were granted. Many other people had similar experiences, resulting in the absence of intellectual property protection for inventors. This failure by the Patent Office not only resulted in an intellectual property mess requiring future correction; it also forced people to keep their best ideas secret.

The government of Japan took just the opposite approach. The New Hydrogen Energy Laboratory (NHE) was created by the Japanese government in 1996 to train their scientists and to discover how the effect could be brought to market. At that time, Fleischmann and Pons were working in France at a laboratory set up by Technova, a subsidiary of Toyota Motor Company. In October 1997, I was invited to Japan for a week to share my understanding of the effect. Here I found a modern laboratory staffed by people from many major universities and companies in Japan. They were using state-of-the-art equipment and occasionally finding modest evidence for the cold fusion effect. Unfortunately, the results were not reproducible and not at a level required to get the attention of industry. Nevertheless, after NHE closed on schedule, many of the staff went back to their institutions and even now some of them continue to advance the field. In that sense, the NHE served its intended purpose. It showed Japanese industry that the effect was real and it encouraged a few scientists to continue their work, which is now bearing fruit. This approach is in sharp contrast to the way the US government handled the problem. Of course, skeptics exist in Japan and several Japanese scientists have been harassed because of their studies. Even so, good progress is being made there, as revealed by three of the twelve ICCF conferences being held in Japan and creation of the Japan CF-research Society[i] which meets regularly. As Hideo Ikegami, at the time with the National Institute for Fusion Science, an organization devoted to hot fusion, explained to me, "Japan has little oil so if there is a 1% chance cold fusion is real, we will explore it. The US has much oil so if there is a 1% chance cold fusion is not real, you will reject it." This sums up the situation rather nicely.

Wired magazine featured "Those Who Dare, A Salute to Dreamers, Inventors, Mavericks, and Leaders" in its November 1998 issue. Twenty-five people, including Mike McKubre and myself, were invited to Los Angeles to attend a party at which all of us mavericks would meet, eat minimalist food, and be given some nice gifts. Once again, the experience was great fun, but had no effect on the history of cold fusion. The same issue of *Wired* contains an excellent article,[51] "What if Cold Fusion is Real?," by Charles Platt.

---

[i] See http://wwwcf.elc.iwateu.ac.jp/jcf/indexe.html,

From 1998 until to 2003, work in my laboratory continued at a steady pace with support from two special people, Charles Entenmann and Jed Rothwell. Fortunately, open-minded people exist who have money and who see a need to help solve some of civilization's problems. Their occasional donations allowed me to explore many kinds of materials and to design and use almost every kind of calorimeter. The result of this effort will be explained in later chapters. In addition, by that time I had collected and catalogued more than 3000 papers published about cold fusion using EndNote[©][j], a very useful computer program designed to handle bibliographic information. This collection was expanded by combining it with one provided on the Internet by Dieter Britz, resulting in a collection now containing more than 3660 citations. Jed Rothwell volunteered to create a website for cold fusion using this bibliography as the starting point. Since then, Jed has expanded the www.LENR-CANR.org site until now it is the main source of scientific information about the subject. Many of the references cited here can be read in full text on the site. More than 640,000 copies of more than 500 full text papers have been downloaded and read by people in many countries—an interest that is growing. A review of the field, "A Student's Guide to Cold Fusion," is also available in English, Spanish and Portuguese. In addition, a book by Jed, available in English and Japanese, entitled "*Cold Fusion and the Future*" can be downloaded for free. The library can be easily searched for keywords using Google or the EndNote[©] version of the database can be ordered and searched in more detail. Thanks to the Internet, facts about the subject are available in spite of conventional publication being made difficult. Two other popular books supporting the field have been published recently. These are "The Rebirth of Cold Fusion; Real Science, Real Hope, Real Energy" by Krivit and Winocur [52], and "Excess Heat, Why Cold Fusion Research Prevailed" by Beaudette[53]. Ignorance is no longer an excuse.

In April of 2003, Lewis Larsen asked me to join Lattice Energy, LLC, as a Senior Scientist, where I remained until April of 2006.

In 2004, thanks to efforts by Peter Hagelstein (MIT), Michael McKubre David Nagel, and Randal Hekman, the DoE was persuaded to again review the subject of cold fusion.[54] A group of 18 scientists[k] responded to an invitation by the DoE to act as a review panel. I was invited at the last minute to the one-day

---

[j] EndNote, Thomson ISI Research, www.endnote.com. This program is available for both the PC and the Mac.

[k] The known reviewers are: Allen Bard (University of Texas), W. Brown (Lehigh University), M-Y. Chou (Georgia Tech.), W. Coblenz (DARPA), G. Hale (LANL), K. Kempar (Florida State University), D. Klepner (MIT), D. Liebenberg (Clemson University), B. Mueller (Duke University), P. Paul (BNL), and J. Smith (former DoE).

meeting on August 23, 2004 in Washington, DC, where we[l] were to be given time to prove the reality of cold fusion. I recognized this as a waste of my time and money. No one can adequately describe cold fusion in such a short time to so many people. Their questions alone would take this much time, leaving little opportunity for an answer. A myth, as described later, was set too firmly in place to be changed so easily. Instead, I declined and sent copies of my extensive review. Not a single person responded to my offer to discuss the subject in more detail and provide more information. As you can see for yourself by reading the reviewer's comments on the LENR-CANR website, only a few of the professional scientists understood the issues[55] and were persuaded to believe anything of importance had been discovered. Subsequent actions by the DoE toward submitted proposals reveals no change in their actual policy. My response to this inadequate review can be read at www.LENR-CANR.org.[56]

Finally, I was honored by being awarded the Giuliano Preparata Medal at ICCF-12 in 2005. This medal has been and will continue to be awarded to a growing number of people who have made significant contributions to the field of LENR[m] (Low Energy Nuclear Reaction).

My experience taught me the futility of trying to understand the LENR effect using a low-level approach. I believe small groups working without adequate funding will not succeed. The region in which nuclear action occurs appears to be located only in what appear to be random spots no bigger than a few microns. Such material can only be studied using proper tools, which are expensive and require special training. Because these small islands of active material are formed by chance, a search for them can be very frustrating, requiring many experimental attempts. So far, no theory has been successful in showing where or when these active islands will form or how they might be created on purpose. However, progress is being made by some of the better funded efforts.

---

[l] The six presenters were Steve Jones, Andrei Lipson, Graham Hubler, Vittorio Violante, Peter Hagelstein, and Michael McKubre in the order of presentation according to the proposed agenda.

[m] Past winners are Yasuhiro Iwamura, Tadahiko Mizuno, Antonella De Ninno, Peter Hagelstein, Yoshiaki Arata, Xing Zhong Li, Michael McKubre, Akira Kitmura, and George Miley. (http://www.iscmns.org/prizes.htm)

CHAPTER 3

Personal Experience Investigating Cold Fusion

3.1 Introduction

All scientists like to see the workings of nature with their own eyes in their own laboratories—and I'm no exception. In fact, if I had not personally witnessed cold fusion myself, I would not have continued my interest. Instead, I might have taken up an equally frustrating and non-reproducible activity, such as golf. However, simply seeing something anomalous is not enough. The behavior must make sense, the observations and explanations must show internal consistency, and other people must witness the effects. Even after these conditions have been met, a trained scientist is frequently torn between accepting "impossible" conclusions and rejecting the whole idea. The challenge is to keep an open mind long enough to learn and understand all of purported evidence, and long enough to be sure the novel explanation is not actually describing a special behavior, before the idea is rejected.

The results published by other people are summarized in the next chapter. This chapter will show some of what I personally observed. Unlike the style presently used in scientific papers and books, which I find inadequate in showing how scientific discovery is actually made, I will describe a very personal experience. Everyone who has worked in this field, or indeed in science in general, has a similar story to tell. Hopefully this example will encourage students to become interested in the process.

3.2 Search for Tritium

The first problem confronting anyone starting out in any new subject is deciding where to start. I considered heat measurement, but quickly abandoned this approach because I knew nothing about calorimetry and did not have time to learn. Fortunately, besides heat production, Fleischmann and Pons claimed they had generated tritium in their electrolytic cell. This nuclear product is easy to detect and its proven presence would add significant weight to their claim. Also, a search for tritium production was the most logical place to start because I was working at the time with some of the world's tritium experts. With financial backing from the DoE, Carol and I set about trying to make tritium. Our approach was to electrolyze heavy-water using various palladium cathodes in a sealed cell. This design allowed us to measure tritium contained in the electrolyte and in the evolving gas. In this way, an accounting could be made of all tritium entering or leaving the cell. Any extra tritium had to result from an abnormal process. Because we planned to do many experiments with limited funding, the cells were designed to be cheap and simple. As you can see in Figure 3, the evolving D_2 and O_2 gases leave the cell through a plastic tube, and pass through a catalyst where they are converted back to D_2O. The resulting D_2O is collected in an IV bag.[a] For a brief time, we were a major user of IV bags in Los Alamos. More than 250 cells were studied using palladium from many different sources containing many different impurities. Usually a half dozen cells would be bubbling together on the table, an arrangement designed to eliminate tritium in the room as an explanation. If only one cell in the group "turned on" and the others did not we would know that the room was not the source of tritium. In addition, for health safety reasons, the tritium content of the air was monitored in the building being used for this study, with no sudden increases noted during this work. Unfortunately, only thirteen cells produced excess tritium and the amount was rather small. To make matters worse, we could not make tritium on purpose no matter how hard we tried. In other words, we did not understand nor could we control the important variables.

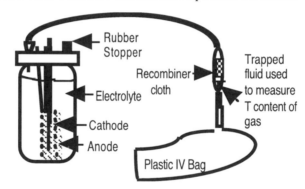

Figure 3. Drawing of the closed cell used to study tritium production.

Typical results obtained from active and inactive cells run at the same time are shown in Figure 4. In this figure, a fraction excess value of 2.0 means the tritium content doubled within the closed apparatus. Note the delay of four days before tritium is produced in cell #73. This delay distinguishes anomalous tritium from cells that might gain tritium from the environment or from tritium that might be dissolved in the palladium, as I will shortly explain. The curve in the figure only connects points and does not reflect the random variation which is ±0.2 in the fraction excess. Note also, the delay is followed immediately by a steady production for about 25 days, with a few bursts being evident. Production stopped after 26 days. It is well known that tritium will be slowly enriched in D_2O by electrolysis, as described in Appendix E. This cannot be the explanation because a complete inventory was made in our studies. Production of bursts followed by gradual reduction in tritium concentration after production stopped

---

[a] An IV bag is used in medicine to administer fluids intravenously.

suggests a mechanism that is able to remove tritium from the liquid and gas. This loss might be caused by tritium dissolving in the palladium cathode, which was not measured and by tritium lost in each sample removed from the cell for analysis. Moments of excessive imagination can even suggest possible nuclear reactions to explain this loss.

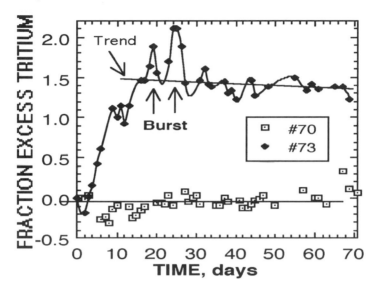

Figure 4. Typical behavior of active and inactive cells in which tritium is measured.

In the process, we made an important discovery. When tritium is produced, it always appears first in the electrolyte, not in evolving gas. This behavior is important because it shows tritium production occurs only at the surface and it leaves the sample before it has a chance to dissolve in the metal. The implications of this behavior will be described in more detail a little later. Being highly controversial, the resulting paper[1] was carefully reviewed and eventually published. Administrators at the Laboratory were in an awkward position. The DoE had concluded such nuclear processes were impossible, yet here they were for all to see. I could almost hear the sigh of relief when our enthusiastic technician threw out the active samples during a mandated cleanup campaign—no samples, no evidence, no need to confront the problem.

Major reasons for rejecting a nuclear source for tritium are based on tritium being in the palladium before the experiment or on it entering the cell from the surrounding environment. Consequently, we set about to test these assumptions. Numerous attempts to find dissolved tritium within the initial palladium and within the environment were made without finding any tritium, similar to the experience of other people. So we chose the opposite approach. Cells were placed in an environment known to contain tritium. In another experiment, a known amount of tritium was dissolved in the palladium cathode. If the behaviors resulting from cold fusion matched those produced by known sources of tritium, then the explanation would be obvious.

When a cell is placed in an environment containing tritium, the tritium content of the electrolyte will slowly increase at a linear rate until diffusion into the cell equals the rate tritium diffuses out. A steady-state concentration will be reached relatively rapidly in our experiments because the wall of the plastic IV bag is thin. This expected behavior, shown in Figure 5, was observed when a typical cell used in this work was placed in an environment known to contain tritium. Note an immediate uptake of tritium followed by steady increase over a period of about 34 days. In contrast, anomalous tritium production was found to occur only after a delay of several days and stopped after about 25 days. In addition, similar cells studied in the same, clean environment showed no change in tritium content while tritium was growing in a neighboring cell. Consequently, the observed behavior of anomalous tritium production in a clean environment and that obtained from a contaminated environment show entirely different patterns of behavior.[2] This comparison eliminates the environment as a source of tritium.

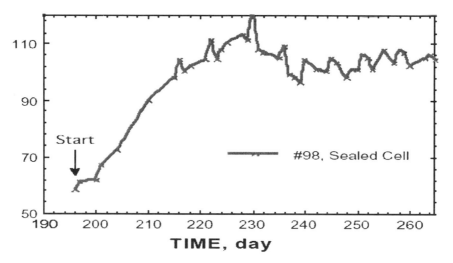

Figure 5. Pickup of tritium in a sealed cell attached to an IV bag located within an environment containing tritiated water vapor. The vertical scale is count/minute in a 1 ml sample. The variations are caused by random scatter in the measurement.

When tritium is dissolved in palladium on purpose, it always appears in the evolving gas, not in the electrolyte. As can be seen in Figure 6, the contained tritium is released immediately after electrolysis starts and it continues at a steadily reduced rate as electrolysis continues. Loss into the gas is a first-order reaction, just as would be expected. Once again, behavior of tritium claimed to be produced by cold fusion and the behavior of tritium known to be dissolved are different. To us at least, this study demonstrated that neither contaminated palladium nor the environment were the source. Skeptics were forced to propose that tritium was not dissolved in the palladium, but was tightly bound to isolated impurities, which somehow caused the observed behavior. They never explained why or how tritium was captured by these special regions, why it would be rapidly released after hours of electrolysis, and how palladium prepared by arc-melting could retain such tritium. Of course, all of these suggested mechanisms made no sense to anyone familiar with the behavior of tritium.

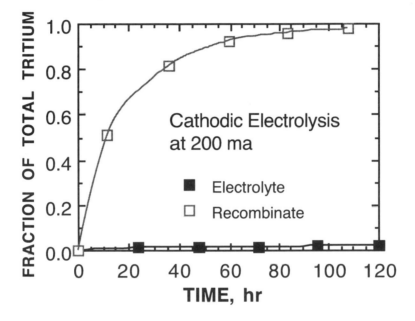

Figure 6. Growth of tritium in the electrolyte and in evolving gas (Recombinate).

Individual studies always have errors. On the other hand, a series of studies using different approaches, but showing the same patterns of behavior, are more difficult to reject. In this case, the observed patterns are completely consistent with anomalous tritium being produced within the cell and not consistent with tritium coming from ordinary sources.

3.3 Effects of Crack Formation

When reacted with deuterium, most palladium forms cracks each time the beta phase forms (See Appendix C for information about the beta phase). During the tritium study, we discovered just how important this crack formation is in determining the local concentration of deuterium in the metal. Loss of deuterium from a palladium surface will be faster from an area where a crack has formed, causing this region to have a lower deuterium content than the rest of the surface. As a result, the surface will be very non-uniform in its deuterium content. Because, as discussed later, a high deuterium content is necessary to achieve a high rate of nuclear reaction, local reductions in concentration can have a large negative effect even when local regions have the potential to be nuclear-active.

This crack concentration can be measured as an increase in physical volume over that calculated using the lattice parameter,[b] shown in Figure 7. As you can see, the size based on the lattice parameter is the same as the size based on physical dimensions at compositions up to about H/Pd=0.6. This means no void space is present in the sample. However, void space starts to grow as the composition is increased in the beta phase.[c] Only a very few samples did not show an increase in excess volume (void space) and they tended to take up deuterium well and to reach a larger H(D)/Pd ratio.

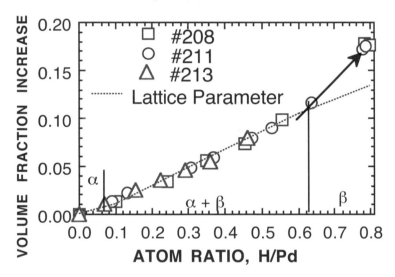

Figure 7. Increase in volume produced by loading palladium with hydrogen. The dashed line is the volume calculated using the published lattice parameter.

By adding tritium to the palladium and using it as a tracer, we were able to quantify the effect of cracks on the loss-rate of deuterium from palladium. In other words, measurement of tritium loss, which is easy to perform, was used to determine the rate of deuterium loss. We found that the larger the excess volume, the faster deuterium left the sample. Therefore, most palladium acts like

---

[b] An atomic lattice consists of planes of atoms and the lattice parameter gives the average distance between these planes without being affected by imperfections or cracks.
[c] Detailed information about the phases in the Pd-D system can be found in Appendix C.

a leaky bucket, with the cracks being a major hindrance in achieving the high deuterium concentration thought necessary.

As can be concluded from Figure 6, when tritium, or deuterium for that matter, leaves palladium, it enters the gas stream rather than dissolving in the electrolyte. This behavior occurs because the dissolved tritium leaves as DT gas that forms on the surface and within cracks. In addition, deuterium leaves as the D_2 molecule. Too little tritium is present to form much T_2 molecule. Once the TD molecule forms, exchange with water is known to be very slow and the solubility of this gas is low in the electrolyte. Consequently, very little of this gas will be found in the electrolyte. Only tritium produced on the surface as an ion can readily enter the liquid, and then only before it forms the DT molecule. This means anomalous tritium, which is always found only in the electrolyte, must have formed on the palladium surface and entered the electrolyte very quickly as T+ ion. Therefore, as other people also concluded, tritium is produced only very near or on the surface of active palladium.

3.4 Anomalous Energy Production

About a year later, I obtained permission and funding to build an isoperibolic calorimeter. This was sealed, contained a recombiner, and was stirred—requirements demanded before claims could be accepted. A drawing of this device is shown in Figure 8. Temperature was measured at two positions within the cell and the deuterium content of the palladium cathode was determined by measuring the change in deuterium pressure within the cell (see Appendix F). For our first study, I was given a piece of palladium by Akito Takahashi from a batch made by Tanaka Kikinzoku Co. (Tanaka 1), which had been shown to produce anomalous energy at the time in Japan. You can imagine my surprise and pleasure when this sample produced excess energy in my calorimeter. In fact, this success seemed too easy, requiring a week of testing to be certain we weren't being fooled. The most dramatic example of excess heat production obtained from this sample is shown as a function of time in Figure 9.

The calorimeter was calibrated before, during, and after the study by applying power to an internal heater while electrolytic current was applied to the cathode. No apparent change in the calibration constant was detected. Excess power was observed only above a critical current and this power was unaffected by periodic reductions in applied current. From what we now know, a critical environment forms on the surface and this reaches the required D/Pd ratio each time current is increased above a critical value. In other words, once the critical environment forms, heat production is controlled entirely by the D/Pd ratio within the critical environment.

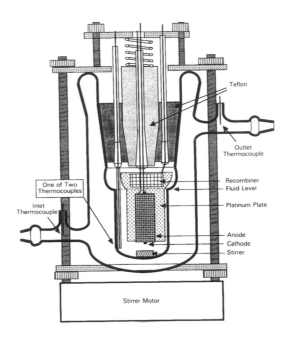

Figure 8. The calorimeter used at LANL.

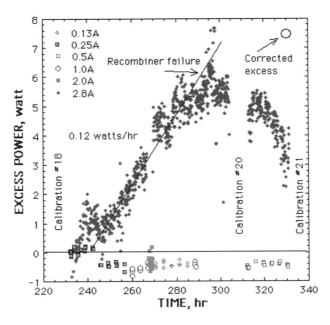

Figure 9. Time variation of excess power production using Tanaka #1 palladium.

In this study, the critical current density lies somewhere between 0.30 A/cm^2 and 0.42 A/cm^2. When the current was reduced below the critical value, the measured zero excess power was constant throughout the study at about -0.5 W even though applied power was varied over a wide range by applying different currents below the critical one. This means a shift in zero had occurred since the calorimeter was last calibrated using a "dead" cathode. Consequently, the measured excess power resulting from this zero-shift has to be added to the plotted values. These tests of zero and repeated tests using the internal heater show that the large and increasing excess did not result from a zero-shift or a change in calibration constant.

Unfortunately, the catalyst used to recombine the D_2 and O_2 gases failed at the time indicated on the figure, causing apparent energy production to drop because the gases were no longer reacting in the calorimeter to form D_2O—seen as increasing pressure within the cell. A "corrected excess" was calculated by adding the energy that would have resulted had the recombiner not failed, as noted on the figure.

A second piece of palladium from a different batch (Tanaka 2), but one claimed to have been made the same way, was delivered from Japan and studied. This piece was found to produce no excess energy, but contained a high concentration of cracks (13.5% excess volume). After informing Takahashi of this fact, a third batch (Tanaka 3) was made and delivered. This time the conditions used during manufacture were closer to those used to make the first batch. This material made excess energy, although less than the first batch. Also, the sample was found to contain a crack concentration slightly greater than the first batch, but much less than the second. The behavior of several samples is compared in Table 1. This study demonstrated two important facts. Cracks are detrimental to anomalous heat production and suitable palladium can be made to produce anomalous heat in independent laboratories using different calorimeters. With no errors being discovered, the work was eventually peer reviewed and published.[3] The results encouraged us to further explore why heat production is so difficult to initiate and what other variables might be influencing success. The work also suggested that pretesting might be effective in eliminating potentially inactive material. Meanwhile, conventional science was taking the opposite approach—saying it was not real because the effect could not be reproduced.

At about this time, I was contacted by Joe Champion[4] from his jail cell in Mexico, with a request to make gold using various alchemical methods. Joe is a modern alchemist who for reasons having nothing to do with cold fusion was locked up for a brief time. He wanted me to mix certain chemicals with gunpowder and ignite the mixture, which he predicted would produce gold. I turned him down. He and an investor then went to John Bockris at Texas A & M with an offer John did not refuse. Although the method seemed to work, the effort got John into trouble.[5,6] In this business, too much willingness to be open-minded can be a danger.

3.5 Study of Palladium

By 1991, study of cold fusion at LANL was meeting increased resistance and paranoia was increasing about possible accidents even when ordinary equipment was used. My new wife, Carol, and I wanted to build a home in Santa Fe and this looked like a good time to retire and start construction. After we finished the house and an adjacent laboratory, I began a study of 90 pieces of palladium furnished by IMRA (Japan). Some of this material produced excess energy, as shown in Figure 10 and listed in Table 1. Notice a critical onset current is required to initiate the effect, as was observed using the Tanaka material. Anomalous power slowly increased while electrolytic current was applied, again like the palladium studied previously. In fact, this behavior is always observed when solid palladium is studied and the necessary measurements are made. Excess power for each run returned to zero when applied current was reduced below the critical value, thereby demonstrating a stable calibration.

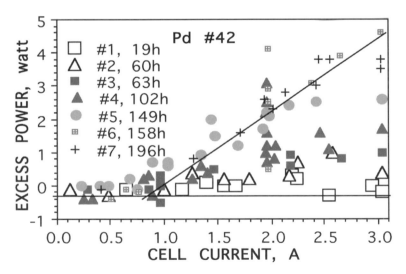

Figure 10. Excess power as a function of applied current after electrolyzing for various times using an unstirred cell.

This same sample was also used to demonstrate where in the cell excess energy was being made by measuring temperature at the top (T) and bottom (B) of the electrolyte, and at the cathode (C), as designated in Figure 11. Behavior of these temperatures was found to be different when excess power was made compared to when it was not. No excess power was detected during Sets #11 and #12, but when excess heat was produced during Set #18, the cathode was found to be warmer than the top and the top was warmer than the bottom of the electrolyte. In other words, extra power came from the cathode. As expected, this extra energy increased the gradient between the top and bottom of the electrolyte. Because the cell was not mechanically stirred, these gradients could produce temperature differences within the cell that changed when excess power was produced.

Table 1. Measured properties and excess power production.

Sample Number	Excess Volume, %	Composition D/Pd	OVC*	Excess Power, W (at 3 A)
Tanaka 1	1.7	0.82		7.5
IMRA #38	2.8	0.875	1.03	3.2
Tanaka 3	2	0.84		2
IMRA #42	1 to 2	0.891	1.25	4.6
IMRA #84	6.7	0.752	1.00	1.5
IMRA #58	4.1	0.833	0.60	0.0
Tanaka 2	13.5	0.75		0.0

* Open circuit voltage compared to an internal Pt reference electrode. See Appendix C.6.

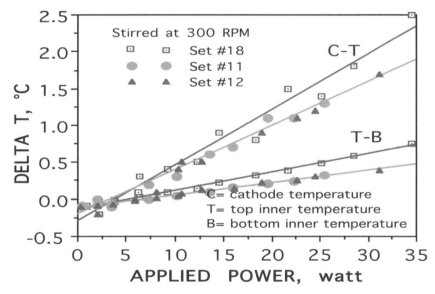

Figure 11. Temperature difference between the cathode and the top of the electrolyte, and between the top and the bottom of the cell.

Sample #42 was found to be unique because the excess volume started very small and did not grow larger upon repeated loading-deloading cycles as was observed to occur when other pieces of palladium were treated in this manner. In other words, only a few cracks were present and the number did not grow upon repeated reaction with deuterium. In addition, energy production was very difficult to kill, returning after a short delay even when the surface was removed by Aqua Regia. Apparently, a sample can reform an active surface after the previous active surface has been removed. This is an important observation because it shows reproducibility can be achieved if certain conditions exist in the metallurgy of material on which an active surface forms and if the required components are available in the electrolyte to allow a new active layer to be electrodeposited. This issue will be discussed in more detail later. Examination using a scanning electron microscope (SEM) showed no obvious characteristics to account for this ability. Clearly, other tools for surface examination are needed.

We started a basic study to show how palladium could be pretested, thereby reducing the growing frustration of working with inactive palladium. This work was published in *Infinite Energy*[7] with many of the figures out of order. Several characteristics, starting with crack formation were examined. Various samples were reacted with deuterium using electrolysis, after which the D/Pd ratio and excess volume were measured. Presence of cracks, as indicated by excess volume, was found to limit the maximum D/Pd, as shown in Figure 12. The wide range of values may be caused by a variable fraction of the cracks reaching the surface, where they can release gas from the sample. Fewer cracks of any kind are an advantage to achieving an average D/Pd ratio known to improve success, as suggested by the least-squares line drawn through the rather scattered data set.

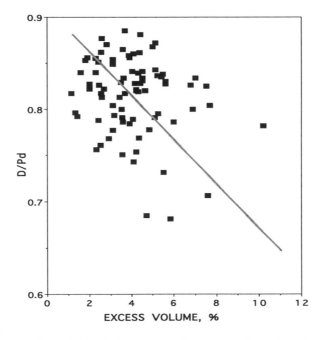

Figure 12. Effect of excess volume on maximum D/Pd.

The study pointed out other variables besides crack formation as being important. These included the deloading rate after current is stopped, open-circuit voltage, and loading efficiency. These variables were later explored in more detail, as described below. Unfortunately, publication of the paper was held up for about a year and then summarily rejected by the then editor of the *Journal of Electroanalytical Chemistry* (JEAC), apparently because cold fusion was mentioned. The information was eventually published in *Journal of Alloys and Compounds*[8] and described at ICCF-7.[9]

3.6 Study of the Loading Process for Palladium

When palladium is reacted with isotopes of hydrogen, competition between two processes determine how much hydrogen can be added. These are the "going-in" (entry) and the "going-out" (exit) processes. These are independent variables with a wide range of values that depend strongly on the nature of the metal and its surface. The "going-in" reaction can be examined using loading efficiency; this quantity being the ratio of the number of atoms of hydrogen presented to the surface by applied current divided into the number that actually dissolve in the metal. The number of atoms presented to the surface per second is equal to the current divided by the Faraday constant and the number actually taken up is based on the composition using the orphaned oxygen method. Figure 13 shows the loading efficiency of an untreated sample and one cleaned with Aqua Regia (HCl+HNO$_3$). Typically, a clean surface reacts with a large fraction of hydrogen presented to the surface by the electrolytic process until the composition has nearly reached the limiting value. A dirty sample will reject some deuterium and show a smaller loading efficiency, even though in this case, both samples reached nearly the same limiting composition.

Once the limiting composition is reached, hydrogen continues to react with the surface, but it leaves just as rapidly, resulting in a constant composition. As a result, a flux of hydrogen continues to pass into and through the sample, with a value that is highly variable. The "going-in" rate is also sensitive to applied current and temperature, being greater for higher applied current and higher temperature. Application of pulsed current or various frequencies superimposed on the DC component will produce additional flux variations. As a result, only the average flux is known. The flux through the region where the nuclear reactions actually occur is unknown and can not be measured. It can only be modified.

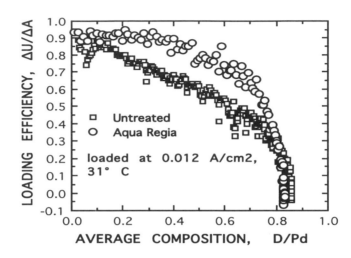

Figure 13. Effect of a surface barrier on loading efficiency. A loading efficiency of 1.0 means that every deuteron presented to the surface dissolves in the palladium lattice.

The "going-out" rate can be studied by measuring the deloading rate after current is turned off. This rate is obtained from a plot of average composition *vs.* square root of time, as shown in Figure 14. In this case, the sample remained in the electrolyte. The initial brief delay seen in the figure has no meaning because it is caused by the composition measuring system. A break in slope is expected when α-PdD forms on the surface. Only the initial slope for β-PdD is used to evaluate the sample. The same behavior is seen when a sample deloads in air. Presumably, deloading is identical in a liquid and in air because D_2 gas molecules form on the walls of cracks, where the surrounding gas or liquid has no influence. Because the environment has little influence, the behavior can be attributed to diffusion being the controlling mechanism for deuterium loss rather than the rate at which D_2 gas forms on the surface. If this is a correct interpretation, the difference in deloading rate between samples is caused mainly by a different concentration of surface penetrating cracks from which D_2 gas can escape. Of course, the higher the composition, the higher the D_2 pressure within unavoidable cracks, so that eventually a D/Pd ratio is reached at which the going-in rate equals the going-out rate and a fixed composition limit is reached during electrolysis.

A more accurate measurement of the composition can be obtained by extrapolating square root of time *vs.* weight, to obtain the composition existing when the current was stopped. Use of this method is important because loss from a sample can be very rapid during the first few minutes after loading stops, especially if a high composition has been achieved. In fact, use of weight is the only unambiguous method to obtain a value for the true average composition and should be used to calibrate other methods, as described in Appendix F.

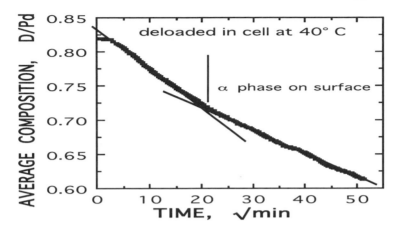

Figure 14. Change in composition *vs.* square root of time after applied current is stopped.

Slopes for numerous samples are compared in Figure 15 as a function of average initial composition. Two types of behavior are seen. Some samples show a rapidly increasing deloading rate as the composition increases. Other samples have a much lower deloading rate, which increases less rapidly as composition increases. The latter behavior is sought. Also shown is the calculated equilibrium D_2 pressure expected to be present within cracks at the indicated compositions. Presumably, the loss rate and the internal pressure deviate at high loss rates because a diffusion gradient forms between the surfaces from which deuterium is lost, *i.e.* the crack wall, and the average composition. This causes the loss rate to be less than expected based on the D_2 pressure, as indicated by the arrow. Two conclusions are clear: the loss rate increases as the average composition is increased, as expected, and some samples have an abnormally low rate even though they have a high average composition.

This same loss is expected to occur while current is being applied. Consequently, loading efficiency, deloading rate, and low excess volume can be used to quickly identify which palladium samples can reach a high composition while subjected active electrolysis.

3.7 Surface Composition Explored

Unfortunately, the average composition, which is easily obtained from measurement (Appendix F), is not the important variable. The important composition is that which exists where the nuclear reactions actually occur. If various places have the potential to be nuclear-active, the region with the higher deuterium or hydrogen concentration will produce the greater reaction rate.

Therefore, in addition to identifying which part of a sample has the ability to be nuclear-active, its location within a very non-uniform deuterium or hydrogen content also must be determined. Just where should the search begin?

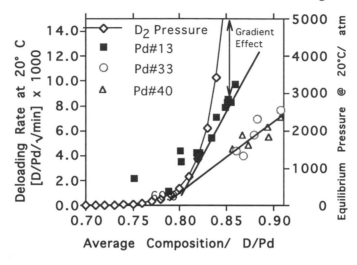

Figure 15. Variation of deloading rate as a function of average composition.

During electrolysis, the surface will have the highest composition. Therefore, if any active sites are present, their reaction rate can be expected to be greatest on the surface. How can this surface composition be determined? Fortunately, changes in deuterium chemical activity at the surface (Appendix C) can be measured using the open-circuit voltage (OCV), a value that is sensitive to the composition of deuterium/hydrogen in the surface. This voltage is measured between the cathode and a reference electrode within the cell when no current flows, *i.e.* when the cell is turned off.[d] This reference electrode can be of various types, each having a different but known voltage. For convenience, the measurements described here use clean, high-surface-area platinum immersed in the electrolyte. The anode cannot be used as a reference because its voltage is strongly influenced by the oxide layer that forms on its surface. Although the exact voltage is uncertain, the relative measurements are accurate and useful. The voltage is measured while current is turned off for about 5 milliseconds or during deloading while current is switched off for the entire time. While current is off, the cell acts like a battery with voltage [E] generated at the cathode created by the $2D^+ + 2e = D_2$ reaction. The value is proportional to hydrogen activity [a] in the surface squared divided by the activity of gas [$P(D_2)$] dissolved in the electrolyte, as described by the equation

---

[d] Voltage is frequently measured in electrochemistry while current is applied. This method is not recommended because too many other sources of voltage are created, which must be estimated and subtracted before the desired value can be obtained.

$E = -[RT/2F] \ln [(a)^2 /P_{D2}]$, where T = temperature, F = Faraday constant, R = gas constant.

Voltage created at the reference electrode is nearly equal to that generated by the cathode when it is first placed in the cell because both electrodes are bathed in the same electrolyte and are exposed to the same pH and deuterium pressure. As a result, the OCV is near zero when the cathode is free of deuterium. The cathode becomes positive with respect to the reference electrode as deuterium is added. If the known activity in the $\alpha+\beta$ two-phase region and the measured voltage are substituted in the equation, the effective D_2 pressure (activity) in the electrolyte can be calculated. Apparently, a very low activity is present even though D_2 bubbled through the solution before current was turned off. This happens because the effective pressures at the cathode and reference electrode surfaces are immediately reduced by reaction at the metal surface with dissolved oxygen, which is also present in the solution. This reaction significantly reduces the local D_2 activity.

Figure 16 shows how the OCV changes as a palladium sample is loaded with deuterium. A reverse of this behavior occurs when a sample is allowed to deload after current is turned off. Occasionally, especially after very high average compositions have been achieved, the OCV measured during deloading shows an interesting behavior, as revealed in Figure 17. This behavior suggests formation of a new phase, which slowly decomposes into normal β-PdD, as deuterium is lost. Such a slow decomposition might cause "life-after-death" to occur, during which anomalous energy continues for a time after current is turned off.[10] A nuclear reaction occurring on the surface would be fed by deuterium moving to the surface from the interior rather than from the electrolyte.

The behavior of the OCV during deloading can be compared to the behavior of platinum and a thin coating of palladium on silver, shown in Figure 18. Platinum dissolves very little deuterium and deloads rapidly without an arrest. This particular sample of platinum had been used previously as a dead cathode and probably contained some dissolved lithium, causing it to deload less rapidly than normal. On the other hand, the palladium deposit formed the two-phase mixture of $\alpha+\beta$ on the surface during deloading, which held the activity constant as long as it was present. Formation of this constant activity allows the OCV to be calibrated with respect to a known deuterium activity. As long as the surface is free of deposited impurities, such as lithium, the composition can be calculated using the measured activity. As will be discussed in Appendix C, the behavior of the OCV can also reveal the presence of impurities and/or reaction with lithium. When the surface becomes nuclear-active, the OCV gives an indication of the deuterium activity required to support this process—hence provides essential information.

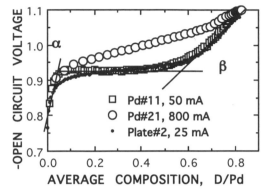

Figure 16. Open circuit voltage measurement during loading at various currents.

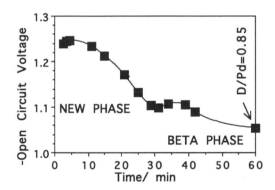

Figure 17. Open circuit voltage during deloading after production of excess energy.

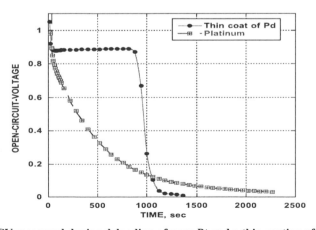

Figure 18. OCV measured during deloading of pure Pt and a thin coating of Pd on silver.

The next question we need to answer is, "What is the actual composition of this high composition phase?" A partial answer can be obtained by studying very thin films of palladium plated on platinum. This measurement was done using the orphaned oxygen technique, as described in Appendix F. As shown in Figure 19, the measured composition of such films is highly variable, but it can achieve a value as high as $D/Pd=1.5$. A D/Pd value above 1.0 can result if the sample is a mixture of β-$PdD_{1.0}$ and a second phase having a greater limiting composition. Large values are obtained here, in contrast to measurements of the average composition, because the large surface composition is not diluted by the low value present in the rest of the sample. A typical average composition of thick palladium is shown as the lower curve. Since this particular film was not energy-active, the average composition of an energy-active surface is probably significantly higher than the one measured here. In addition, deloading from a surface is very non-uniform, as can be seen by examining bubble production. Therefore, the maximum composition of such a thin film, or indeed any palladium surface, is well above the average value, in addition to being above $D/Pd=1.5$. This observation means that all theories based on the properties of β-PdD are riding the wrong horse.

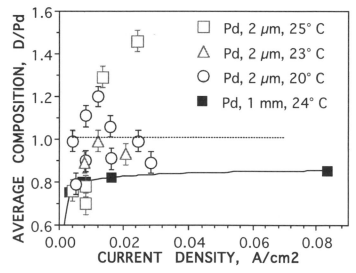

Figure 19. Measured average composition of thin films of Pd plated on Pt after being subjected to different current densities.

Based on these studies, the actual energy-active phase might be γ-PdD_2. This idea was further developed in a paper given at the Asti conference in 1995[11] and later published in *Infinite Energy*.[12] As discussed in Appendix C, this compound contains various impurities, any one of which might be essential to its behavior, but which are not acknowledged by theory. The present approach would be like trying to develop a theory of superconductivity while ignoring the elements present in the sample or developing a recipe for soufflé without acknowledging eggs are required.

3.8 Writing Reviews

The results of my work were used by ENECO to encourage the Patent Office to grant the Fleischmann and Pons patent. This and similar efforts by other people to obtain patent protection for their ideas were completely ignored. To help in this effort, I wrote a review to bring all of the better data under one roof so that everyone could see the bigger picture. This was published in the *Journal of Scientific Exploration*[13] with a brief version in an earlier volume.[14] Again, no change occurred at the Patent Office or in any other government agency. In 1997, the *International Journal of Modern Physics* rejected one of my reviews because it discussed cold fusion, even though one of the editors requested the paper. This work was eventually published in *Infinite Energy*.[15] In 2001, another review was submitted to four journals, one after another. These journals are *Physics Review B*, *Review of Modern Physics*, *Chemical Review*, and *Journal of Electroanalytical Chemistry*. Rejection was based on the usual myth or outright hostility. To be certain I did not miss the point, the editor of *Reviews of Modern Physics*, Dr. George Bertsch, rejected the paper with the promise, "Cold fusion is a classic example of pathological science. I will certainly not publish articles supporting its disproven claims." This review was even rejected recently by arXiv, the physics preprint archive, even after Prof. Brian Josephson sponsored it. You can decide for yourself whether this paper deserves such treatment. A full text is available at www.LENR-CANR.org as "Cold Fusion-An Objective Assessment." My most recent review "A Student's Guide to Cold Fusion," which gives a relatively brief overview, is on the same website and has been very popular to general readers. I mention these examples of rejection, which are similar to what many people have experienced, to demonstrate how information about the subject is managed by the scientific press. Recent experience suggests many avenues for publication previously available no longer provide the service because reviewers have become even less willing to give cold fusion the benefit of doubt. Hopefully, this situation will improve as the facts become better known.

3.9 Trip to the NHE Laboratory (Japan)

Dr. Naoto Asami sent some palladium for testing, which had been made at great expense for the New Hydrogen Energy Laboratory (NHE) in Japan. Unfortunately, they had detected no excess energy from this material. As expected, my tests showed these samples to be flawed—material I would expect to be completely inert. After hearing this, Dr. Asami invited me to Japan in

1997. The visit started with some anxiety because the person sent to meet me at the airport waited at the wrong location. Once this problem was solved, the trip was filled with the hospitality for which the Japanese are famous. After discussing the palladium problem in some detail, we discovered impurities were being introduced during manufacture as well as during subsequent annealing. Unfortunately, too little time and money were available to change the method of manufacture. As a result, NHE closed without trying pretested palladium.

3.10 Exploration of Errors in Calorimetry

Calorimetry, although simple in concept and old in application, is beset by subtle errors, as some people are more than willing to point out. Even today, many people making such measurements are unaware of some real potential problems while other people imagine problems where none exist. Therefore, I undertook the challenge to explore the source of various errors, rather than to rely on speculation. I found, as many other people discovered, the expected temperature gradient within the electrolyte is not important when power is applied by electrolysis, in contrast to what has been claimed.[16] As can be seen in the Figures 20 and 21, a small electrolytic current added during Joule heating reduces the gradient to insignificant values. This behavior, which is discussed in Section 7.9, makes an accurate calibration possible using an internal resistor only when electrolytic current is applied at the same time.

Mechanical stirring will reduce error caused by a gradient, but at the same time it will change the calibration constant because the stagnant layer of fluid at the cell wall will be disrupted. This layer influences the overall thermal conductivity of the wall and is very sensitive to how fast the fluid moves. Mechanical stirring reduces the effect of this barrier, but it does not eliminate it completely, as can be seen in Figure 22. Notice how sensitive the calibration constant is to small disturbances in fluid-flow near zero stirring rate. This demonstrates that, in the absence of mechanical stirring, even small changes in the convection currents, caused by changes in bubble generation rate or by changes in power generation, will affect the calibration constant. The effect does not become constant even at high stirring rates, requiring use of a constant stirring rate at all times even though all temperature gradients have been eliminated at the higher stirring rate.

Based on this work, measurement of small amounts of heat using unstirred isoperibolic calorimeters (see Section 7.9.2 for additional description) will probably contain more error than is claimed if an internal Joule heater is used for calibration without applying electrolytic current. Fleischmann and Pons avoided this problem by frequently calibrating their cell while applying electrolytic current. Nevertheless, the effect may still have some influence on their data, especially at the lowest claimed excess power levels. In my work, calibration was based only on electrolytic calibration using a newly cleaned platinum cathode, with Joule power used only to determine whether the calibration constant had changed while electrolysis was on going.

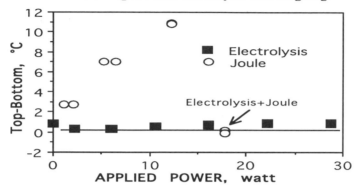

Figure 20. Gradient between the top and bottom of the electrolyte as a function of applied power.

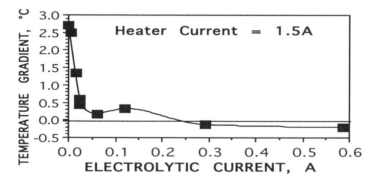

Figure 21. Reduction in gradient as electrolysis current is increased at a fixed heater current.

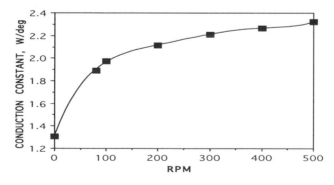

Figure 22. Effect of stirring rate on calorimeter constant.

3.11 Experience with Flow Calorimetry

These problems caused me to explore flow calorimetry. The initial design was intended to measure heat production using both flow and isoperibolic methods.[17] However, the dual method did not work well because, at the low flow rates needed for the flow mode, the reference temperature for the isoperibolic mode, provided by the flowing fluid, was not sufficiently stable. This calorimeter, shown in Figure 23, had a heat recovery of 98.7% and no more than 1.2% difference between calibrations using an internal heater and a dead Pt cathode. Because these two methods generate heat at different locations and local heating produced by D_2+O_2 recombination is not possible during Joule heating, this agreement demonstrates the calibration errors proposed by Shanahan[18] are absent. The cell contained several glass covered thermistors, which were only used to determine the temperature of the cathode; a recombining catalyst to insure no gas left the cell; a Teflon stirring bar to reduce thermal gradients within the cell; and a Pt reference electrode. The cell was contained within a vacuum Dewar and the whole assembly was contained in a box in which the temperature was held at 20°± 0.01°C. Consequently, the calorimeter was well isolated from the outside temperature, which was kept constant in any case. Water flow was measured using a container located on a balance, which filled and then emptied automatically. Orphaned oxygen was measured by weighing oil that was displaced from a reservoir (Appendix F), which allowed the D/Pd ratio to be measured at all times. I used this design until it was replaced by much better Seebeck calorimeters, as described in Section 7.9.6 and Appendix B.

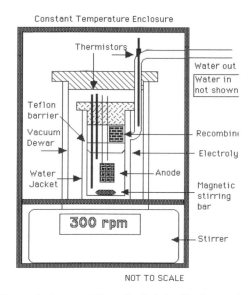

Figure 23 Drawing of the flow calorimeter. The cathode is inside the anode and not visible.

3.12 Surface Deposits

The flow calorimeter was used to study electroplated palladium on platinum because electrodeposition appeared to have much greater success compared to the use of bulk palladium and it is less expensive. In addition, an unexpected event occurred that required a re-examination of my approach using palladium sheet, as I will describe. During previous studies, calibration was normally made using a clean platinum cathode. This metal was flame cleaned before use and was always found to be completely "dead". However, on one occasion, the platinum cathode was not cleaned after it had been used. Much to my surprise, the expected "dead" cathode suddenly produced excess energy, as plotted in Figure 24, thereby proving once again that nature sometimes rewards sloppy work. Subsequent examination using a scanning electron microscope revealed a complex surface with a visible crack, as can be seen in Figure 25. This crack resulted because a layer of PdD, which had deposited on the platinum surface, shrank when it lost deuterium after the study. Presumably, the layer resulted when Pd^{++}, that formed by dissolution of Pd-Li alloys from previously studied palladium cathodes, subsequently electrodeposited on the platinum. The surface is proposed to be an alloy of palladium, platinum, and lithium, with most of the platinum in the EDX spectrum resulting from the underlying platinum substrate. Which part of the surface was active is impossible to determine. In addition to palladium, the surface also contained copper (Cu), iron (Fe), and a little oxygen (O), as can be seen in the EDX spectrum provided as Figure 26. Copper and iron are not expected to be present in the cell, although this belief cannot be supported by available data.

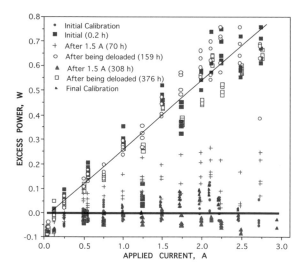

Figure 24. Excess power generated by a Pt cathode on which an active layer had been deposited during previous "normal" electrolysis using a Pt anode and a Pd cathode.

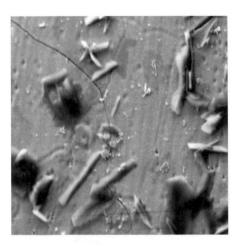

Figure 25. SEM picture of the active platinum surface. A crack is visible along with deposited material.

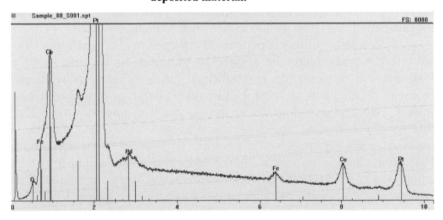

Figure 26. EDX spectrum of a coated Pt surface using 20 kV electrons.

Energy production from this sample was explored in several different ways. Applied current had the same effect on power production as had been seen before when bulk palladium was studied, except power was produced without the need for a critical current. A critical current is not needed, in contrast to bulk palladium, presumably because platinum does not allow significant diffusion from the backside of the active surface deposit, hence extra deuterium is not needed to compensate for this loss. In other words, a small current can achieve the critical surface composition because the deuterium has nowhere to go. The behavior was duplicated after deuterium was removed and reapplied several times. In other words, the sample showed reproducible behavior. Various treatments were found to reduce excess power and other treatments caused it to return to consistent values, as indicated by the line on the figure. Scatter of the calibration points around zero gives a measure of random scatter in the data, which is ±50 mW.

This observation opened my eyes to the possibility of palladium being simply an inert material on which an active deposit slowly forms, with this layer being difficult to distinguish from the inert Pd substrate. Slow deposition of this active material would provide one reason why activation of a cathode traditionally takes so long, although the delay in achieving a high average D/Pd ratio would add additional time. Use of palladium that is capable of reaching a high average deuterium content would only be required to reduce loss of deuterium from the active film into the bulk metal. Also, the reverse situation might be true, *i.e.* the required active layer might prevent loss of deuterium, hence allow a higher average D/Pd ratio to be retained by the bulk material. Platinum, copper, gold, nickel, iron, silver or even carbon might be far superior as basic cathode materials on which an active layer is applied because these materials diffuse deuterium only very slowly. This would avoid the difficult and time consuming loading process required when bulk palladium is used. However, deposition of an active surface on such material may not occur under the same conditions that are successful when palladium is used.

As a result of this experience, I changed my approach to a study of how an active surface might be applied to platinum or to other inert elements. Platinum was chosen as the best substrate for subsequent studies because it can be easily cleaned and recycled. Unlike the co-deposition method used by Szpak et al.,[19] I applied a coating before the sample was placed in the calorimeter. In this way, the coating could be characterized beforehand and the chemical reactions in the calorimeter would not be complicated by the deposition reaction. Various methods were used including deposition of palladium from a solution in which nanosized particles of various substances were suspended. This method incorporates suspended particles of potentially active material into the growing film of deposited palladium, where they can be held while being exposed to deuterium during subsequent electrolysis. Excess energy was occasionally found, as shown in Figure 27, for carbon particles incorporated in palladium. In addition, this study demonstrated anomalous energy production using surface layers of many different materials applied along with palladium. However, anomalous energy cannot be attributed to these additional materials alone because they might have changed the structure of the palladium to an active form or the active form might have been produced by pure coincidence. For example, palladium layers, when deposited by electrodeposition without suspended particles, sometimes become active. In other words, this work gave no insight into what might be the active material. It only showed, once again, that an active surface could be applied to otherwise inert material.

3.13 Experience with Seebeck Calorimetry

Thanks to generous support by Charles Entenmann, I was able to purchase a Seebeck calorimeter and lease another from Thermonetics, Inc. in early 2001. In addition to allowing energy production to be measured with greater accuracy, I was also able to study several active cathodes using both the flow calorimeter and the Seebeck. Figure 27 compares the results of one such comparison. Although the amount of excess energy is small, it is clearly visible in both instruments. The designation "constant" refers to application of constant current for a significant time, while "sweep" describes application of a fixed current only long enough for the calorimeter to reach steady-state, after which the current is changed. This process moves the current up and then down in value over the full current range. During application of constant power, points were taken every 6 minutes after the calorimeter had stabilized. However, power production had not stabilized in this time and continued to increase, as shown by the steady increase at each current. This effect is also visible in the sweep points because the power values are lower when current is increased compared to values taken as current is reduced during the following sweep cycle.

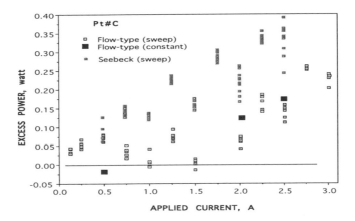

Figure 27. Effect of applied current on excess power production using a cathode made by co-plating carbon particles with Pd on Pt.

A wide range of conditions for deposition of an active surface on platinum was used. Most attempts failed. Nevertheless, the critical role of surface characteristics became increasingly apparent. Eventually the Thermonetics calorimeter was phased out in favor of my own custom-built designs, one of which is described in a paper given at ICCF-10 and in Chapter 7.[20] The first instrument of this design had an uncertainty of less than ±30 mW in the normal range of a Fleischmann and Pons study or ±0.2% of applied power with a power range up to 60 watts. Another calorimeter with an improved design is described in Appendix B.

3.14 Attempts to Replicate the Case Effect

After McKubre and co-workers reproduced the Case effect[21] at SRI, I undertook to do the same. Jed Rothwell supplied funds to build a system needed to purify the Case catalyst and to load it with very pure deuterium. Cells were constructed in which the catalyst could be heated to 200° C while being exposed to 4 atm of D_2. A temperature gradient up to 100° could be created across the catalyst bed, a condition thought to be important. A Seebeck calorimeter was built to hold the pressurized cell. Catalyst was supplied from several sources, including from the batch used by Case. In addition to obtaining some of the original Case catalyst, I manufactured my own using the method described by United Catalyst, Inc., the supplier of the active material. About a year was spent on this effort without seeing any unambiguous excess energy.

As is common in this field, the ability to make active material was lost. In this case, the drum of charcoal used by United Catalyst as the substrate was thrown out during cleanup, perhaps explaining why the catalyst they made later did not work. Because the unique characteristics required to make the material become nuclear-active were not determined at the time, it is impossible to manufacture more active material according to known specifications.

3.15 Replication of the Letts-Cravens Effect

Dennis Letts and Dennis Cravens[22,23] produced excess energy when they applied low-power laser light (680 nm) to a specially prepared cathode while it was being electrolyzed in D_2O + LiOD. Letts published a complex procedure[24] for treating the palladium before the required gold layer was applied. This pretreatment of the palladium was found to be unnecessary. Active layers could be made more easily, as described at ICCF-10.[20] Apparently, the Letts-Cravens effect is just as sensitive to the nature of the surface as is the Fleischmann and Pons effect, although each is sensitive to different characteristics of the surface. Figure 28 shows the relationship between excess power and application of various amounts of laser light.

Fleischmann-Pons heat was also produced without laser light being applied, as shown in Figure 29 based on measurements using the Seebeck calorimeter. In other words, excess energy was generated after the laser was turned off. The laser added no additional energy, beyond that supplied by the laser itself (30 mW), when the current was 0.5 A. However, when applied current was increased to 1.0 A, the amount of Fleischmann-Pons heat as well as the extra heat from the laser increased. Excess energy gradually decayed away after 3000 minutes. The Letts-Cravens effect does not generate tritium[25] and the nature of the reaction that generates extra energy is, as yet, unknown. Although Letts demonstrated the effect at SRI, additional efforts to replicate the effect have proven difficult.

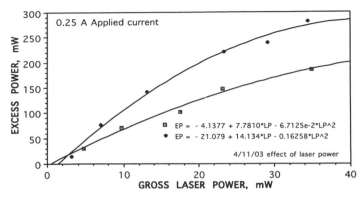

Figure 28. Excess power produced when laser power is applied to a 2 mm² spot on the cathode surface while using an isoperibolic calorimeter.

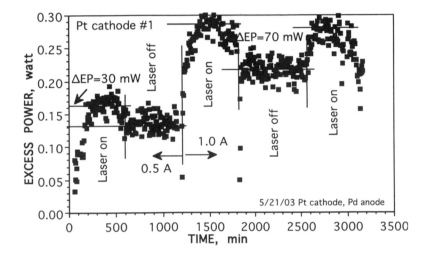

Figure 29. Effects of applying laser light (30 mW) to produce excess energy.

3.16 Development of Better Seebeck Calorimeters

I constructed a better Seebeck calorimeter, as described in Appendix B, thanks to support by Lattice Energy, LLC. As a result, uncertainty in heat measurement is greatly reduced while the calorimeter has the additional advantage of being very simple to use.

3.17 Conclusion

My work persuaded me to accept that tritium and anomalous heat can be made in an electrolytic cell. Both novel products are very sensitive to what has deposited on the cathode surface. Because the effect is rare and controversial, measurements must be made carefully and with a good understanding of the tools being used. Like Fleischmann and Pons, I chose heat measurement as the main tool and set about understanding and constructing good calorimeters. The same approach needs to be applied to future work so that suspicion of error being the cause of claimed anomalous energy production can be eliminated.

Many years of study revealed one important fact. Because the effect is very difficult to initiate, more information is gained about what does not work than what does. Failures would fill a small library. While such information is important, it is not very interesting to read. As a result, such experience does not get published or otherwise made available. However, several general patterns emerge from these failures. Small particles are not the only requirement, use of ^{6}Li gives no benefit, and high average loading alone does not always work. In fact, some of the best results have been obtained using poorly loaded material, based on the average composition. Active material may, nevertheless, have a high surface composition even when the average is low. Various impurities dissolved in the palladium, such as boron, carbon, oxygen, lithium, calcium, platinum, neodymium, or gold, by themselves, do not ensure success. In fact, none of the conditions predicted by the various theories produce success more often than can be expected from chance. A common experience is to obtain an active sample, an active batch of material, or discover a successful method only to have the sample stop working; to have the batch used up and found to be impossible to replace; or for the method to give good results only for a short time. For the most part, Nature has seemed content to give just enough encouragement to keep a person interested without allowing the effect to be understood. Nevertheless, some of the better-funded efforts are making progress.

Next we will examine what other people have reported and continue to observe. Clear patterns of behavior are starting to guide the studies and progress is being made as funding becomes available.

CHAPTER 4

What is Known or Believed?

4.1 The Myth of Cold Fusion

Replication is the gold standard of reality. If enough people are able to make an effect work, the consensus of science and the general public accept the effect as being real and not error or figment of imagination. With that goal in mind, this chapter lists the many effects attributed to cold fusion and gives a partial history of how this information was obtained in laboratories all over the world. More detail is provided in later chapters to show how these observations are related. Many cited papers are available at www.LENR-CANR.org and sources for the conference proceedings, in which many papers are located, are listed in Appendix D. If hundreds of replications are not sufficient and you insist on 100% reproducibility, an easy demonstration, or a commercial device, you will be disappointed and need read no further. If you decide to continue, a distinction needs to be made between what is known and what is believed about the subject.

A Myth has formed about cold fusion not being duplicated, being based on error, and being an example of "pathological science",[1] *i.e.* wishful thinking. None of this description is correct. The basic claims have been duplicated hundreds of times and continue to be duplicated by laboratories all over the world, although success is difficult to achieve. Fleischmann and Pons were wrong in a few minor ways[2,3] and were certainly wrong about how easy the claims would be to replicate. Nevertheless, their measurements of anomalous energy have been evaluated[4-7] and the real or imagined errors are shown to be unimportant.[8-10] Additional evaluations found their calorimetry to be sufficiently correct to justify their claim for anomalous energy.[11-13] Since then, replication of heat production has been growing and achieving increased levels of power production. While no study is ever completely free of error, the question always is, "Do the measurements justify the conclusion?" In this case, reasonable people can differ as to whether they would have published the data of Fleischmann and Pons if this had been their own work. However, after studying the accumulated evidence presented in this book, I do not believe reasonable people can ignore what has been discovered since then.

Skeptics love to apply the pejorative "pathological science" to the work of other scientists they do not believe. Of course, some claims richly deserve this insult because even a good scientist can be deceived by wishful thinking on occasion. But wishful thinking works both ways. Wishing a claim to be unreal and based on error without providing evidence for these assumptions can be described just as accurately as pathological skepticism. Other defining characteristics of such an approach can be suggested. Pathological skeptics often assume the people who have a viewpoint different from their own are ignorant of possible error and are being controlled by delusion rather than by facts. This approach is especially clear in politics, but happens less often in science. Most scientists take great pains to find errors and to test for prosaic possibilities. In contrast, pathological skeptics tend to propose imagined errors and frequently ignore real evidence in their haste to cast doubt. Another characteristic is revealed when a scientist fails to duplicate another person's work and then concludes, without a doubt, the fault is in the original work rather than in his own. For example, failure to detect expected nuclear products in an experiment is frequently used to show that Fleischmann and Pons were wrong. More likely, no nuclear reaction was made to happen in the first place because the conditions were not correct. Ideally, believers and skeptics should engage in a dialogue with both sides being open to new ideas, respectful of each other, and willing to explore the issues, an approach not uncommon in many other fields of science.

If this source of ideal energy were not so important, people interested in cold fusion could take comfort in being treated just as badly as are many other people who have new ideas.[14] This luxury is no longer possible because the stakes are so high. The growing energy shortage is causing increased conflict between nations and a growing stress on economic health by the increased cost of energy. Warming of the earth is being helped by a rising concentration of CO_2 in the atmosphere and the environment in some regions is being degraded by accumulation of radioactive by-products from fission reactors. We can no longer afford to pretend pathological skepticism is just another quirk of human nature, like a lazy family member who sits around doing nothing while complaining about what everyone else is doing.

4.2 Why was Cold Fusion Rejected?

Many kinds of nuclei can be made to fuse together provided the positive charge existing on each can be overcome, thereby allowing each to get sufficiently close. Because hydrogen and its isotopes (protium, deuterium, and tritium), have only a single positive charge, mutual repulsion can be overcome with relative ease. Even a spark produced by a few thousand volts in deuterium gas will cause a few deuterons to fuse, as indicated by the few detected neutrons. Conventional theory can explain these reaction rates. A problem arises when the observed rates become much larger.

Fleischmann and Pons, and Jones[15] (BYU) both claimed nuclear reactions could be made to occur in electrolytic cells. This claim, by itself, should attract attention because this environment is not expected to cause any kind of nuclear reaction, no matter how small. More important, the Fleischmann and Pons observation was unique because the high rate might allow their process to become a competing source of energy and a threat to conventional theory, rather than being simply a scientific curiosity. A mere scientific curiosity can be accepted without a battle; an economic and intellectual threat cannot.

The huge difference in reaction rates between the Fleischmann and Pons work and what Jones observed needs to be understood and appreciated. As an example, one watt of power produced by fusion, as claimed by Fleischmann and Pons, generates about 10^{12} neutrons/second according to conventional theory, which would kill anyone nearby. On the other hand, a few neutrons/second, as Jones found, represent insignificant energy. This huge difference makes neutron emission, on which Jones based his claim, important only to theoreticians. While low neutron emission was acceptable to most scientists when evaluating the Jones' results, the absence of a high neutron flux was used to reject the Fleischmann and Pons claims. This expectation would be true and offer a reason to reject the claims if the nuclear process were identical to the one operating at high energy, *i.e.* during hot fusion. However, the new phenomenon discovered by Fleischmann and Pons is not caused by hot fusion. The assumption that hot fusion is operating is the basic flaw in the argument used to reject their claims.

Recent theories have explored how neutrons might be involved without actually being emitted. This approach has raised the basic question, "Do neutrons play any role at all in generating energy or causing transmutation in the cold fusion environment?" This possibility is being hotly debated and will be discussed later. For this and other reasons, the absence of neutrons is looking less and less like a good reason to reject the claims, while, at the same time, gives a good reason to be grateful for not having to deal with this potentially deadly radiation.

Indeed, absence of conventional nuclear products of any kind, except perhaps for tritium, made the high rate claimed for heat production suspicious in many people's eyes. While this was a legitimate concern, rejection did not wait until the proper measurements were made. Only now, 18 years later, have the nuclear products been detected on a sufficient number of occasions and with sufficient care to be credible. In addition, sufficient time was not given to find plausible explanations. Instead, what was known about nuclear interaction at the time was used with great certainty to prove the impossibility of the novel effects. Later examination of conventional theory shows that many conclusions opined to be absolutely true are not true at all.

4.3 Excess Power Production

Excess power is the term used to describe power in excess of that being applied to the reaction cell. Most experimental methods apply power as electrical energy and the resulting heat is measured using a calorimeter. When more heating power is measured than is applied, this extra power is being generated in the reaction cell. The fundamental question is, "What is causing this extra power?"

When the amount of power is sufficiently large, no prosaic or trivial source is plausible. Possible sources are discussed in Chapter 5.

Table 2 lists nearly 200 published claims for anomalous (excess) power. Excess energy, which is not tabulated, is accumulated while the listed power is produced and the amount depends, in part, on how long power was produced or on how much patience the experimenter had. Many cells are turned off before energy production stops of its own accord. Examples of excess power between 5 mW and 183 watts published between 1989 and 2005 are listed, almost all of which are well in excess of the claimed uncertainty of the measurement. While many papers report success using several samples, only the most productive result is listed.

In addition to these reported successes, many failures have been published, including purported negative results obtained at Harwell, UK,[16] California Institute of Technology,[17] and MIT,[18] each of which played a major role in the early rejection. Of course, published negative reports are overshadowed by the many failures not reported. Accepting these negative studies as evidence against anomalous power being real would be like having many groups each collect random rocks from a beach, have the samples carefully tested for diamonds, and then when only a few diamonds are found, conclude that diamonds do not exist anywhere in nature because the observations could not be reproduced when other random rocks were examined. Such an approach would be considered absurd in any other field of study, yet it was applied to claims of Fleischmann and Pons. Even when anomalous energy is reported, many examples of normal behavior are also observed during the same study, thereby showing the cause is not a universal error. To make this point clearer, people have many opportunities to test their apparatus for error by running "identical" experiments, only a few of which give positive results. Error would be expected to affect all studies preformed the same way using the same apparatus. Whether this low success rate is because heat production is an imaginary event based on error, as the skeptics believe, or because special conditions required to initiate the novel process are not present, will be explored in this chapter and throughout the book.

Table 2 provides information about the general type of calorimeter, with the designations explained at the end of the table. Each type will be described in greater detail in Chapter 7. An electrolytic cell can be either open or closed. If it is open, it does not contain a catalyst, hence generated gases can leave the cell, usually through a bubbler to protect the cell from invasion of air. Closed cells contain a catalyst to combine the D_2 and O_2 gases and reform D_2O, thereby reducing potential error. Methods not using electrolysis do not require such a catalyst and are designated "NA." All methods use a material on which the nuclear reactions are proposed to occur, which is designated "Substrate." However, this is not the nuclear-active-environment (NAE); instead it is the material on or into which the NAE forms. Surrounding the substrate is an "Environment" consisting of gas, liquid or plasma. If this environment is simply heated gas, the method is designated "Ambient." If the environment is a liquid with a few volts applied, the "electrolytic" method is being used. If voltage in excess of about 100 volts is applied to either a liquid or gas, a large number of ions will form, thereby creating the "plasma" method. These methods are discussed in more detail in Chapter 6. The listed power is extracted from a range of values listed in the publication and is the maximum reported value. Values depend on the method used, size of the active region, applied temperature, accuracy of the calorimeter, and luck of the day. A question mark is used when excess power is reported, but without information needed to convert the reported value to watts from values reported as watt/cm^2, watt/cm^3 or power efficiency.

Table 2. Studies reporting anomalous power production.

Reference	Calorimeter	Closed-Open	Method	Substrate	Electrolyte	Max. Watt
Dardik et al.[19]	DW Iso.	open	electrolytic	Pd	LiOD+ D$_2$O	1.8
2004						
Strinham[20,21]	flow	NA	sonic	Pd	D$_2$O	40
Savvatimova and Gavritenkov[22]	flow	NA	plasma	Ti	D$_2$ gas	?
Mizuno et al.[23,24]	flow	open	plasma	W	K$_2$CO$_3$+ H$_2$O	?
Tian et al.[25]	Iso.	open	electrolytic	Ni	K$_2$CO$_3$+ H$_2$O	?
Szpak et al.[26]	Iso.	open	electrolytic	Pd(co-deposition)	ND$_4$Cl +D$_2$O+ PdCl$_2$	0.24
Campari et al.[27]	Iso.	NA	ambient	Ni	H$_2$ gas	25
Dash and Ambadkar[28]	Iso.	closed	electrolytic	Pd	H$_2$SO$_4$+ D$_2$O	0.93
Dardik et al.[29]	DW Iso.	open	electrolytic	Pd	LiOD+ D$_2$O	33
2003						
Wei et al.[30,31]	Iso.	open	electrolytic	Case type	?+ D$_2$O	0.45
Tsvetkov et al.[32]	Iso.	NA	fused salt	Ti	LiF+LiD+ KCl+LiCl	0.35
Swartz and Verner[33]	DW Iso.	open	electrolytic	Pd	D$_2$O pure	1.5
Storms[34]	Seebeck	closed	electrolytic	Pd,Au (laser)	LiOD+ D$_2$O	0.30
Miles[35]	Iso. dual	open	electrolytic	Pd particles	LiOD+ D$_2$O	0.25

Reference	Calorimeter	Closed-Open	Method	Substrate	Electrolyte	Max. Watt
Li et al.[36]	Iso.	NA	diffusion	Pd	D_2 gas	7
Letts and Cravens[37,38]	Iso	closed	electrolytic	Pd,Au (laser)	LiOD+ D_2O	0.7
Karabut[39,40]	flow	NA	plasma	Pd Sputtered	D_2 gas	12
De Ninno[41]	cathode temperature	open	electrolytic	Pd	LiOD+ D_2O	0.02
Dardik et al.[42]	flow	NA	Plasma, superwave	Pd	D_2 gas	2.9
Celani et al.[43]	flow, resistance	open	electrolytic	Pd	C_2H_5OD+ D_2O+ $Th(NO_3)_4$	1.4
2002						
Warner et al.[44]	Seebeck	open	electrolytic	Ti	H_2SO_4+ D_2O	0.51
Tian et al.[45]	Iso.	NA	ambient	Pd	H_2 gas	49 0.004
Tian et al.[46,47]	Seebeck	NA	diffusion	Pd-Ag	D_2 gas	8
Swartz et al.[48]	DW Iso.	open	electrolytic	Ni	D_2O+ H_2O(pure) LiOD	0.36
Sun et al.[49]	Iso.	open	electrolytic	Ti	D_2O	76.5
Storms[50]	Seebeck, flow	closed	electrolytic	Various Deposited	LiOD+ D_2O $ND_4Cl +$	0.45
Miles et al.[51]	Iso.	open	electrolytic	Pd	D_2O	0.27
Li et al.[52]	complex	NA	diffusion	Pd	D_2 gas	0.44
Kirkinskii et al.[53,54]	Iso.	NA	diffusion	Pd-black	D_2 gas	0.3
Karabut[55]	flow	NA	plasma	Pd	D_2 gas	15
Fujii et al.[56,57]	flow	open	electrolytic	Pd coated	Li_2SO_4+ H_2O	7.8
Del Giudice et al.[58,59]	Iso.	open	electrolytic	Deposited Pd wire	LiOD+ D_2O	0.02
Chicea[60]	Iso.	open	electrolytic	Ni	Li_2SO_4+ H_2O	0.3
Castano et al.[61]	Iso.	open	electrolytic	Ni.Pd thin film	Li_2SO_4+ H_2O	0.3[b]
Isobe et al.[62]	flow	closed	electrolytic	Pd	LiOD+ D_2O	2.6
Arata and Zhang[63]	flow	NA	sonic	Pd,Ti,Au	D_2O, H_2O	?
2001						
	Seebeck		current flow	U		1.3
Dufour et al.[64]	Iso., flow	NA	AC plasma	Pd	H_2	8.6
2000						
Zhang et al.[65]	Seebeck	open	electrolytic	Pd	$LiNO_3+$ D_2O	0.025

Reference	Calorimeter	Closed-Open	Method	Substrate	Electrolyte	Max. Watt
Warner and Dash[66]	Seebeck	closed	electrolytic	Ti	H_2SO_4+ D_2O	0.4
Storms[67]	flow	closed	electrolytic	Pt coated with Pd	LiOD+ D_2O	0.8
Mizuno et al.[68,69]	flow	open	plasma	W	K_2CO_3+ H_2O	40
Miles[70-72]	Iso.	open	electrolytic	Pd-B,Pd-Ce	LiOD+ D_2O	0.2
Miles[73]	DW Iso.	open	electrolytic	Pd	LiOD+ D_2O	0.09
McKubre et al.[74]	flow	NA	ambient	Pd	D_2 gas	10
				U	H_2	4.0
Dufour et al.[75]	flow	NA	AC plasma	Pd	H_2	8.6
Campari et al.[76]	Iso.	NA	ambient	Ni	H_2 gas	70
Isobe et al.[77]	?	closed	electrolytic	Pd	LiOD+ D_2O	2
Bernardini et al.[78]	Iso.	open	electrolytic	Ti	K_2CO_3+ D_2O	1
Arata and Zhang[79-82] **1999**	flow	NA	ambient	Pd	D_2 gas	12
Szpak et al.[83] **1998**	Iso.	open	electrolytic	Pd	LiOD+ D_2O	0.4
Takahashi[84,85]	Iso.	closed	electrolytic	Pd	LiOD+ D_2O	5
Stringham et al.[86]	Iso.	NA	sonic	Ti,Ag,Cu, Pd, Pd-Ru,Pd-Ni,Pd-Pt-	D_2O	17
Savvatimova and Korolev[87]	complex	NA	plasma	W	D_2 gas	?
Oya et al.[88-90]	flow	open	electrolytic	Pd, Pd-B	LiOD+ D_2O	4
Ohmori and Mizuno[91,92]	Iso.	open	plasma	W	K_2CO_3, Na_2SO_4+ H_2O	183
Mengoli et al.[93,94]	Iso.	open	electrolytic	Ni	Na_2CO_3 or K_2CO_3+ H_2O	0.8
Mengoli et al.[95]	complex	open	electrolytic	Pd	K_2CO_3+ D_2O	0.8
Lonchampt et al.[96]	complex	open	electrolytic	Pd,Pt(?)	Li_2SO_4+ D_2O	?
Lonchampt et al.[97]	flow	open	electrolytic	Ni beads	Li_2SO_4+ H_2O	0.25
Li et al.[98]	Iso.	NA	ambient	Pd	D_2 gas	25.9
Iwamura et al.[99-101]	flow	closed	electrolytic	Pd+CaO	LiOD+ D_2O	3.2

Reference	Calorimeter	Closed-Open	Method	Substrate	Electrolyte	Max. Watt
Gozzi et al.[102]	flow	open	electrolytic	Pd	LiOD+ D$_2$O	10
Focardi et al.[103]	Iso.	NA	ambient	Ni	H$_2$	38.9
Cain et al.[104]	flow	open	electrolytic	Pd	LiOH+ D$_2$O+ H$_2$O	?
Bush and Lagowski[105]	Seebeck	open	electrolytic	Pd	LiOD+ D$_2$O	0.06
Biberian et al.[106]	Iso.	open	solid	La$_{0.95}$AlO$_3$	D$_2$ gas	0.05
Arata and Zhang[107-109]	flow	NA	ambient	Pd	D$_2$ gas	24
1997						
Swartz [110]	DW Iso.	open	electrolytic	Ni	Pure H$_2$O	2
Ohmori et al. [111]	Iso.	open	electrolytic	Au	Na$_2$SO$_4$, K$_2$CO$_3$, K$_2$SO$_4$+ H$_2$O	0.937
Dufour et al. [112]	flow	NA	Plasma(AC)	Pd	H$_2$ D$_2$	9.5 13.5
Numata and Fukuhara[113]	complex	NA	ambient	PdD	D$_2$ gas	6
Mengoli et al.[94]	complex	open	electrolytic	Ni	K$_2$CO$_3$+ H$_2$O	1
Focardi et al.[114]	Iso.	NA	ambient	Ni	H$_2$ gas	20
Cammarota[115]	flow	NA	ambient	Ni	H$_2$ gas	1.2
1996						
Kopecek and Dash[116]	Iso.	closed	electrolytic	Ti	H$_2$SO$_4$+ D$_2$O	1.2
Li et al.[117]	Iso.	NA	ambient	Pd	D$_2$ gas	0.639
Yasuda et al.[118]	flow	closed	electrolyte	Pd	LiOD+ D$_2$O	5
Celani et al. [119]	Iso.	open	electrolyte	Pd	LiOD+ D$_2$O	100
Roulette et al.[120]	Iso.	open	electrolyte	Pd	LiOD+ D$_2$O	101
Preparata et al.[121]	Iso.	open	electrolyte	Pd	LiOD+ D$_2$O	30
Oyama et al.[122]	Iso.	closed	electrolyte	Pd, Pd-Ag	LiOD+ D$_2$O	0.6
Oya et al.[123]	Iso.	open	electrolyte	Pd	LiOD+ D$_2$O	2.5
Oriani[124]	Seebeck	NA	solid	SrCeYNb O$_3$	D$_2$ gas	0.7

Reference	Calorimeter	Closed-Open	Method	Substrate	Electrolyte	Max. Watt
Niedra and Myers[125]	Iso.	open	electrolyte	Ni	K_2CO_3+ H_2O	11
Mizuno et al.[126]	Iso.	NA	solid	SrCeYNb O_3	D_2 gas	1.5
Miles and Johnson[127,128]	DW Iso.	open	electrolyte	Pd	LiOD+ D_2O	0.05
Lonchampt et al.[129]	complex	open	electrolyte	Pd	LiOD+ D_2O	0.3
Kamimura et al.[130]	Iso.	closed	electrolyte	Pd	LiOD+ D_2O	0.6
Iwamura et al.[131]	flow	closed	electrolyte	Pd	LiOD+ D_2O	1
Isagawa and Kanda[132]	Iso.	open	electrolyte	Pd	LiOD+ D_2O	6.3
Dufour et al.[133]	Iso.	NA	plasma(AC)	Pd	D_2, H_2	14 10
De Marco et al.[134]	flow	?	electrolyte	Pd	LiOD+ D_2O	11
Cellucci et al.[135]	flow	open	electrolyte	Pd	LiOD+ D_2O	10
Arata and Zhang[136,137]	flow	NA	ambient	Pd	D_2 gas	20
1995						
Zhang et al.[138]	Seebeck	closed	electrolyte	Pd	NaOH+ H_2O, D_2O KOH, LiOH+ D_2O,	?
Takahashi[139]	Iso.	open	electrolyte	Charcoal	H_2O	?
Takahashi et al.[140]	Iso.	open	electrolyte	Pd	LiOD+ D_2O	3.5
Samgin et al.[141]	Iso.	NA	solid	Sr-Ce-O	D_2 gas	2.5
Ota et al.[142]	flow	closed	electrolyte	Pd, Pd-B	LiOD+ D_2O	0.35
Ogawa et al.[143]	Iso.	?	electrolyte	Pd	LiOD+ D_2O	?
Noble et al.[144]	Iso.	closed	electrolyte	Pd	H_2SO_4+ D_2O	?
Miles[145]	DW Iso.	open	electrolyte	Pd	LiOD+ D_2O	0.4
Karabut et al.[146]	flow	NA	plasma	Pd	D_2 gas	2.8
Isagawa et al.[147]	Iso.	open	electrolyte	Pd	LiOD+ D_2O	6.8
Hasegawa et al.[148]	Iso.	closed	electrolyte	Pd	LiOD+ D_2O	?
Gozzi et al.[149]	Iso.	open	electrolyte	Pd	LiOD+ D_2O	19

Reference	Calorimeter	Closed-Open	Method	Substrate	Electrolyte	Max. Watt
Dufour et al.[150]	Iso.	NA	plasma(AC)	Pd	H_2 gas	5.5
Cravens[151,152]	flow	open	electrolyte	Ni,Pd,Ni bead	Li_2SO_4+ H_2O	1.7
Celani et al.[153]	Iso.	open	electrolyte	Pd	LiOD+ D_2O	5
Biberian[154]	Iso.	NA	solid	$AlLaO_3$	D_2 gas	0.5
Bertalot et al.[155]	flow	open	electrolyte	Pd	LiOD+ D_2O	11
1994						
Storms[156,157]	Iso.	closed	electrolyte	Pd	LiOD+ D_2O	2
Notoya et al.[158]	Iso.	open	electrolyte	Ni	K_2CO_3+ H_2O	0.9
Miles and Bush[159,160]	Iso.	open	electrolyte	Pd	LiOD+ D_2O	0.06
McKubre et al.[161]	flow	closed	electrolyte	Pd	LiOD+ D_2O	1
Focardi et al.[162]	Iso.	NA	ambient	Ni	H_2 gas	50
Bush and Eagleton[163]	Iso.	closed	electrolyte	Ni	Rb_2CO_3+ H_2O	?
Bockris et al.[164,165]	Iso.	open	electrolyte	Pd	LiOD+ D_2O	18
Arata and Zhang[166]	Iso.	NA	ambient	Pd	D_2 gas	28
1993						
Zhang et al.[167]	Iso.	open	electrolyte	Ti	NaOD + D_2O	?
Storms[168,169]	Iso.	closed	electrolyte	Pd	LiOD+ D_2O	7.5
Ramamurthy et al.[170]	Iso	open	electrolyte	Ni	Li_2CO_3, K_2CO_3+ H_2O, D_2O	0.8
Pons and Fleischmann[171,172]	adiabatic	open	electrolyte	Pd	LiOD+ D_2O	0.8
Ota et al.[173]	flow	closed	electrolyte	90Pd+10Ag	LiOD+ D_2O	1.3
Okamoto et al.[174,175]	flow	open	electrolyte	Pd	LiOD+ D_2O	6
Ohmori and Enyo[176,177]	Iso.	open	electrolyte	Sn	K_2CO_3+ H_2O	0.9
Mizuno et al.[178]	Iso.	NA	solid	Sr(CeYNb)O_3	D_2 gas	50
Miles et al.[160,179,180]	DW Iso.	open	electrolyte	Pd	LiOD+ D_2O	0.06

Reference	Calorimeter	Closed-Open	Method	Substrate	Electrolyte	Max. Watt
Hugo[181]	flow	closed	electrolyte	Pd-Ag	LiOD+ D_2O	2.6
Hasegawa et al.[182]	Iso.	closed	electrolyte	Pd	LiOD+ D_2O	0.5
Gozzi et al.[183]	Iso.	open	electrolyte	Pd	LiOD+ D_2O	19
Fleischmann and Pons[184]	Iso.	open	electrolyte	Pd	Li_2SO_4+ D_2O	144
Dufour et al.[185-187]	Iso.	NA	plasma(AC)	Pd	H_2, D_2	2.07
Criddle[188]	Iso.	open	electrolyte	Ni	K_2CO_3+ H_2O	?
Celani et al.[189,190]	flow	open	electrolyte	Pd	LiOD+ D_2O	?
					K_2CO_3,	1.1
					RbOH,	0.7
					Rb_2CO_3,	
Bush and Eagleton[191]	Iso.	closed	electrolyte	Ni	Cs_2CO_3 +H_2O	
Bertalot et al.[192-194]	flow	open	electrolyte	Pd	LiOD+ D_2O	3
Bazhutov et al.[195]	Iso.	open	electrolyte	Ni	Cs_2CO_3+ H_2O	?
Aoki et al.[196,197]	Iso.	open	electrolyte	Pd	LiOD+ D_2O	27
1992						
Yuan et al.[198]	Iso.	NA	electrolyte	Pd	LiCl+ KCl+ LiD fused	1080
Wan et al.[199]	Iso.	open	electrolyte	Pd	LiOD+ D_2O	?
Takahashi et al.[200,201]	Iso.	open	electrolyte	Pd	LiOD+ D_2O	15
Srinivasan et al.[202]	Iso.	open	electrolyte	Ni	K_2CO_3+ H_2O	1
Ray et al.[203]	Iso.	open	electrolyte	Pd	LiOD+ D_2O	?
Oyama et al.[204]	Iso.	closed	electrolyte	Pd	LiOD+ D_2O	0.008
Ota et al.[205]	flow	closed	electrolyte	Pd-Ag	LiOD+ D_2O	1.3
Notoya and Enyo[206]	Iso.	open	electrolyte	Ni	K_2CO_3+ H_2O	8
Noninski[207]	Iso., dual cells	open	electrolyte	Ni	K_2CO_3+ H_2O	?
Mizuno et al.[208]	Iso.	closed	electrolyte	Pd	LiOD+ D_2O	?
Miyamaru and Takahashi[209]	Iso.	open	electrolyte	Pd	LiOD+ D_2O	2

Reference	Calorimeter	Closed-Open	Method	Substrate	Electrolyte	Max. Watt
McKubre et al.[210]	flow	closed	electrolyte	Pd	LiOD+ D_2O	1.2
Kunimatsu et al.[211,212]	Iso.	closed	electrolyte	Pd	LiOD+ D_2O	?
Kobayashi et al.[213]	Iso.	closed	electrolyte	Pd	LiOD+ D_2O	?
Karabut et al.[214,215]	complex	NA	plasma	Pd	D_2 gas	30
Isagawa et al.[216]	Iso.	open	electrolyte	Pd	LiOD+ D_2O	30
Gozzi et al.[217]	Iso.	open	electrolyte	Pd	LiOD+ D_2O	9
Celani et al.[218]	flow	open	electrolyte	Pd	LiOD+ D_2O	4
Ohmori and Enyo [177]	Iso.	open	electrolyte	Sn	K_2SO_4+ H_2O	0.907
Bush[219]	Iso.	open	electrolyte	Ni	K_2CO_3+ H_2O	4
Bush and Eagleton[220]	Iso.	open	electrolyte	Pd-Ag	LiOD+ D_2O	3
1991						
Yun et al.[221]	Seebeck	open	electrolyte	Pd	LiOD+ D_2O	0.26
Will et al.[222]	Iso.	closed	electrolyte	Pd	D_2SO_4+ D_2O	0.010[a]
Bush et al.[223]	DW Iso.	open	electrolyte	Pd	LiOD+\ D_2O	0.52
Szpak et al.[224]	Iso.	open	electrolyte	Pd	LiOD+ D_2O	?
Norinski and Norinski[225]	Iso.	open	electrolyte	Pd	K_2SO_4+ D_2O	2.6
Eagleton and Bush [226]	Flow+ Iso.	closed	electrolyte	Pd	LiOD+ D_2O	6.8
Mills and Kneizys[227]	Iso.	open	electrolyte	Ni	K_2CO_3+ H_2O	?
McKubre et al.[228]	flow	closed	electrolyte	Pd	LiOD+ D_2O	0.5
1990						
Liaw et al.[229,230]	Iso.	NA	fused salt	Pd	LiCl+ KCl+ LiD	25.4
Zhang et al.[231]	Seebeck	open	electrolyte	Pd	LiOD+ D_2O	0.017
Scott et al.[232-234]	flow	open-closed	electrolyte	Pd	LiOD+ D_2O	3
Schreiber et al.[235,236]	DW Iso.	closed	electrolyte	Pd	LiOD+ D_2O	?

Reference	Calorimeter	Closed-Open	Method	Substrate	Electrolyte	Max. Watt
Pons and Fleischmann [10,237,238]	Iso.	open	electrolyte	Pd	Li_2SO_4+ D_2O	2.8
Yang et al. [239]	Iso.	open	electrolyte	Pd	LiOD+ D_2O	9
Guruswamy and Wadsworth [240]	Iso.	open	electrolyte	Pd	LiOD + D_2O	7.6
Lewis and Sköld [241]	flow	open	electrolyte	Pd	LiOD + D_2O	1
Oriani et al. [242]	Seebeck	open	electrolyte	Pd	Li_2SO_4+ D_2O	3.6
Miles et al. [243,244]	DW Iso.	open	electrolyte	Pd	LiOD+ D_2O	0.14
Hutchinson et al. [245-247]	Iso.	open	electrolyte	Pd	LiOD+ D_2O	3
McKubre et al. [248]	flow	closed	electrolyte	Pd	LiOD+ D_2O	1.25
Appleby et al. [249,250]	Seebeck.	open	electrolyte	Pd	LiOD+ D_2O	0.046
Kainthla et al. [251]	Iso.	open	electrolyte	Pd	LiOD+ D_2O	1.08
Beizner et al. [252,253]	DW Iso.	open	electrolyte	Pd	LiOD+ D_2O	1
1989						
Kainthia et al. [251]	Iso.	open	electrolyte	Pd	LiOD+ D_2O	1.08 Pd-1.54
Santhanam et al. [254]	Iso.	open	electrolyte	Ti, Pd	NaCl+ D_2O	Ti-0.31
Fleischmann and Pons [255]	Iso.	open	electrolyte	Pd	LiOD+ D_2O	26.8

a- Corrected for apparent typographical error. Value given as 10 W in the paper.

b- Neutral potential correction was made incorrectly. 2.01 V is used rather than 1.48 V.

Iso. = Isoperibolic Calorimeter where temperature is measured between the source of heat and a fixed reference temperature.

DW Iso. = Double wall isoperibolic calorimetry where temperature is measured across a barrier separated from the source of heat.

Flow = Fluid carries heat away from a source and the amount of power is based on temperature change of the fluid.

Seebeck = The source of heat is surrounded by a wall that is able to convert temperature across the wall to a voltage that is proportional to the rate of heat flow through the wall.

Complex = A method using a combination of temperature measurements.

Closed-Open = An open cell does not contain a catalyst to recombine the gases being generated by electrolytic action. A closed cell contains such a recombiner so that the cell can be closed to the atmosphere and the cell can be considered to be a chemically closed system.

4.3.1 Heavy Hydrogen (Deuterium)

This large collection of experience is impractical to discuss in detail. However, a few studies are particularly interesting in showing unique or unexpected behavior. More detail will be provided in the following sections.

Table 3 lists the original data reported by Fleischmann and Pons,[238,255] which triggered all the work reported here. Surprisingly, the worldwide response, both positive and negative, was far in excess of what the few published numbers would justify in any other field of science. Normally, a few measurements, as were reported initially, would be ignored in other fields until more work had been done. As Fleischmann and Pons readily acknowledged, this was only preliminary work largely paid for out of their own pockets, which they had not intended to make public. Unfortunately, events spiraled out of their control and started the process prematurely. Nevertheless, the excessive response encouraged intense studies in many laboratories and a willingness of a few scientists to acknowledge anomalous results. Without this over-reaction, such unexpected behavior would have been completely ignored as error. Instead, people were encouraged to report behavior thought to be impossible—behavior that now has been witnessed hundreds of times in dozens of laboratories.

Table 3. Initial data reported by Fleischmann and Pons.

Cathode	Current, mA/cm^2	Excess Power, W
0.1 cm dia.x10 cm rod	8	0.0075
	64	0.079
	512	0.654[a]
0.2 cm dia.x10 cm rod	8	0.036
	64	0.493
	512	3.02[a]
0.4 cm dia.x10 cm rod	8	0.153
	64	1.751
	512	26.8[a]
0.2x8x8 cm sheet	8	0
	1.2	0.027
	1.6	0.079
1x1x1 cm cube		exploded

a = rescaled from 1.25 cm length
error: <1%, ~±0.002 W

Their excess power was obtained using a Pyrex electrolytic cell containing a palladium cathode, a platinum anode, and 0.1M LiOD + D_2O electrolyte. Excess power was found to increase when applied current was increased and when the size of the sample was increased, effects that have now been seen many times by other people. They attributed the size effect to the volume, which instead is now known to be related to surface area. Nine more successful results were published in 1990,[238] based on electrolytes containing 0.5

M Li_2SO_4 or 0.45 M Li_2SO_4 + 0.1 M LiOD, along with 16 additional studies using 0.1 M LiOD. Fourteen "blanks"[a] were also described in this later paper. As Pons points out,[237] much of this data was submitted as a letter to *Nature* to counter complaints of their failure to use "blanks", but the letter was refused by the Editor. This rejection by a respected journal was an early and often repeated example of bias toward information supporting the claims.

One of the first replications of heat production was provided by Prof. Appleby *et al.*,[249,250] working at Texas A & M. This active research program was started very soon after the Fleischmann and Pons announcement[b] using various sizes of palladium wire, all of which produced heat when electrolyzed for many hours in 0.1 M LiOD, but not in LiOH or NaOD. An example of one result is shown in Figure 30. Maximum excess power was 14% of applied power and well within the accuracy of the calorimeter (<1 microwatt, ±0.7% of applied power). Because this result was exactly as claimed by Fleischmann and Pons, everyone was encouraged to expect easy replication, which did not happen. What made this work different from the many failed efforts? The unique feature of this study was the use of a stainless steel container to hold the electrolyte. Such materials were avoided by everyone else because the cathode could be contaminated by elements such as Fe, Ni and Cr from the metal alloy. Bockris and students,[251] also working at Texas A & M, produced excess power on 3 out of 10 occasions when they electrolyzed cells using Ni wire attached to the electrodes as well as a Ni anode during initial loading—conditions expected to transfer Ni to the cathode surface. Some excess power was also observed by Lewis and Sköld[241] (Uppsala University) when a stainless steel cooling coil was present in their cell. In contrast, Bertalot *et al.*[194] obtained no excess heat when nickel was used as the anode and Jow *et al.*,[256] working at the U.S. Army Electronics Technology Laboratory and Devices Laboratory, failed to replicate the expected excess power even when a stainless steel cell similar to Appleby's cell was used. Apparently, having nickel or stainless steel present is not sufficient.

Similar studies made at about the same time using Pyrex cells without nickel failed to produce detectable excess power.[16-18,257-269] Many similar failures to replicate the Fleischmann-Pons success have been reported over the years even though the gross conditions have been closely duplicated. On the other

---

[a] A "blank" is a measurement expected to produce no anomalous effects and is made in order to reveal errors that might be missinterpreted as anomalous behavior.

[b] Additional detail about the cell and Pd samples can be found in the paper by Bertalot et al.[194] (See: Bertalot, L., Bettinali, L., De Marco, F., Violante, V., De Logu, P., Dikonimos, T., and La Barbera, A., Analysis of tritium and heat excess in electrochemical cells with Pd cathodes, in *Second Annual Conference on Cold Fusion, "The Science of Cold Fusion"*, Bressani, T., Del Giudice, E., and Preparata, G. Societa Italiana di Fisica, Bologna, Italy, Como, Italy, 1991, pp. 3.)

hand, many attempts have been successful. Why are some studies successful and others failures, even though the conditions appear identical? Does material deposited on the cathode play a role? At the time most of these studies were made, these questions were not recognized as being important, so answers were not sought. Only later, after much effort was expended, was the presence of a very small amount of impurity on the cathode surface found to be very important to success. This experience shows the importance of having a critical understanding of a process before claims are evaluated or rejected.

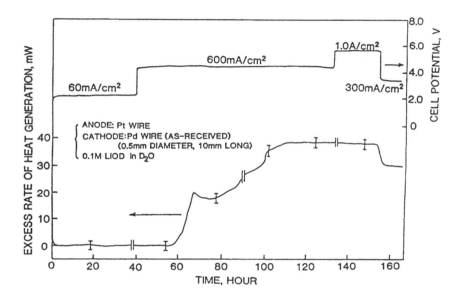

Figure 30. Excess power produced by 0.5 mm diameter x 10 mm long Pd wire electrolyzed in 0.1 M LiOD using a Pt anode. The area of the Pd is 0.157 cm^2. (Appleby et al.)

In an attempt to improve success, several variations have been explored. Oriani et al.,[242] at the University of Minnesota, achieved success by attaching a piece of palladium to the anode and Bertalot et al.[193] (ENEA, Frascati, Italy) used a palladium anode. I had occasional success using a palladium anode as well.[270] In each case, a thin layer of palladium is deposited on the palladium cathode, but this time without nickel being present. Szpak and co-workers,[51,83,224,271,272] at the Naval Ocean Systems Center (CA), electroplated palladium using $PdCl_2$ dissolved in the electrolyte, which applied a thick coating of palladium to a copper or nickel cathode within the electrolytic cell. Occasionally, this method has made excess energy very quickly while generating X-rays. Others researchers[273-275] also have used this co-deposition method with success.

Starting in 1990, other methods besides aqueous electrolysis were explored. Liaw et al.,[229,230] working at the University of Hawaii, demonstrated excess power production at much higher temperatures than can be obtained using a water-based electrolyte. They electrolyzed a cell containing a fused salt of LiCl+KCl held near 370° C, in which was dissolved a little LiD. The anode, to which deuterium goes in such a cell, was palladium and the cathode was aluminum. Figure 31 compares the total amount of applied power to the total amount of power measured by the calorimeter. Excess power, which reached 25.4 W, is the difference between the two plotted curves. Excess production stopped when the small amount of dissolved deuterium was exhausted from the electrolyte. A small but significant amount of helium was retained by the palladium.[276] This study was replicated by Yuan et al.[198] at the National Tsing Hua University (Taiwan) while measuring low level neutron emission. Although the excess power was real in this study, its value is uncertain.

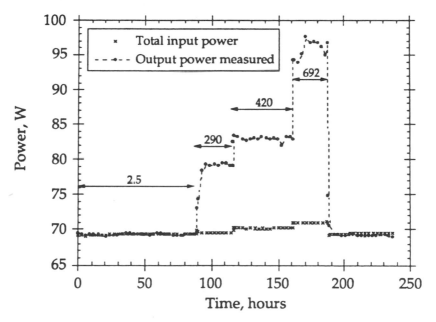

Figure 31. Power from a fused salt cell containing LiCl+KCl+LiD. The numbers on the figure are applied current density, mA/cm^2. (Liaw et al.)

As expected, the D/Pd ratio is low at such high temperatures, being 0.03 at 430° C and 300 mA/cm^2.[277] In addition, palladium quickly forms an alloy with lithium and potassium under these conditions, which would significantly alter the physical and chemical environment of the anode surface with unknown consequences. As a result, this success is difficult to reconcile with current models of behavior.

A program based on gas discharge was started at the Scientific Industrial Association LUTCH (Russia) and has provided a steady stream of information. When sufficient voltage (above 100 V) is applied across a gap between two electrodes bathed in low-pressure deuterium gas, plasma forms. Deuterium ions, having energy provided by the applied voltage, bombard the cathode. Karabut et al., in a series of papers,[39,40,55,146,214,215,278-296] describe how they were able to produce excess power greater than 30 watts, along with emission of charged particles and X-rays. Transmutation products were also observed and studied, as discussed later (Section 4.5).

A plasma can also be created in a liquid when sufficient voltage is applied—a technique tested by Nakamura et al.[297] and developed by Mizuno and Ohmori (Hokkaido University, Japan) in a series of papers.[23,24,68,69,91,92,298,299] Cirillo and Iorio[300] (Laboratorio M. Ruta, Italy) used the same method to duplicate many of the results reported by Mizuno and Ohmori. The method is described in detail in Section 6.2.5. Excess heat, transmutation products, and radiation have been detected, although again the behavior is not consistent with conventional expectations.

In 1991, a very successful program was started at SRI under the direction of Michael McKubre, made possible by initial funding from the Electric Power Research Institute (EPRI). A first-class laboratory was created for the sole purpose of answering the growing list of questions. Even though the phenomenon was replicated many times and several important variables were identified, funding was eventually terminated. The Institute for Applied Energy (Japan) and DARPA (USA) kept the work going at a much lower level. McKubre and co-workers[74,161,210,228,301-309] removed any doubt about the production of excess power from an electrolytic cell by using an advanced flow calorimeter. These measurements were unique because they were made at constant electrolyte temperature, which can not be achieved in other calorimeters, because application of increased current normally causes the electrolyte temperature to increase. As a result, temperature and applied current change at the same time, with unknown interaction. They were able to show that anomalous heat production is related to the average composition of the palladium cathode, that the characteristics of the palladium are important, and that a relationship exists between energy and helium production. The helium-heat studies are discussed later in Section 4.4.2.

The average deuterium content of the cathode is based on changes in the resistance of the metal. Figure 32 shows the defining relationship between resistance ratio[c], which is measured, and the composition, which is calculated from this curve. For the purpose of showing the effect of composition on heat

---

[c] The resistance ratio is resistance of the sample containing deuterium divided by the resistance before it has been reacted with deuterium.

production, the curve in Figure 32 is divided into three regions. Nineteen samples in the region from D/Pd = 0.7 to 0.9 never made any excess power. In the region between 0.90 and 0.95, nine samples made no power and six produced excess power. When the average D/Pd ratio was above 0.95, all fifteen samples reaching this composition made excess power. In other words, excess power production requires a large average deuterium content in bulk palladium, a condition difficult to achieve.

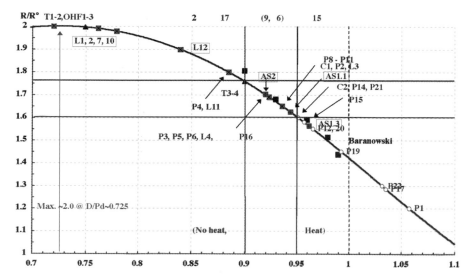

Figure 32. Relation between D/Pd and occasions when excess heat was produced. (McKubre *et al.*)

The relationship between excess power and average composition of a single active sample is shown in Figure 33. Similar behavior has been reported by other researchers.[88,130,208] However, not all such measurements produce the same relationship between average D/Pd and excess power because sample size and shape affect the relationship between the average and surface compositions[d]. This behavior provides one more reason why so many attempts to make excess energy failed. In this case, failure resulted because the required average D/Pd

---

[d] When discussing the average composition, it is important to realize that this easily measured composition (see Appendix F) is not the deuterium content of the surface where the NAE is presumed to be located. The surface composition will be influenced by the average composition, but the exact relationship between the two is unknown. In part, this relationship will be influenced by the diffusion constant of the cathode material, the applied current, and the number of surface penetrating cracks that are present.

ratio for the particular conditions was not achieved. More is said about this requirement in later Chapters.

While these studies were under way in the US, work in Japan and in Italy was progressing nicely. Takahashi and co-workers at Osaka University were looking for the relationship between excess heat and neutron emission using active metal made by Tanaka Kikinzoku Kogyo. The procedure adopted to initiate neutron production, using low applied electrolytic current followed by high current, proved successful in also generating excess heat.[200,201,209] This method was adopted in Italy by Celani (INFN-LNF, Frascati) with similar success. They carried the approach one step further by applying high pulsed current to thin Pd wires to which various surface "poisons" had been applied. An increase in the D/Pd ratio was achieved.[153,189,190,218]

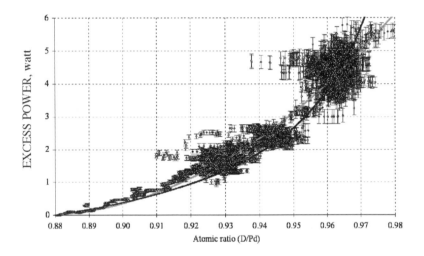

Figure 33. Relationship between the average D/Pd ratio of a palladium cathode and excess power.[210]

Excess energy can be generated without even using electrolysis or plasma. Simply exposing finely divided palladium powder, and perhaps other metals as well, to deuterium gas has been found to work. After exploring various ways to load palladium, Arata[e] and Zhang,[310-313] working at Osaka University, developed a clever way to load palladium powder with pure deuterium gas under pressure, as described in a series of papers.[63,79-82,107-109,136,137,166,314-325] A tube of palladium metal is filled with palladium-black[f] and sealed. This capsule is

---

[e] Dr. Yoshiaki Arata was awarded the highly prestigious Emperor Prize for "The Order of the Sacred Treasure for Science and Technology" in Japan.
[f] Palladium-black is very finely divided palladium metal. The material used by Arata and Zhang had an initial particle size near 5 nanometers.

electrolyzed as the cathode in a normal electrolytic cell containing $LiOD+D_2O$ in order to generate deuterium that diffuses through the wall and builds up a pressure of very pure D_2 gas on the inside. Presumably, high-pressure D_2 gas obtained from a conventional metal tank could be used to achieve the same result. A drawing of the flow calorimeter and capsule cells can be seen in Figure 34. Certain batches of palladium-black were discovered to generate excess energy, helium, and occasionally tritium once the pressure had reached a critical value. Production of energy continues as long as pressure is maintained. In fact, an active capsule showed energy production each time it was loaded with D_2 gas, even after remaining unloaded for years. An example of energy production while the capsule is being electrolyzed is shown in Figure 35. Each prepared sample produces excess power with a different pattern over time, but all active material is found to produce significant excess energy as electrolysis continues. For example, one 5-gram sample of palladium-black generated 50 MJ in 800 hours. A method has been found to control excess power production so that a fixed and predictable amount can be obtained.[79] Replication of the method has been accomplished at SRI[74,307,326] using a flow calorimeter of a design different from that used by Arata. Excess power production by two electrolytic cells, one containing D_2O and another one containing H_2O, are compared in Figure 36. Such a simple method cries out to be developed into a practical device as Arata has proposed to do.[327]

Energetics Technologies in Israel is actively exploring various methods using a novel approach. Dardik and co-workers[19,29,42,328] apply power as a complex wave, called a super-wave. A graphic example of excess power production by the electrolytic method can be seen in Figure 37 where the total power, based on the use of an isoperibolic calorimeter, is compared to the amount of applied power. The amount of applied power is about 0.72 W, while the amount of generated power reaches 33 watts while generating 1.1 MJ of excess energy, which is significantly greater than the expected error of ~2%.

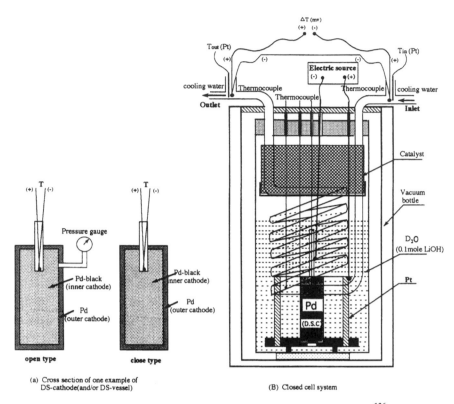

(a) Cross section of one example of
    DS-cathode(and/or DS-vessel)

(B) Closed cell system

Figure 34. Drawing of the flow calorimeter and capsule cell. [136]

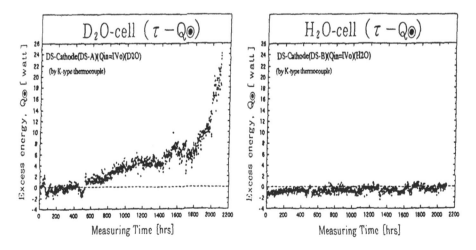

Figure 35. Palladium capsule containing palladium-black showing excess power production when deuterium is present but none when a capsule contains hydrogen.[108]

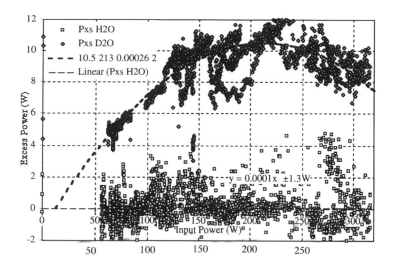

Figure 36. Replication of heat production at SRI using an Arata cell. (McKubre *et al.*)

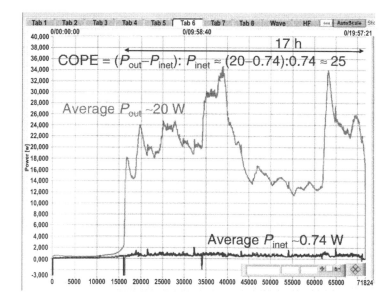

Figure 37. Total power compared to applied power for an electrolytic cell as a function of time using a super-wave superimposed on the electrolytic current. (Dardik *et al.*)

4.3.2 Light Hydrogen (Protium)

Initially, cells containing H_2O instead of D_2O were assumed to be dead, hence suitable as a "blank". Indeed, when H_2O is added to the D_2O in working cells, excess power stops. You can imagine the surprise and shock when Mills and Kneizya[227,329] reported making excess power using a nickel cathode with an electrolyte of K_2CO_3 and H_2O. Efforts to replicate this claim were quickly undertaken. Bush and Eagleton[220] (California State Polytechnic University) reported excess power using LiOH in the electrolyte and a Ni cathode. Later efforts by these workers[191] produced excess power using either RbOH, Cs_2CO_3 or K_2CO_3 in the electrolyte. In addition, the combination Ni/K_2CO_3 generated enough calcium in the electrolyte to be roughly consistent with the amount of measured excess energy.[219] A small amount of copper seemed to improve excess heat production. The effect of applied current on excess power production is shown in Figure 38 for two cells containing the Ni/$K_2CO_3+H_2O$ combination.[191] Note that a critical current, as observed when the Pd/D_2O combination is used, is not evident. Unlike the behavior when D_2O is used, some studies have reported a drop off in excess power as current is increased above a critical value. Later work by Bush[163] using Ni/Rb_2CO_3 produced a radioactive product with a half-life of about 3.8 days. In contrast to the predictions of Mills, use of Na_2CO_3 in the electrolyte does appear to produce excess energy.[93] According to Swartz et al.,[48] even pure H_2O used as the electrolyte produces excess power and the amount is increased when 3.7 % D_2O is added.

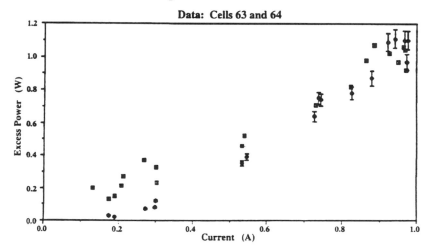

Figure 38. Effect of applied current on excess power from a Ni/K_2CO_3 + H_2O cell. (Bush and Eagleton)

Notoya and Enyo[206] (Hokkaido University, Japan) also detected additional Ca in the electrolyte when Ni/K_2CO_3 was used and a radioactive product was obtained when Ni/Na_2CO_3 was used instead.[330] Use of other cathode materials besides Ni and electrolytes besides K_2CO_3, were explored by Ohmori and Enyo[177] (Hokkaido University, Japan). No excess was detected using the Ni/Na_2CO_3, Ni/Na_2SO_4 or Ni/Li_2SO_4 combinations, but a large excess was obtained using Sn/K_2SO_4. A large program in India at the Bhabha Atomic Research Centre, under the direction of Srinivasan,[202] observed excess energy and tritium production when Ni/K_2CO_3 was used. Excess energy increased when D_2O was added to the H_2O. Swartz[48] and Bush[219]also noted a positive effect of adding D_2O. Through the years, many studies supporting energy production have been published.[56,93,94,97,110,111,138,139,151,152,158,170,176,191,195,330-336] However, Shkedi et al.[337,338] point out that some reported heat might result from uncertain corrections for Faraday efficiency in some studies. See Appendix A for more information about this potential error.

One of the very few patents granted in the US was given to James Patterson[339,340] for producing energy using beads covered with Ni and Pd during electrolysis in 0.1 M Li_2SO_4 + H_2O. This claim was even demonstrated in public.[341] Most attempts at replication have failed.

Light hydrogen has been found to produce extra energy when specially treated metal is simply exposed to the gas—in this case the metal is nickel while exposed to hot H_2. This technique has been patented in Italy and is being explored by Focardi et al.[27,76,103,114,162,342,343] Their work gives evidence for nuclear reactions being the source of the anomalous energy. However, success depends on using specially treated nickel, because this element does not readily react with hydrogen. Even then, the surface hydride does not penetrate very far into the bulk material. Mengoli et al.[93] have studied the hydriding process and recommend activation by first cleaning with HCl and then electrolyzing the metal as the anode in a $K_2CO_3+H_2O$ electrolyte until a black nickel oxide forms. This oxide is subsequently reduced and converted to the hydride when the nickel is exposed to hot hydrogen gas or used as a cathode. A more pronounced reaction can be obtained by using sintered nickel. Because the hydriding process is mainly diffusion limited, the greater the number of grain boundaries in sintered material the faster the penetration rate. Also, repeated loading and deloading generates cracks that give access to an increased volume of the metal. More information can be obtained by consulting Baranowskyi.[344]

4.3.3 General Behavior

Success has been mixed, and many experiments fail. Some failures occur because the conditions are frequently deficient in ways that are important to success. For example, the essential role of small amounts of material on or within the cathode surface is still not fully understood or controlled. Consequently, conditions are not truly duplicated. True replication will not be achieved until the essential conditions are replicated on the surface in each experiment. Because this variable is uncontrolled, everyone finds some batches of palladium work well and others are completely inert. An example of this experience is provided by Miles[145] in a long running study done at the Naval Air Warfare Center Weapons Division, China Lake (CA) and is summarized in

Table 4. The amount of excess power is small because the samples were small. Nevertheless, some batches were clearly more successful than others.

Table 4. Success ratio reported by Miles.

Source of Cathode Material	Maximum Excess Power, W	Success Ratio
Johnson-Matthey	0.4	9/14
Fleischmann-Pons	0.06	2/2
Johnson-Matthey	0.04	1/1
Tanaka Kikinzoku Co. (Japan)	0.06	1/3
Johnson-Matthey	0	0/1
IMRA (Japan) Pd-Ag alloy	0	0/1
Naval Research Laboratory	0	0/4
John Dash	0	0/2
Pd/Cu	0	0/2
Wesgo	0	0/6
Co-deposition	0.15	2/34

As the experimental evidence shows, heat production has been achieved many times using heavy-water (D_2O) and less often using light-water (H_2O). Other methods, besides electrolysis, have been found to work. Success depends on the nature of the cathode material, not on the method. Palladium normally is used with deuterium and nickel is active with hydrogen. The reverse combinations have too few examples to eliminate them as being unworkable, especially because the deposited surface layer seems to be more important than the underlying material. Within this large collection of data, some studies are totally wrong, some might be good but are too poorly described to be analyzed and believed, and some reported anomalous power might have resulted from processes having nothing to do with cold fusion. As an additional issue, it is not important whether the reported values might be uncertain by a few percent, which is to be expected. The important issue is whether all results are in error by 100%. Potential errors are explored in Chapter 7. Meanwhile, general patterns of behavior associated with heat production are examined here in more detail.

The range of 156 published studies can be seen in Figure 39, with 111 resulting from the electrolytic method. All studies report the maximum error in their published values as being well below 2 watts. If the results were solely caused by random error, all values would lie within the first cell of the histogram. More detail is provided in Figure 40 where a comparison is made between all data and that obtained using the electrolytic method at the low end of the range. Clearly, a large number of values fall well outside of the stated error range of most calorimeters. This wide range of values is to be expected because the amount of excess power is influenced by conditions within the cell. For example, the size and treatment of the substrate, as well as temperature and applied energy are important variables. The amount of active material and the concentration of reactants also play critical roles. These variables do not have the same values in all studies. Nevertheless, significant amounts of heating power have been reported by many researchers, well in excess of the error expected in the measurements. This is not a small effect near the limit of detection.

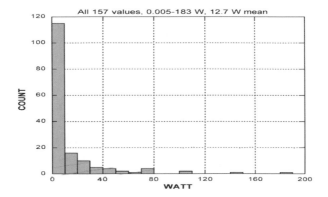

Figure 39. Histogram of excess power measurements.

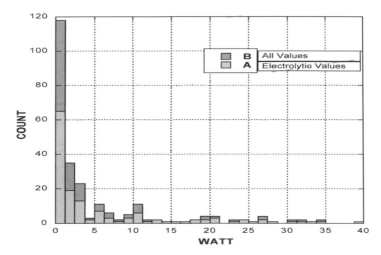

Figure 40. Comparison between all data and those obtained using the electrolytic method.

Evidence of how the current applied to the cathode affects the behavior of anomalous heat production can be seen when excess power density (W/cm^2) is plotted as a function of current density (mA/cm^2) in Figure 41. Each observation is characterized by very little or no excess power produced at low applied current, followed by a rapid increase as applied current is increased. Although a maximum shown by the solid line appears to exist for conditions used during these studies, recent work has produced more excess heat for the same applied current than the maximum would indicate. In other words, anomalous power is getting larger rather than going away as people improve the methods, quite the opposite of the behavior expected if the observations were caused by error. This typical behavior is shown more clearly in Figure 42 where a linear comparison is provided. This universal relationship between applied current and excess power, based on many studies using different calorimeters, provides evidence for operation of a common process, in spite of a wide range of values being reported. In other words, the whim of nature produces a wide variation in values, but a similar pattern of behavior occurs because the behavior is caused by the same mechanism.

The especially large values reported by Liaw et al.[229] were obtained using fused salt electrolysis at 450°C. Other studies are based on $LiOD+D_2O$ electrolysis at "normal" temperatures. In general, the larger the temperature, the more excess energy is reported, even when aqueous electrolysis is used. Other universal behaviors are also seen. For example, when the average D/Pd ratio of the palladium cathode is measured and compared to generated anomalous power, the relationship obtained by McKubre[210] and his group at SRI International is obtained, as shown clearly in Figure 32. Everyone who has made suitable measurements has seen this general behavior. Of course, the amount of heat generated will also depend on how much nuclear-active environment (NAE) is present, a highly variable condition. In addition, forcing palladium to take up a large amount of deuterium is not easy. In other words, the amount of power depends on the amount of NAE and the amount of deuterium or hydrogen present, quantities that are completely independent of each other and frequently not under the control of the researcher. Other variables are discussed in later chapters.

It is important to realize that different combinations of calorimeters, methods and environments are used and each variation produces anomalous energy on occasion. In addition, both deuterium and hydrogen are found to be active, although not under the same conditions. Even though palladium is used in most studies, it is not the only metal on which the NAE can form. Consequently, observations are consistent with how nature is expected to behave, i.e. the same behavior is seen no matter who looks, which tool is used, and how the behavior is initiated. Once again, if error is the explanation, we have to assume that many unknown and large errors are operating under a variety of conditions.

If nuclear reactions do occur, nuclear products should be present and detectable. Many kinds of nuclear products have been detected as described next.

4.4 Helium and Tritium Production

While heat production was being explored and debated, the necessary nuclear products were actively sought. Unfortunately, the information demanded by skeptics came too late to stall the rush to judgment. Now 18 years late, many of the expected products have been seen, although not all nor in the expected amounts.

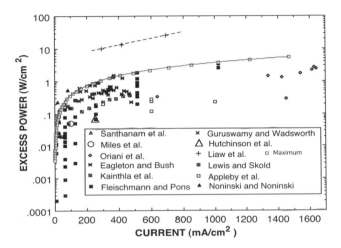

Figure 41. A sample of heat measurements compared as a function of current density and log excess-power density.[345] Cited references are listed in Table 2.

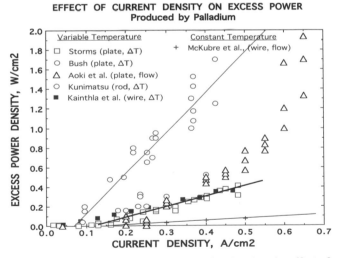

Figure 42. Comparison between several studies showing the effect of current density on power density. Cited references are listed in Table 2.

When conventional fusion occurs, the reaction can take five paths depending whether tritium (t) or normal hydrogen (p) are present, both of which are common impurities in deuterium (d). In addition, deuterium is a common impurity (~6000 ppm) in ordinary hydrogen and must be considered as a possible reactant when H_2 and H_2O are used. The possible reactions result in distinctive and easily detected products, as listed in Table 5. Each of these reactions produces significant amounts of energy, noted as million electron volts (MeV)[g] per event for each of the energetic products.

Table 5. Reactions involving deuterium in a fusion reaction.

$d + d > {}^3He(0.82 \text{ MeV}) + n(2.45 \text{ MeV})$
$d + d > p(3.02 \text{ MeV}) + t(1.01 \text{ MeV})$
$d + d > {}^4He + gamma (23.5 \text{ MeV})$
$d + t > n(14.01 \text{MeV}) + {}^4He(3.5 \text{ MeV})$
$d + p > {}^3He + gamma (5.5 \text{ MeV}$

As can be readily appreciated, this energy source makes burning coal and oil look trivial. Even fission energy obtained from the same weight of uranium is small by comparison. This difference becomes especially apparent when the energy used in the extraction and concentration of the active uranium isotope (^{235}U) is taken into account. In addition, the generated radioactive material resulting from fission must be stored for centuries. Cold fusion does not suffer from any of these serious problems.

Of the nuclear products, neutron emission is the easiest to detect and was expected to occur in large amount. However, neutron emission is seldom found even though over 500 papers have described efforts to detect them. Even when neutrons are occasionally detected, the emission rate is too small and too infrequent to give insight into the heat-producing reaction. However, it is important to point out that any neutron emission supports claims for an unexpected nuclear reaction whether the rate is trivial or not. While this is a factor worth considering, the efforts to find neutron emission are not discussed in this book.

---

[g] 1 MeV = 2.3×10^{10} cal/mol = 9.6×10^{10} J/mole = 2.7×10^4 kW-hr/mole. As an example of the huge energy available, consider the following personal example. One MeV produces 2.3×10^7 kcal when 6.02×10^{23} (1 mole) events occur. Consequently, when the deuterium in 20 g (18 ml or 0.61 oz) of D_2O is caused to fuse to make ^4He, 5.4×10^8 kcal or 6.3×10^5 kw-h would result from a 100% efficient process. My house uses about 10^4 kw-hr in a year, so this amount of heavy-water would supply my energy needs for about 63 years. With a little care, a fluid ounce of D_2O could last a person for their entire lifetime. Even the D_2O contained as an impurity in the same amount of ordinary tap water would supply my home with electric power for about 4.5 months.

Tritium production is also too small and too infrequent to provide much information about the major processes. Nevertheless, because the amount detected can be at least 10^5 greater than the number of neutrons produced, tritium production will be discussed because it offers much better proof for an anomalous nuclear reaction than does neutron production.

Helium production is unexpected because it is rarely made by hot fusion and always is accompanied by gamma emission, which is not detected during cold fusion. Finding helium production to be the main source of excess energy was a major surprise and a frequent reason used by a few people to reject the whole idea.

4.4.1 Tritium

Tritium (^{3}H or t) is a radioactive isotope of hydrogen present in the environment only in a very small amount. Sixty-one claims for tritium production between 10^6 and 10^{16} atoms are listed in Table 6 along with the neutron/tritium (n/t) ratio. Some studies are so poorly done, they deserve to be ignored. On the other hand, a few are so well done, they are hard to reject because tritium is easy to detect and, in sufficient quantity, can not be explained by environmental tritium (contamination) or by error. Possible errors are discussed in Section 7.3 along with an evaluation of a few studies in which errors were largely eliminated.

Table 6. Summary of selected tritium measurements.

Source	Method	Substrate	Environment	Amount	n/t
Dardik et al.[29] 2003	electrolysis	Pd	LiOD+ D_2O	7.5 x background	
Romodanov et al.[346] 2002	plasma	W	D_2+H_2	10500 x background	
Violante et al.[347]	electrolysis	Pd	LiOD+ D_2O	10 x background	
Celani et al.[348] 2001	electrolysis	Pd	C_2H_5OD + D_2O	2.2 x background	
Clarke et al.[136,326] 2000	ambient	Pd	D_2 gas	$1.8x10^{15}$ atoms	
Yamada et al.[349]	deload	Pd, MnO_2	D_2 gas	$3x10^6$ atoms	
Romodanov et al.[350] 1998	Ambient(hot)	Ta, Nb	H_2 gas D_2 gas	$2.4x10^7$ a/sec $8.9x10^7$	
Szpak et al.[351]	electrolysis	Pd	LiOD +D_2O	$2.6x10^{12}$ atoms	
Romodanov et al.[352]	ambient	Fe-Cr+Ni-Ti	D_2+H_2 gas	$1x10^{11}$ atoms	
Romodanov et al.[353]	plasma	Nb	D_2+H_2 gas	$4x10^{10}$ atoms	

Source	Method	Substrate	Environment	Amount	n/t
Romodanov et al.[354]	ambient	Fe	D_2+H_2 gas	7×10^{11} atoms	
Claytor et al.[355]	plasma	Pd-Rh-Co	D_2 gas	400 x background	
1996					
Sankaranarayanan et al.[356]	electrolysis	Ni	$Li_2CO_3+H_2O$, D_2O	3 x background	
Romodanov et al.[357]	plasma	Zr	D_2 gas	10^9 atoms/sec	10^{-7} to 10^{-9}
Itoh et al.[358,359]	ambient	Pd	D_2 gas	?	
Claytor et al.[360,361]	plasma	Pd	D_2 gas	1400 x background	
1995					
Sankaranarayanan et al.[362,363]	ambient	Ni	H_2 gas	777 x background	
Lipson et al.[364]	ambient	$YBa_2Cu_3O_7+D_y$	D_2 gas	2×10^9 atoms/g	
1994					
Notoya et al.[158,365]	electrolysis	Ni	$K_2CO_3+D_2O$	13 x background	
Notoya et al.[158,365]	electrolysis	Ni	$K_2CO_3+H_2O$	26 x background	
Aoki et al.[197]	electrolysis	Pd	$LiOD+D_2O$	5×10^9 atoms	
1993					
Will et al.[366,367]	electrolysis	Pd	$D_2SO_4+D_2O$	5.8×10^{11} atoms	
Stukan et al.[368]	electrolysis	Pd	$Li_2CO_3+D_2O$	2×10^8 t/sec	10^{-5}
Ramamurthy et al.[170]	electrolysis	Ni	$Li_2CO_3+D_2O$	10 x background	
Chien and Huang[369]	electrolysis	Pd	$LiOD+D_2O$	1000x background 10^7-10^9 atoms/sec	
Notoya[370]	electrolysis	Ni	$K_2CO_3+H_2O$	10-100 x background	
Gozzi et al.[183]	electrolysis	Pd	$LiOD+D_2O$	2×10^{14} atoms	
1992					
Takahashi et al.[200]	electrolysis	Pd	$LiOD+D_2O$	3 x background	10^{-5} to 10^{-6}
Mengoli et al. [371,372]	electrolysis	Ti	$NaOD+D_2O$	2×10^{11} atoms	
Stella et al.[373]	electrolysis	Pd	$LiOD+D_2O$	2 x background	10^{-7}
Sevilla et al.[374]	electrolysis	Pd	$LiOD+D_2O$	2.5×10^{10} atoms	
Srinivasan et al.[202]	electrolysis	Ni	$K_2CO_3+D_2O$, H_2O	339 x background	
Srinivasan et al.[202]	electrolysis	Ni	$Li_2CO_3+H_2O$	145 x background	
Matsumoto et al.[375]	electrolysis	Pd	$D_2SO_4+D_2O$	10 x background	10^{-4}

Source	Method	Substrate	Environment	Amount	n/t
Lee and Kim[376]	electrolysis	Pd	LiOD+D$_2$O	2 x background	
Gozzi et al.[217,377-379]	electrolysis	Pd	LiOD+D$_2$O	17 x background	3x10^{-6}
Claytor et al.[380]	plasma	Pd	D$_2$ gas	7 x background	
Clarke and Clarke[381]	ambient	Ti	D$_2$ gas	9x10^7 atoms	
Chien et al.[382-384]	electrolysis	Pd	LiOD+D$_2$O	10^{15} atoms	
Celani et al.[218]	electrolysis	Pd	LiOD+D$_2$O	1.4 x background	
1991					
Will et al.[222]	electrolysis	Pd	D$_2$SO$_4$+D$_2$O	9x10^{10} atoms	
Szpak et al.[224,385]	electrolysis	Pd	Li$_2$SO$_4$+D$_2$O	3 x background	
Mengoli et al.[372,386]	electrolysis	Pd	LiOD+D$_2$O	10 x background	
Lanza et al.[387]	ambient	Ti,Zr,Hf,Ta	D$_2$ gas	5 x background	3x10^{-6}
Kochubey[388]	ambient	Pd organic complex	D$_2$ gas	5 x background	
De Ninno et al.[389]	ambient	Ti	D$_2$ gas	10 x background	
Calytor et al.[390]	plasma	Pd-Si	D$_2$ gas	2.2 x background	4x10^{-9}
1990					
Yang et al.[239]	electrolysis	Pd	LiOD+D$_2$O	10 x background	
Wolf et al.[391]	electrolysis	Pd	LiOD+D$_2$O	10^{16} atoms	10^{-7}
Storms and Talcott[392,393]	electrolysis	Pd	LiOD+D$_2$O	4 x background	
Srinivasan et al.[394,395]	ambient	Ti	D$_2$ gas	58 x background	
Sona et al.[396]	electrolysis	Pd	LiOD+D$_2$O	1.6 x background	10^{-7} to 2x10^{-6}
Matsumoto et al.[397]	electrolysis	Pd	D$_2$SO$_4$+D$_2$O	10^2 atoms/sec	10^{-6}
Fernández et al.[398]	electrolysis	Ti	Li$_2$SO$_4$	10^{10} atoms	10^{-6}
Claytor et al.[399]	plasma	Pd	D$_2$ gas	7x10^{12} atoms	1-3x10^{-9}
Kraushik et al.[395]	ambient	Ti	D$_2$	1.4x10^{16} atoms	
Chêne and Brass[400]	electrolysis	Pd	LiOD+D$_2$O	10^{10} atoms	
1989					
Bockris et al.[401-403]	electrolysis	Pd	LiOD+D$_2$O	2x10^5 x background	10^{-8}
Iyengar and Srinivasan[404,405]	electrolysis	Ti,Pd,Pd-Ag	LiOD,NaOD, Li$_2$SO$_4$	8x10^{15} atoms	10^{-6} to 10^{-9}
Adzic et al.[406]	electrolysis	Pd	LiOD+D$_2$O	49 x background	
Sánchez et al.[407]	electrolysis	Ti	Li$_2$SO$_4$+D$_2$O	1.2x background	

Attempts to produce tritium were undertaken at many laboratories immediately after people learned of the Utah claims. Prof. Bockris at Texas A & M was one of the first to start a study and was the first to report seeing tritium[402] in five cells. This success quickly unleashed a controversy at Texas A & M[408-411] leading to accusations of fraud by Gary Taubes, a contrived article in *Science*[h], and an internal investigation, which cleared Bockris of all charges. The observation that started this unwarranted series of events is shown in Figure 43. The production behavior is similar to that observed at LANL (Figure 4). Perhaps the use of nickel anodes, in contrast to the use of platinum, was the reason for this early success. Later,[369,382] another active sample having detected copper on its surface and significant H_2O in the D_2O generated an amount of tritium that was 100 times background. Production could be stopped by simply shaking the cell or adding D_2O. The production rate could even be changed by changing the cell current (voltage). This unusual behavior is shown in Figure 44 where the tritium concentration can be seen to increase in a linear manner and then change its production rate when cell current was increased at about 90 hours.

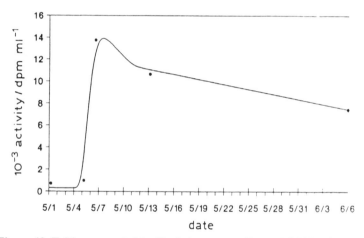

Figure 43. Tritium reported by Bockris group at Texas A&M in 1989.

In 1989 and 1990, a large and successful program was under way in India at the Bhabha Atomic Research Centre.[404,412] Tritium was produced using a variety of methods and chemical environments, as well as being measured when neutron emission was detected. Later, Will and co-workers[366] (National Cold Fusion Institute, Utah) undertook a very complete study using closed cells, a $D_2SO_4+D_2O$ electrolyte, and measurement of the D/Pd ratio. It is important to

---

[h] The accusation made by Taubes in his article, published in the June 1990 issue, was countered by a letter to the editor I wrote to Science on June 25, 1990. The known behavior of tritium in such cells made the charge very unlikely to be correct. This letter was ignored, thereby allowing the false accusation to remain unchallenged.

note that the electrolyte slowly acquired H_2O and the Pd cathode was plated with nanocrystals of Pd before the study. Their high D/Pd ratios are biased on the high side because the H/D ratio would have been greater than they expected, leading to an error when all of the dissolved element was assumed to be deuterium. Nevertheless, they found significant tritium, up to $5.7x10^{11}$ atoms and up to 57 times background, in three cells using D_2O and none in companion cells using H_2O. The frequent observation of H_2O in tritium producing cells raises the question, "Does tritium production require significant protium in the NAE?"[413]

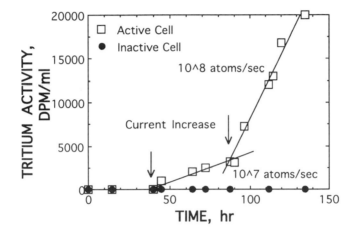

Figure 44. Effect of increase in cell current on the production rate of tritium. (Bockris *et al.*)

By 1993, Fritz Will and co-workers at the National Cold Fusion Institute had laid to rest the possibility of tritium being present in palladium. They carefully analyzed their palladium as well as 90 other pieces from various sources and found no tritium.[414,415] Many other claims for tritium production made since 1993 have eliminated prosaic sources as plausible explanations for most observations.

Evidence that tritium is produced along with neutrons, based on gamma measurement, was provided by Sánchez *et al.*[407] (Figure 45). Although the ratio of n/t was not given, the two nuclear products are clearly coupled. Daniele Gozzi *et al.*[378] (Universitá La Sapienza) found the same correspondence between neutron emission and tritium production in a Pd/D_2O cell. Quantitative measurements of the n/t ratio observed during extended production are listed in Table 6. In addition to neutron emission being frequently associated with tritium production, ~3 MeV protons, as normally accompanies tritium production, was reported by Aiello *et al.*[416] when $PdD_{0.7}$ was deloaded under vacuum.

The frequent association between conditions favoring crack formation and production of neutron and tritium raises the question, "Are neutrons and tritium always produced together during certain types of crack formation, but with an abnormal ratio favoring tritium production?" While this process may account for the occasional bursts of tritium and neutrons, the process would have no relationship to the production of the other nuclear products attributed to cold fusion because crack formation occurs at such a small rate and it is not a steady process as required to make steady observed power. As a result, we have additional evidence for unexpected nuclear reactions, but no insight into the source of detected energy.

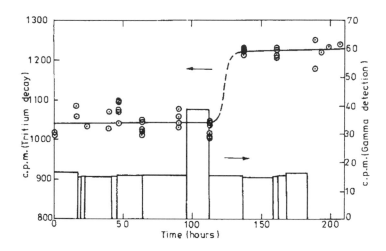

Figure 45. Neutron emission, based on gamma detection, associated with tritium production in a Ti/D$_2$O electrolytic cell.[407]

Tritium is claimed to be produced in cells containing ordinary light-water when nickel electrodes are used.[158,170,202,334,347,356,362,363,365] This observation is totally unexpected and still defies satisfactory explanation.

Tritium can be produced at a higher applied energy than available in an electrolytic cell by using gas discharge in low-pressure D$_2$ gas. A few studies are especially interesting. Claytor and co-workers (LANL) subjected D$_2$ gas to a voltage discharge, first in contact with layers of Si and Pd,[399,417] and later using various alloys of palladium as a cathode in a gas discharge device.[355,360,361,418] A typical result from the latter work is shown in Figure 46. The presence of tritium was determined by measuring the ion and electron current generated in D$_2$ gas by the beta decay. The results of this measurement were confirmed by converting the gas to water and measuring tritium using the scintillation method. (See Section 7.3 for more detail about these methods.) Simply applying a voltage of 2500 V, which is the maximum voltage used, is not sufficient to generate tritium at the measured rates. They found success to be very sensitive to the nature of the palladium or palladium alloy used as the cathode. Because the cathodes are very small, the tritium production rate is small, being about 10^{13} atoms/cm^2-hr. As mentioned in Chapter 2, this work passed through extensive review at LANL during which no prosaic explanation nor error could be found. A similar technique has been used by other people to make an amount of tritium much too large to be consistent with present theory,[187,346,357,419,420] as listed in Table 6.

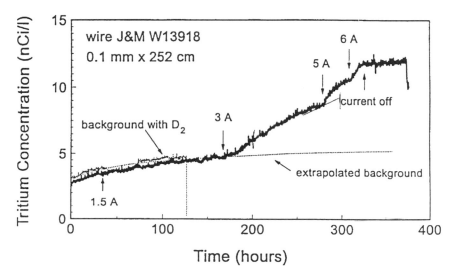

Figure 46. Tritium produced by subjecting palladium wire to brief pulses of 1500-2500 V in D$_2$ gas.(Claytor *et al.*[361])

Other environments have been explored and found to generate tritium. The superconductor, YBa$_2$Cu$_3$O$_7$D$_2$, was found by Lipson and co-workers[364] (Inst. of Phys. Chem., Russian Academy) to produce significant tritium when cooled to its Curie temperature (88-93K) in deuterium gas. Tritium is produced[421] while a mixture of titanium and D$_2$O is pulverized. This method also generates neutrons when other materials as well are pulverized in the presence of deuterium.[422-424] Such mechanical methods produce cracks from which neutrons and tritium, in equal amounts, would be expected to result from the hot fusion process. However, the amount of tritium is far in excess of the amount predicted. Several other unexpected conditions also generate tritium without involving mechanical destruction of a lattice. For example, bombardment of an alloy containing Fe-Cr-Ni-Ti by H$^+$ is found to produce tritium.[425] Even a fine powder of palladium generates tritium[326] along with heat and ^{4}He when exposed to D$_2$ gas.[80,307,308]

Like every aspect of the cold fusion process, tritium production seldom occurs, is highly variable when it is detected, and is dependent on the nature of the substrate. Even harder to believe is its production when certain metals are simply exposed to deuterium. The amount of tritium is always much too small to account for measured energy or to be a health hazard. On the other hand, much more tritium is found than should accompany the few emitted neutrons, being between 10^5 and 10^9 tritium atoms for each neutron. In other words, tritium production can not account for anomalous energy but, at the same time, the amount is too great to be easily ignored and its production rate is not consistent with the rate of neutron emission based on the experience using hot fusion. Although observed tritium reveals an incomplete conventional theory, it is not the whole story for cold fusion. What other nuclear reactions can be proposed?

4.4.2 Helium

Although helium (^{4}He) can form when two deuterons fuse together, the reaction is seldom detected even during the hot fusion process. When it does occur under conventional conditions, simultaneous gamma ray emission is required and is detected. Although gamma rays are occasionally detected when heat is generated during the cold fusion process, the energy and quantity are not consistent with what would be expected if the observed helium resulted from the hot fusion process. Therefore, skeptics rejected the claims because this expected radiation is absent. Instead of basing an evaluation of cold fusion on how hot fusion behaves, the questions should be, "Are the measurements sufficiently accurate and internally consistent to demonstrate a helium-energy connection?" If so, "What is the source of the helium other than the obvious one?" Answers to these questions are explored below and in later chapters.

Measurement of helium is a challenge because air contains enough helium (5.24 ppm) to make the small detected amount appear to be the result of an air leak or diffusion through the walls of the apparatus. In addition, very few laboratories have access to tools needed to measure small helium concentration with required accuracy. In spite of this limitation, on at least seven occasions at laboratories in three countries,[41,74,105,109,149,159,223] helium has been found in amounts consistent with energy production. Of these efforts, four deserve special discussion because great care was taken and the data are presented in a form permitting evaluation.

Miles, at Naval Weapons Center, China Lake (California), and his team first explored the relationship between energy and helium production, using a double-wall isoperibolic calorimeter, and they reported the work in a series of publications,[159,179,180,223,426-428] as summarized by Miles in a recent review.[429] This investigation was debated in a series of papers between Miles[430-432] and Jones[433,434] in which Miles successfully defended his work. Several arguments can be used to support the results. First, 12 studies produced no extra energy and produced no extra helium. Second, out of 21 studies producing extra energy, 18 produced extra helium with an amount consistent with the amount of excess energy. The exceptions were one sample having a possible error in heat measurement and two studies using a Pd-Ce alloy. Miles calculates the chance occurrence of this result as being 1 in 750,000. The data based on nine values is listed in Table 7 and compared in Figure 47 to the known energy for the fusion reaction of 23.82 MeV/fusion or 2.6×10^{11} He atoms/watt-sec. One point that is obviously in error is omitted in this data analysis and a few typographical errors are corrected. Table 7 also lists three measurements made by Bush and Lagowski (University of Texas working at SRI) using an all-metal system with a Seebeck calorimeter. These results are in excellent agreement with the Miles study and have less scatter because the helium background was much lower.

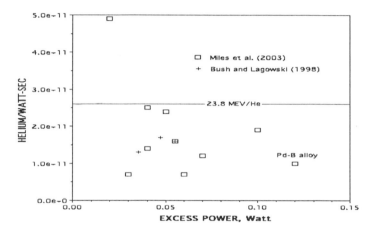

Figure 47. Helium production compared to excess power production.

What additional insight can be obtained from the behavior of the alloys containing boron or cerium? Excess energy produced by the Pd-B[i] alloy is consistent with values obtained from samples not containing boron, suggesting that boron does not enter into a nuclear reaction to make energy or helium. However, the presence of boron apparently does improve power production. Absence of helium when the Pd-Ce alloy was studied might have resulted because all active deuterium reacted with cerium to create energy by a transmutation reaction without producing helium. Although this possibility is difficult to justify based on theory, it must be considered in view of the published observations by Iwamura and co-workers (Table 10).

The measured helium values are expected to have a negative bias because some unknown amount will be retained by the palladium. The values obtained by Miles *et al.* indicate 46% was retained in their study, a very reasonable amount if half of the emitted alphas went in the direction of the bulk material and were captured, while the other half went into the solution and were detected. In addition, some extra energy might result from other reactions, such

---

[i] A two-phase mixture of Pd and $Pd_{16}B_3$ has shown occasional success in producing excess energy in spite of loading poorly and seldom reaching even $PdD_{0.5}$.

as transmutation without helium being produced. The values reported by Bush and Lagowski are consistent with 42% of the helium being retained by the metal—a reasonable amount in good agreement with the Miles' value. All values listed in Table 7 are compared in Figure 47 to the theoretical value for the 2D = He, which is shown by the horizontal line.

Table 7. Summary of selected samples studied by Miles et al.[429]

Power, W	Helium atoms/500 ml**	He/watt-sec
0.100	1.34×10^{14}	1.9×10^{11}
0.050	1.05×10^{14}	2.4×10^{11}
0.020	0.97×10^{14}	4.9×10^{11}***
0.055	1.02×10^{14}	1.6×10^{11}
0.040	1.09×10^{14}	2.5×10^{11}
0.040	0.84×10^{14}	1.4×10^{11}
0.060	0.75×10^{14}	0.7×10^{11}*
0.030	0.61×10^{14}	0.7×10^{11}*
0.070	0.90×10^{14}	1.2×10^{11}
0.120	1.07×10^{14}	1.0×10^{11}
(background)	0.51×10^{14}	Average = 1.4×10^{11}
		STDEV = 0.7×10^{11}

* decimal point error corrected
** not corrected for background
*** eliminated from average

Summary of Samples Studied by Bush and Lagowski [105]

0.047		1.7×10^{11}
0.035		1.3×10^{11}
0.055		1.6×10^{11}
		Average = 1.5×10^{11}
		STDEV = 0.2×10^{11}

While the relationship between heat and helium production was being explored in the US, successful studies were performed in Italy and reported by Gozzi et al.,[102,149,183,435,436] at Universitá La Sapienza. Figure 48 shows an example from this study of helium production while heat was being made in a Fleischmann-Pons type cell. The figure compares energy being measured by calorimetry to that calculated from the measured helium using the value of 23.8 MeV/He. Heat and helium are apparently being made at the same time.

The relationship between energy and helium production found by McKubre[74] and co-workers (SRI International), using palladium deposited on carbon (a "Case"[437] catalyst), is shown in Figure 49. Two methods were used to calculate power production, differential and gradient, with good agreement between the two. In contrast to the work described above, the sample consisted of finely divided palladium deposited on coconut charcoal, which was exposed to D_2 gas and then heated. The helium content of the cell increased over a period of 45 days and exceeded the concentration in air after 15 days. This event eliminates an air leak or diffusion of helium as an explanation. The line through the data is consistent with about 25% of the helium being retained by the sample if all helium resulted from a fusion reaction. In this case, the alpha particles would have been stopped in the carbon substrate where they would be retained less effectively compared to when they are stopped in palladium. It would be an amazing coincidence if this consistency with expectation and previous work were caused by helium being released from the charcoal. This possibility is further reduced by no helium being detected before the charcoal was exposed to deuterium.

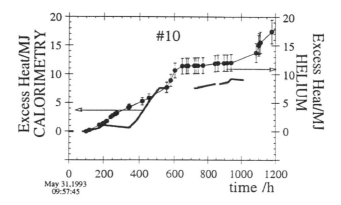

Figure 48. Heat and helium produced as a function of time and compared as energy from each source. (Gozzi et al.[183])

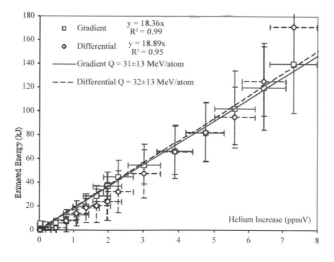

Figure 49. Relationship between energy and helium production from finely divided palladium on carbon heated in D_2 gas. (McKubre et al.)

A single very careful measurement made at SRI is reported by Peter Hagelstein and co-workers.[438] A solid rod of palladium produced two bursts of heat (Figure 50) from which helium was captured in an all metal system. After heat production stopped, the sample was deloaded and reloaded several times, which resulted in extraction of additional helium from the palladium. The total amount of helium obtained gives 24.8±2.5 MeV/He.

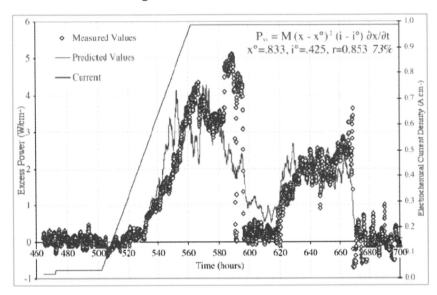

$$P_{xs} = M (x - x°)^2 (i - i°) \partial x/\partial t$$
$$x° = .833, i° = .425, r = 0.853 \ 73\%$$

Figure 50. Excess power produced by a palladium cathode while helium was collected (SRI). The predicted amount of power, based on an equation involving the flux, is also plotted.

If the gross values are combined, an upper limit of <43±12 MeV/He is obtained. If 50% of the helium is assumed to be retained by solid palladium, when this was used, this value is reduced to 21±12 MeV/He. The value of 24.8±2.5 MeV/He, obtained after an effort was made to extract all helium, is in excellent agreement with the corrected value. By combining all measurements, a value of 25±5 MeV/^{4}He is proposed to be the amount of heat produced by formation of each helium atom using the cold fusion process, whatever that process might be. Although this value is consistent with d-d fusion being the source of energy and helium, other reactions may also be consistent, as discussed in Chapter 8.

Fifteen additional studies[62,77,136,183,276,281,383,435,436,439-444] found unexpected helium in metal cathodes or in the cell after energy had been made under cold fusion conditions, but the energy-helium relationship was not reported.

Helium-3 (^{3}He) is seldom detected but when it is seen, it apparently results from tritium decay.[326,381,445-447] On the other hand, Mamyrin *et al.*[448] measured the ^{3}He/^{4}He ratio in many materials and found a wide variation they could not explain. Even so, the amount of ^{3}He is very small and only observed because the sensitivity of the measurement is so high.

Once again we are faced with good work being done by independent laboratories producing an "impossible" result. To reject this work, we have to assume that errors in helium measurement and errors in heat measurement both conspire to produce a similar ratio regardless how or by whom the measurement is made. In addition, we need to assume these errors only operate when anomalous heat is actually detected. If the data are accepted, we also need to accept that somehow helium and energy are apparently being created at the same time without generating gamma emission. Or this information can be simply ignored, as it was by many members of the DoE panel[449,450] convened in 2004 to evaluate cold fusion.[451] To avoid this conflict between experiment and theory, several attempts are made in Section 8.3 to explain helium production without the need for conventional fusion and the resulting gamma emission.

4.5 Transmutation as a Source of Nuclear Products

Helium and tritium are not the only unexpected nuclear products being reported. Evidence for transmutation reactions is also accumulating. These are nuclear reactions in which one element is converted into another. This can happen in several ways. For the first method, an isotope of hydrogen might fuse into the nucleus of a heavy element, such as a proton entering into the nucleus of potassium to form calcium[25,206,219] or several deuterons fusing with the nucleus of barium to make samarium.[452] The second method might occur when two heavy elements fuse together, a process involving an even greater Coulomb barrier. For example, Sundaresan and Bockris[453] (Texas A & M) propose that iron found[454-457] after an arc is formed under water between carbon rods is made by carbon fusing with oxygen. Finally, an element can split into two smaller elements, a process called fission. Fission of uranium is common, but lighter atoms, such as palladium, are not known to experience fission. Nevertheless, this possibility has been suggested to occur when an element is placed in the NAE,[458,459] especially after having acquired one or more deuterons[40] or photons.[460] The source of these unexpected nuclear products is still being hotly debated.

Because many nuclear reactions in addition to fusion are observed, the phenomenon is now referred to as Low Energy Nuclear Reactions (LENR) or Chemically Assisted Nuclear Reactions (CANR). The entire field is now called Condensed Matter Nuclear Science (CMNS). For the sake of consistency and habit, all of these reactions are called cold fusion in this book even though this is not technically correct. Indeed, simple d+d fusion may not even make a significant contribution.

Reported transmutation products have a range of atomic numbers. One example of this range can be seen in Figure 51, based on the data provided by

Miley *et al.*[461] Miley and co-workers electrolyzed a complex Ni-Pd cathode in $Li_2SO_4+H_2O$ and found a collection of elements with maximum production rates having atomic numbers near 12 (Mg, Si), 30 (Fe, Zn, Se), 48 (Ag, Cd) and 82 (Pb). A distribution of elements with similar peaks and valleys, shown in Figure 52 was found on a palladium cathode after it had been subjected to plasma electrolysis in D_2O by Mizuno *et al.*[462] Other studies are summarized in Table 8.

Pure nickel heated in hydrogen gas shows a different pattern, with most of the detected elements at atomic numbers less than nickel and clustering in a region about 1/2 the atomic number of nickel, presumably as a result of fission. This result is less compromised than the electrolytic method because fewer impurities are present.

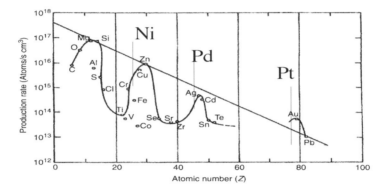

Figure 51. Production rate of elements found on the surface of a Ni+Pd cathode after electrolysis using a $Li_2SO_4 + H_2O$ electrolyte. (Miley *et al.*)

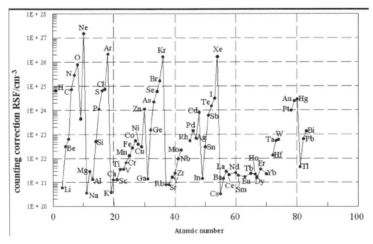

Figure 52. Elements formed on a Pd cathode after being exposed to plasma electrolysis in D_2O. (Mizuno *et al.*[462])

Table 8. Reported transmutation products.

Source	Substrate	Environment	Method	Detected
Wang et al.[481]	Pd	H_2SO_4+D_2O	electrolytic	Ag,Ni,Fe,Ti,S,Pt
Wang et al.[481]	Ti	H_2SO_4+D_2O	electrolytic	Ag,Ni,Fe,Ti,S,Pt
Szpak et al.[272,482]	Pd	LiOD+D_2O	electrolytic	Si,Mg,Zn,Ca,Al
Savvatimova et al.[22]	Ti	D_2 gas	plasma	Al,Mg,Br,Sr,Rb,S,F,O,Ni,Cr, Fe,Sn (isotope ratio change)
Mizuno et al.[23]	W	K_2CO_3+H_2O	plasma	Ca,Fe,Zn
Lochak and Urutskoev[483]	Ti	H_2O	fuse	Na,Mg,Al,Si,K,Ca,V,Cr,Fe, Ni, Cu,Zn
Karabut[40,288]	Pd	D_2 gas	plasma	Li,C,N,Ne,Si,Ca,Fe,Co,Zn, As, Ag,Cd,In(isotope ratio change)
Focardi[342]	Ni	H_2 gas	ambient	Cr,Mn
Cirillo and Iorio[300]	W	K_2CO_3+H_2O	plasma	Re,Os,Au,Hf,Tm,Er,Y
Celani et al.[484]	Pd	C_2H_5OD+ Th,Hg	electrolytic	Cu,Zn,Rb,Cs,Pb.Bi
Campari[27]	Ni	heated H_2 gas	ambient	Na,A,Si,S,Cl,K,Ca,Fe,Zn
Yamada et al.[485]	Pd	H_2 gas	diffusion	Ti,Cr,Mn,Fe,Ni,Cu,Ag
Violante et al.[463]	Ni	D_2O	electrolytic	Cu (isotope ratio change)
Passell[486]	Pd	D_2 gas	plasma	Pd isotope change, Co,Zn,Au,Ir
Ohmori et al.[464]	Re	K_2CO_3+D_2O, H_2O	plasma	K (isotope ratio change)
Celani et al.[43]	Pd	C_2H_5OD+ Th,Hg,Sr	electrolytic	Sr→Mo(isotope ratio change)
Violante et al.[347]	Ti	D_2O+Li_2SO_4	electrolytic	Zn,Cu,Ag (isotope ratio change)
Yamada et al.[487]	Pd	H_2O+Na_2CO_3	electrolytic	Li,B,Mg,Al,K,Ca,Ti,Cr,Mn, Fe, Co,Ni,Cu,Zn,Ba,Pb(isotope ratio change)
Warner et al.[44]	Ti	D_2O+H_2SO_4	electrolytic	Au
Vysotskii et al.[488]	Cs		biological	Ba
Matsunaka et al.[443]	Pd	D_2O	electrolytic	Fe,Zn
Karabut[55]	Pd	D_2 gas	plasma	C,Ca,Ti,Fe,Co,Zn,As,Ag,Cd (isotope ratio change)
Iwamura et al.[475,489]	Pd	D_2 gas	diffusion	Cs→Pr, Sr→Mo
Goryachev[490]	Ni	27 MeV electron	bombard	Ni→Rh
Di Giulio et al.[491]	Pd	PdD	laser	Ca,Fe,S,Zn,Ti,Cu,Cr
Arapi et al.[492]	Pd	D_2 gas	plasma	Li,Be,Fe,Ni,Cu,Ba
Yamada et al.[493]	Pd	D_2 gas	plasma	Fe,Cu
Warner and Dash[66]	Ti	D_2O+H_2SO_4	electrolytic	Cr
Wang et al.[494]	TiH	H^+	bombard	He^4

Source	Substrate	Environment	Method	Detected
Vysotskii et al.[495]		Na, P	biological	Na+P→Fe
Passell and George[496]	Pd	D_2 or D_2O	electrolytic	Zn
Nassisi and Longo[497]	Pd	PdD	laser	Zn
Mizuno et al.[69]	W	$K_2CO_3+H_2O$	plasma	Al,Si,Ca,Ti,Cr,Fe,Ni,Zn,Ge, Pd,Ag,In
Li et al.[498]	Pd	$Pd+D_2O$		Ni
Iwamura et al.[499]	Pd	D_2 gas + C	diffusion	Mg,Si,S
Iwamura et al.[499]	Pd	$LiOD + D_2O$	electrolytic	F, Al, Si
Hanawa[454]	C	H_2O	plasma	Si,S,Cl,K,Ca,Ti,Cr,Mn,Fe,Co, Ni,Cu,Zn
Dufour et al.[75]	Pd	H_2 gas	plasma	Mg,Zn,Fe
Castellano et al.[500]	Pd	PdD	laser	Na,Mg,Al,P,S.Cl.Ca,Ga,Fe, Ni,Zn,Cu,Sn
Campari et al.[76]	Ni	heated H_2 gas	ambient	F,Na,Mg,Al,Si,P,S,Cl,K,Ca, Cr, Mn,Fe,Cu,Zn
Bernardini et al.[78]	Ti	$K_2CO_3+D_2O$	electrolytic	Sc (radioactive)
Ransford[457]	C	H_2O	plasma	Fe,Cr
Ohmori and Mizuno[501]	W	Na_2SO_4+ H_2O	plasma	Cr,Fe,Ni,Re,Pb
Focardi et al.[114]	Ni	heated H_2 gas	ambient	C,O,Mg,Si,K,S,Cl,Al, Na,Fe,Cu
Klopfenstein and Dash[502]	Ti	$D_2SO_4+D_2O$	electrolytic	Al.S,Ca,Fe (Ti isotope change)
Qiao et al.[471,503]	Pd	D_2	ambient	Zn
Ohmoi et al.[465]	Au	Na_2CO_3 or $Na_2SO_4 + H_2O$	electrolyte	Hg,Kr,Ni,Fe,Si,Mg (isotope change)
Ohmori and Mizuno[92]	W	$K_2CO_3 + H_2O$	plasma	Ni,C,Fe,Cr,Pb (isotope change)
Notoya et al.[504]	Ni	$K_2CO_3 + H_2O$	electrolytic	Os,Ir,Pt,Au,K
Nassisi[505,506]	Pd	H_2, D_2 gas	XeCl laser	**Al**,Au,C,Ca,Cl,Cr,**Fe**,K,Mg, Na, Nd,Ni,V,Zn,O,S,**Si,** delayed n
Jiang et al[507]	Pd	$NaOD + D_2O$	electrolytic	Mg,Al,Si,Fe,Cu,Zn,Pt
Jiang et al.[456]	C	H_2O	plasma	Fe
Iwamura et al.[100,101]	Pd	$LiOD +D_2O$	electrolytic	Ti,Cu,Fe (isotope change)
Nakamura et al.[508]	Ni, (C anode)	$(NH_4)_2MoO_4+$ H_2O	plasma	radioactivity
Ohmori et al.[111]	Au	Na_2SO_4,K_2SO_4 ,K_2CO_3,KOH+ H_2O	electrolytic	Fe (isotope change)
Qiao et al.[509]	Pd	H_2	ambient	Zn,Tb
Kopecek and Dash[116,510]	Ti	$H_2SO_4+D_2O$	electrolytic	S,K,Ca,V,Cr,Fe,Ni,Zn
Ohmori and Enyo[335]	Au, Pd	$Na_2SO_4+H_2O$	electrolytic	Fe (isotope change)

Source	Substrate	Environment	Method	Detected
Yamada et al.[511]	Pd	D_2 gas	plasma	C
Karabut et al.[287]	Pd	D_2	plasma	Na,Mg,Br,Zn,S, Mo.Si
Miley et al.[336,461,512]	Ni	$Li_2SO_4+H_2O$	electrolytic	Major elements: Cr,Fe,Mn,Cu,Zn,Se,As,Cd, Ag (isotope change)
Savvatimova and Karabut[283]	Pd	H_2, D_2 gas	plasma	As,Br,Rb,Sr,Y,Cd(isotope change)
Notoya[330,334,513]	Ni	$H_2O+Cs_2SO_4$ solid electrolyte	electrolytic	Ba
Mizuno et al.[514,515]	Pt	SrCeNbY oxide	electrolytic	Pt(radioactive), Al,Ca,Mg,Bi, Sm,Gd,Dy
Sundaresan and Bockris[453]	C	H_2O	plasma	Fe
Singh et al,[455]	C	H_2O	plasma	Fe
Mizuno et al,[462,467]	Pd	$LiOH+D_2O$	electrolytic	Ti and Cr(isotope change), Ca,Mn,Fe,Co,Cu,Zn,Cd,Sn, Pt,Pb
Dash et al.[516,517]	Pd	$H_2SO_4+D_2O$	electrolytic	Ag
Matsunoto[518]	Pd	$K_2CO_3+H_2O$	plasma	Ni,Ca,Ti,Na,Al,Cl,Cd,I
Bush and Eagleton[163,519]	Ni	$Rb_2CO_3+H_2O$	electrolytic	Sr(radioactive)
Savvatimova et al.[281]	Pd	D_2 gas	plasma	Li,B,V,Cr,Fe,Ni,Cu,Sr,Zr,Na ,Al, Si,Ti,Nb,Mo,Ag,In (isotope ratio change)
Notoya[331]	Ni	$K_2CO_3 + H_2O$	electrolytic	K→Ca
Komaki[520]		H_2O	biological	Na→K, Na→Mg, K→Ca, Mg→Ca
Dillon and Kennedy[521]	Pd	$LiOD +D_2O$	electrolytic	Zn,Cu,Cr,Fe
Bush and Eagleton[522,523]	Ni	$Rb_2CO_3 + H_2O$	electrolyte	Rd→Sr
Ohmori and Enyo[177]	Ni	$K_2CO_3 + H_2O$	electrolytic	K→Ca
Rolison and O'Grady[524]	Pd	$Li_2SO_4 +$ D_2O,H_2O	electrolytic	Rh,Ag
Williams et al.[16]	Pd	$LiOD +D_2O$	electrolytic	Li,Cu,Zn,Fe,Pb,Si,Pt
Divisek et al.[261]	Pd	$LiOD + D_2O$	electrolytic	Pb, Cu
Greber[525]	Pd	$LiOD + D_2O$	electrolytic	Pb,Hg,Bi,Zn

Electrolyte = solution through which current is passed to initiate a Faraday-type reaction.

Plasma = Sufficient voltage is applied to either a gas or liquid to form gaseous ions as an arc or spark.

Laser = Laser light is applied in order to stimulate nuclear reactions.

Diffusion = Deuterium or hydrogen is diffused through palladium from the gas phase.

Fuse = Metal is rapidly melted by high current while under water.

Ambient = Metal substrate is placed in the indicated gas.

Bombard = Substrate is bombarded with the indicated charged particle.

Biological = Transmutation products are made in the presence of living organisms.

The general experience summarized in Table 8 is compared in Figure 53 where the number of occasions certain elements were reported on the cathode surface when palladium was used as the cathode in a D_2O-based electrolyte are plotted as a function of atomic number. Note that only a few detected elements have a greater atomic number and weight than the most abundant element present, i.e. Pd, while most have a smaller atomic number and weight. This pattern is consistent with the behavior shown in Figures 51 and 52 and suggests a fission reaction might have occurred. Of the anomalous elements reported, iron (Fe), copper (Cu), and zinc (Zn) are most frequently found in high concentration and frequently with an abnormal isotopic ratio. When suitable measurements are made, these products are also found along with heat production and other indications of anomalous activity.

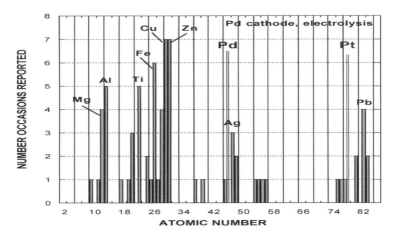

Figure 53. Occasions when elements are reported as transmutation products. Locations of Pd and Pt are noted for convenience.

Some of the reported elements are normal impurities made visible after being concentrated on the surface. However, many of the elements, especially those having abnormal isotopic ratios listed in Table 9 cannot be explained this way. Isotope abundance is characteristic of each element and can not be easily changed, especially for the heavier elements. For example, the normal ratio of $^{63}Cu/^{65}Cu$ is 2.24, but Violante and his group[463] (ENEA, Italy) measured values of 1.37 and 0.97 for copper deposited on a nickel cathode after electrolyzing it in a copper-free electrolyte of $Li_2SO_4+H_2O$. Ohmori and Mizuno[464] (Hokkaido University, Sapporo, Japan) formed a plasma within an electrolyte of $K_2CO_3 + H_2O$ or D_2O using a rhenium cathode. Regardless of the hydrogen isotope used in the electrolyte, the potassium isotope ratio, $^{41}K/^{39}K$, changed from the natural value of 0.072 to between 0.27 and 0.56. In another study, they found a large amount of iron on the cathode, with the isotope ^{57}Fe increased by about a factor of ten, after gold was electrolyzed in $Na_2SO_4 + H_2O$.[465] Savvatimova and

Gavritenkov[22] (LUTCH, Russia) reported a large change in the distribution of deposited calcium isotopes after subjecting titanium to glow discharge in deuterium gas. These are only a few examples provided after people began using a mass spectrometer for analysis.

Table 9. Examples of isotope enrichment.

Source	Method	Isotope Change	Remarks
Donohue and Petek[466]	Electrolysis, D_2O	Pd	no change
Savvatimova et al.[281,283]	Plasma, Pd in D_2 gas	$+^{54}Fe$, $+^{57}Fe$, $+^{11}B$, $+^{51}V$, $+^{53}Cr$, $+^{61}Ni$, $+^{63}Cu$, $+^{87}Sr$, $+^{90}Zr$	Many elements produced
Mizuno et al.[462,467]	Plasma, Pd in D_2O	100% ^{63}Cu, $+^{57}Fe$, $-^{56}Fe$, $+^{53}Cr$, $-^{52}Cr$, $-^{39}K$, $-^{64}Zn$, $+^{91}Ir$, $-^{93}Ir$, $+^{185}Re$, $-^{187}Re$	Many elements produced.
Ohmori and Enyo[111,335]	Electrolysis, Pd and Au in H_2O	$+^{54}Fe$, $+^{57}Fe$	Fe increased as excess energy increased.
Savvatimova et al.[468]	Plasma in Ar+Xe gas	$+^{104}Pd$	Many elements produced.
Miley[458]	Electrolysis, Ni in H_2O	$+^{107}Ag$, $-^{109}Ag$, $+^{63}Cu$, $-^{65}Cu$	Many elements produced.
Iwamura et al. [99]	Electrolysis, Pd in D_2O	$+^{57}Fe$,	Pd-CaO-Pd cathode
Ohmori et al. [91]	Plasma, W in H_2O	$+^{56}Fe$, $+^{52}Cr$, $+^{206}Pb$, $-^{208}Pb$	Cr and Fe found together on the W
Karabut [55]	Plasma, Pd in D_2 gas	$+^{57}Fe$, $+^{110}Cd$	Many elements produced
Celani et al. [43]	Electrolysis, Pd in $D_2O+C_2H_5OD$	$+^{63}Cu$, $+^{39}K$,	Many elements produced
Ohmori et al. [464]	Plasma, Re in $H_2O/D_2O + K^+$	$+^{41}K$	
Violante et al. [463 469]	Electrolysis, Ni in H_2O	$+^{65}Cu$	Laser light used
Kim and Passell [470]	Various methods	$+^7Li/^6Li$	
Savvatimova and Gavritenkov [22]	Plasma, Ti in D_2 gas	$+^{40}Ti$	Many elements produced

+ = indicates increase in concentration
- = indicates decrease in concentration

A variation on gas loading has been used by Qiao et al.[117,471,472] Palladium wire was exposed to H_2 or D_2 gas in a cell in which was located a heated tungsten wire. Ions were formed on the wire and these were accelerated to the palladium cathode where they increased loading to a high level—a method initially described by Oates and Flanagan.[473] Helium was found to have accumulated in the cell after a year of this treatment. During this time, excess energy production was detected as well. In addition, when the palladium used as a "blank" in a cell containing H_2 was examined by EDX, the palladium surface revealed significant zinc and terbium along with aluminum, silicon, lead, calcium, copper, and iron—elements that were not present initially. Is it possible, once again, that hydrogen is not a good blank?

All transmutation reactions must build upon elements initially present on the surface, sometimes called the target or "seed" element. This requires a connection be made between the seed and the product if the type of nuclear reaction is to be determined. A few clear cut examples of transmutation resulting from fusion between a single proton and the cation present in the H_2O electrolyte have been reported.[163,177,219,331,474] In contrast, a similar transmutation has not been reported when D_2O is used in an electrolytic cell, although such a reaction is reported to occur in biological systems. Beyond these few examples, imagination is required, which is made difficult because fusion increases the atomic number while fission will cause the atomic number of the product to be a fraction of the seed. The amount of change in atomic number depends on how many deuterons or protons were added to the seed and whether particles were emitted. So many potential seeds are present that the outcome can be very complex.

The problem was simplified somewhat by Iwamura and his team at Mitsubishi Heavy Industries (Japan). Deuterium was caused to diffuse through a layer consisting of 400 Å thick palladium, containing the target elements on the front surface, and alternating layers of calcium oxide and palladium, all applied to a palladium substrate (Figure 54). This method caused the target elements to fuse with various numbers of deuterons to give identifiable products as summarized in Table 10.[452,475-477] The method has even been patented[478] in Japan. Amazingly, the applied target elements were found to take up deuterons in units of two and the number of such units depended on the element studied. A connection between the seed and product element was revealed by a decrease in concentration of the seed and an increase in the amount of product, an example of which is shown in Figure 55.

Table 10. Summary of reactions reported by Iwamura et al.

1.	Ba + 6d = Sm + ? Q = 67.6 MeV
2.	Cs + 4d = Pr + ? Q = 50.5 MeV
3.	Sr + 4d = Mo + ? Q = 53.4 MeV
4.	Cs + 2d = La (radioactive, EC, no γ) + ?, Q = ~24 MeV

? = unknown and undetected emission

Reaction #1 has been replicated at Kobe University in Japan by Kitamura,[479] reaction #3 has been replicated in Italy by Celani and his group using electrolysis,[43] reaction #2 has been replicated by Higashiyama et al. at Osaka University (Japan),[480] and attempts to replicate the general method are under way at the Naval Research Laboratory (NRL) in Washington DC.

The generated energy/event, which is calculated from the mass defect produced by the indicated reaction, is very large compared to a fusion reaction. Nevertheless, only about 13 mW/cm^2 can be calculated to have resulted from this study. This energy must be communicated to the lattice by some mechanism. If these were normal reactions, emission of a gamma ray would be expected because only one element atom is produced, which is forbidden by conventional models (see Section 8.2.4). Another product, indicated by the question mark, should be sought. No matter which process helps carry away energy and conserve momentum, it must occur without causing a change in atomic number or in gross mass, which severely limits the possibilities.

This observation raises an additional issue. The "seeds" were contained in a palladium lattice. Why was palladium not affected? Such issues severely challenge all theories.

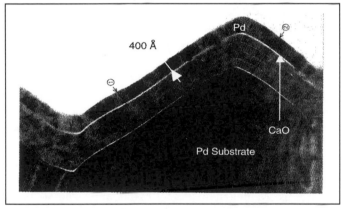

Figure 54. Cross-section of the layer used by Iwamura *et al.*

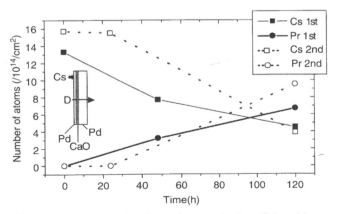

Figure 55. Example of reduction in seed concentration (Cs) and increase in product concentration(Pr). (Iwamura *et al.*)

A little irony is added to the history by David Williams and co-workers[16] at Harwell (UK). Their failure to find anomalous heat, based on what at the time was considered careful work, was used to discredit Fleischmann and Pons, although later analysis indicated some anomalous heat was actually produced.[526-528] Largely ignored was their reported finding of Fe, Cu, and Zn on the cathode surface—elements considered normal impurities at the time. In view of more recent studies, their work might have provided the first evidence for transmutation products.

An interesting aspect of the newly observed transmutation reactions, aside from providing ammunition to skeptics, is its relationship to alchemy,[529,530] the ancient origin of chemistry. According to claims provided by alchemists, metals could be transformed into other metals, generally gold, when mixed with a special substance they called the Philosopher's Stone. This material has the same abilities that are attributed to the nuclear-active environment (NAE). Of course their explanation was much different from the one accepted today, yet evidence for their success remains in objects and texts. Could we have stumbled upon a process observed over 500 years ago before the existence of elements was even recognized? Roberto Monti[531,532] and Joe Champion[533] have explored this aspect of the phenomenon by following the ancient recipes. It was one such recipe provided by Champion that started Bockris down this unexpected path and opened the door to the possibility of transmutation reactions being part of the cold fusion phenomenon. People were detecting unexpected elements in their cold fusion cells since the beginning of the field, but these elements were generally ignored because they were thought to be contamination. Once the group at Texas A & M began to detect what appeared to be transmutation products using the "gun-powder" method provided by Champion, interest picked up in both a negative and positive way. While Bockris was subjected to increased attacks on his reputation by his fellow professors, Miley, at the University of Illinois, began to take an interest. Gradually, with reluctance in some quarters, the idea gained support. Two conferences organized by Bockris helped in this process. Proceedings of the Second International Low Energy Nuclear Reactions Conference, held on September 13-14, 1996 in Collage Station, TX, were published as Volume 1, #3, of the *Journal of New Energy*. As Table 8 demonstrates, this is now a growing branch of the cold fusion field—thanks to alchemy.

4.6 Emissions as Nuclear Products

When nuclear reactions occur in nature, various kinds of radiation are expected and observed. These emissions can be electromagnetic radiation (EMR) (gamma or X-rays) or can be energetic particles such as electrons (beta emission), positrons (positive electrons), or helium (alpha emission). Proton and triton emission might also occur if the nuclear reaction involves fusion. Emission can occur immediately (prompt) or can extend over a period of time (radioactive) as the nuclear products lose energy. Modern methods leave little doubt about the types and energy of conventional radiation being emitted, although identification can be a challenge if the radiation is too weak or if can not leave the apparatus at all. Emission of prompt radiation produced by cold fusion is now being explored as suitable detectors are placed within the apparatus and as costly radiation and particle detectors are used. Some of the results are listed in Table 11.

Table 11. Reported studies of radiation.

Source	Method	Environment	Type	Detection	Energy	Other Behavior
Lochak and Urutskoev[483]	Ti melted	Arc in H_2O	S	film	?	transmutation
Karabut[40]	plasma	D_2, <1.77 kV	X-ray	TLD,film, SD	1.5-2.5 keV	heat, transmutation bursts
Lipson et al.[534]	electrolytic	He implant, Ti,Pd	α	Si-SSD,CR-39	9-16 MeV	
Lipson et al.[534]	ambient	PdD/PdO/Au (deload)	p (or)		1.7 MeV	
Lipson et al.[534]			d		2.8 MeV	
Kowalski et al.[535]	ambient	TiDx, PdD (deload)	?	CR-39	?	
Focardi et al.[27,342]	ambient	H_2,Ni	γ	NaI, Ge	744 keV[j]	heat, transmutation
Oriani and Fisher[537-539]	electrolytic	Pd,Ni, Li_2SO_4+ D_2O,H_2O	?	CR-39	?	
Miles[427,429]	electrolytic	LiOD,Pd	?	GM counter	?	heat, helium
Keeney et al.[540]	ambient	TiD$_x$ (deload)	p	SD	2.6 MeV	

[j] The reported value of 661.5 keV was corrected by later measurements to 744 keV according to Takahashi[536.] (See: Takahashi, A., Progress in condensed matter nuclear science, in *12 th International Conference on Cold Fusion*, Takahashi, A. World Scientific Co., Yokohama, Japan, 2006.)

Source	Method	Environment	Type	Detection	Energy	Other Behavior
Lipson et al.[295,541]	electrolytic	LiOD,Pd	α	CR-39	11-16 MeV	heat
			p	CR-39	1.5-1.7 MeV	
			X-ray	TLD	?	
	plasma	D₂,Ti,<2.5 kV	α	CR-39	13.0 MeV	
			p	CR-39	3.0 MeV	
			d	CR-39	2.8 MeV	
			X-ray	SD, TLD	1.1-1.4 keV	
	laser	TiDx	α	CR-39	13.0 MeV	
Cecil et al.[542]	plasma	TiD	α	Si-SSB	6.8 MeV	
Afonichev[543,544]	ambient	TiD (deform)	RF	antenna	?	tritium
Lipson et al.[545,546]	electrolytic	LiOD, Pd	α	CR-39	11.0-16.0 MeV	heat
			p		1.7 MeV	
Violante et al.[347]	electrolytic	Pd,Ni, Li₂SO₄+ D₂O,H₂O	X-ray	Ge	6.2, 7.2, 7.4, 8.8 keV	transmutation, tritium
Tian et al.[46]	ambient	Pd-Ag, D₂	X-ray	TLD	?	heat
Yamada et al.[493]	plasma	PdD, D₂	γ	film, NaI	106 keV	autoradiograph
Lipson et al.[547,548]	ambient	PdD/PdO/Au (deload)	p	Si-SSD	2-3 MeV	
Roussetski[549] {Roussetski, 1998} #1348,550	ambient	PdD/PdO/Au (deload)	p	CR-39	3 MeV	
			t		1 MeV	
			³He		<0.8 MeV	
					890 keV	
Bernardini et al.[78]	electrolytic	Ti	γ	Ge	1123 keV 665 keV,	Radioactive, heat
Campari et al.[76]	ambient	Ni,H₂	γ	NaI, Ge	412 keV	heat, transmutation

Source	Method	Environment	Type	Detection	Energy	Other Behavior
Savvatimova [282,551]	ambient	PdD,TiD	beta + S	film, electro-meter	?	Radioactive
Iwamura et al. [101]	electrolytic	Pd, D_2O	X-ray	NaI	10-100keV	Heat, neutron transmutation
Qiao et al. [471]	ambient	Pd,H_2O, D_2O	?	CR-39	?	heat, transmutation
Szpak and Mosier-Boss [552]	electrolytic	$PdCl_2$+ D_2O	X-ray	Ge	12 keV 22 keV	heat Co-deposition
Lin and Bockris [553]	gunpowder	Hg_2Cl_2 + PbO+FeS	β	Proportional counter	?	17.7 h half-life transmutation
Notoya et al. [554]	electrolytic	Ni,Pt + various salts	γ	Ge	Many values	
Rout et al. [555,556]	ambient	PdH,PdD (deload)	S		?	autorad.
Roussetski [550]	ambient	PdD/PdO/Ag (deload)	p?	CR-39	1.7, 2.7, 4 MeV	
Cellucci et al. [135]	electrolytic	Pd,LiOD	X-ray	film	89 keV	heat, helium
Mizuno et al. [514]	Solid electrolyte	Sr+Ce+Y+Nb, oxide, Pt	γ	Ge-Li	76.8 keV	radioactive Tritium, radioactive decay
Itoh et al. [359]	ambient	PdD (deload)	X-ray	NaI	21 keV	
Matsumoto [557,558]	plasma	metal, K_2CO_3+ H_2O	S	film	?	transmutation
Karabut et al. [146]	plasma	Pd,D_2	γ, β	Ge ?	0.1-3MeV 0.01-2MeV	Radioactive, heat, transmutation
Karabut et al., [284]	Plasma 200-600 V	Pd,D_2	γ parti-cles	Ge, film Si-SSD	100-300keV 5-6 MeV	Radioactive decay
Manduchi et al. [559]	ambient	Pd, H_2, D_2	?	CR-39	?	neutrons
Koval'chuk et al. [560]	electrolysis	Pd, Ni, D_2O,H_2O $LiClO_4$	EMR	Photo-multiplier	380-420 nm	
Taniguchi [561,562]	electrolysis	Pd, LiOD+ D_2O	?	Si-SSD	4-10 MeV	
Matsumoto [563]	electrolysis	Ni, H_2O+ K_2CO_3	S	film	?	
Matsumoto [564]	electrolysis	Pd, D_2O+ NaCl	S	film	?	
Mo et al. [565,566]	ambient	TiD, PdD (deload)	α	Si-SSD	5 MeV	

Source	Method	Environment	Type	Detection	Energy	Other Behavior
Long et al. [567]	Plasma 17 kV	Various + D_2	γ	NaI	10s keV, 3.4MeV, 5.8MeV	neutrons
Uchida et al.[568]	electrolysis	Pd,Ti, D_2O (loading)	?	GM counter	?	bursts
Bush and Eagleton[220]	electrolysis	Pd-Ag, LiOD+ D_2O	X-ray	NaI	characteristic	Decay, heat
Karabut et al.[214]	Plasma 100-500 V	Pd, D_2	X-ray	Ge, film	15-20 keV	Heat, decay
			α	Si-SSD, CR-39	2-4 MeV	decay
			γ	Ge	0.2 MeV	
Bush et al.[223]	electrolysis	Pd,LiOD+D_2O	?	film	?	heat, helium
Wang et al.[569,570]	ambient	Ti, Pd, D_2	α	CR-39	?	
Jin, et al.[571]	ambient	PdD (deload)	?	CR-39	?	
Dong et al.[572,573]	ambient	PdD (deload)	?	CR-39	?	
Taniguchi and Yamamoto[574,575]	electrolysis	Pd-Ag, LiOD+ D_2O	p	Si-SSD	2-3 MeV	
Matsumot[576,577]	electrolysis	Pd, LiOD+ D_2O	S	film	?	
Jones et al.[578]	ambient	PdD (deload)	p?	scintillator	2.3-3.0 MeV	burst
Cecil et al.[579]	ambient	TiD (deload)	α or t	Si-SSD	>4.5 MeV	burst
Celani et al. [580]	electrolytic	PdD (load)	γ	NaI	?	burst
Fleischmannn et al. [3]	electrolytic	PdD	γ	NaI	2.8 MeV[k]	+heat

p= proton (hydrogen)
α = alpha (helium)
t= triton (tritium)
d = deuteron (deuterium)
γ = gamma (photon)
β = beta (electron)
S = strange
Burst = Occasional large signal over a relatively brief time.

---

[k] The value of 2.496 MeV given by Fleischmann and Pons was challenged and corrected by Petrasso et al. to give 2.8 MeV. However, the statement that this peak has an abnormal width, hence an artifact, is not true and is based on a misinterpretation by Petrasso et al.[3.] (See: Fleischmann, M., Pons, S., Hawkins, M., and Hoffman, R. J., Measurements of gamma-rays from cold fusion., *Nature (London)* 339 (622), 667, 1989).

Radiation and radioactivity result from the cold fusion processes. Most early studies did not find these expected products, which added doubt to a nuclear process being involved. However, as can be seen in the table, later studies found many kinds of radiation when the proper detectors were used and the required conditions were present. Fortunately, none of this radiation is a health hazard nor is it easy to detect outside of the apparatus, which makes the process safe to study and safe as an eventual source of energy. To make these observations more significant, such energetic radiation and radioactivity are never produced by the mild chemical environments typical of many cold fusion studies and can not be dismissed as error. In other words, regardless of the relationship to cold fusion, such radiation is not expected to result from these low-energy environments. Even more amazing, radiation is detected occasionally that does not produce tracks in film like any known radiation, which is designated "S" in Table 11. As people look more carefully, they continue to see behavior for which no conventional explanation exists.

4.6.1 Prompt X-ray Emission

Attempts to measure X-rays resulting from the slowing down of emitted particles (Bremsstrahlung) were unsuccessful early in the field's history.[269,581-584] However, in no case was evidence provided to show that cold fusion actually occurred during these measurements. In contrast, Bush and Eagleton[220] measured a rich assortment of characteristic X-rays while heat was being made by a Pd-Ag alloy cathode electrolyzed in LiOD (Figure 56). Miles and Bush[427] detected X-rays using a GM (Geiger-Müller) detector and X-ray-sensitive film when a cell was making energy. Evidence of radiation using a GM detector also is reported by Uchida et al.[568] and images on X-ray film cast by the internal structure of cells caused by X-ray emission is reported by Gozzi et al.,[102] Cellucci et al.,[135] Antanasijevic and co-workers,[585] (Institute of Physics, Yugoslavia) and Szpak et al.[224] (Naval Ocean Systems Center). An example in Figure 57 of X-ray exposure obtained by Szpak et al. shows an image of the nickel cathode structure on which palladium was being electrodeposited. Several point sources of X-ray are revealed by the multiple images. Violante and co-workers[463] (EURATOM-ENEA, Italy) also detected X-rays (20-25 keV) when Ni was electrolyzed in D_2O after which the unexpected copper detected on the surface of the cathode was found to have an abnormal isotopic ratio. Similar X-ray emissions were obtained by Itoh and co-workers[359] (Mitsubishi Heavy Industries, Japan) when $PdD_{0.83}$ was coated with copper and deloaded in vacuum while neutron and tritium production were observed to occur. X-ray bursts were seen using a NaI (sodium iodide) detector when palladium was electrolyzed in D_2O.[586] Lead (Pb) was found on the surface after the study. Karabut[289,296](FSUE "LUTCH", Russia) has detected X-ray emission that, instead of being emitted in all directions, was focused in well defined beams. More is said about this observation in Sections 5.2 and 6.2.9.

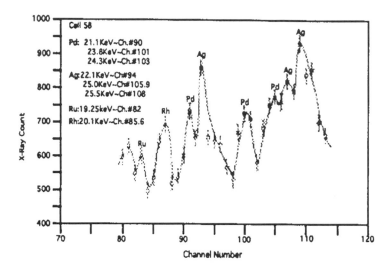

Figure 56. X-ray emission from a F&P-type cell while energy was produced using a Pd-Ag cathode. The number of counts in various channels is connected by a hand-drawn line.[220]

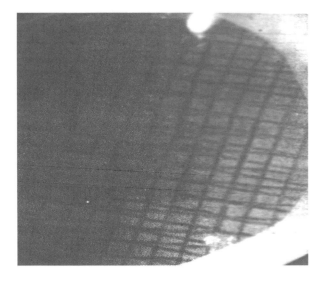

Figure 57. Example of X-ray exposure of film located outside of a Fleischmann and Pons-type cell. (Szpak *et al.*)

Only a few examples of such emission are available because suitable detection methods are seldom in place when the process starts.

4.6.2 Prompt Gamma Emission

Many efforts were made to detect the expected gamma emission from the D+D = He fusion reaction, but without success. Of course, many of these efforts were made when no cold fusion reaction was actually occurring. Nevertheless, this radiation has never been detected even when energy was being produced. As an example, Figure 58 shows only a small increase in gamma emission between 2.64-3.14 MeV and no change above 3.14 MeV while 2.5±0.2 W of excess power is produced, as reported by Scott et al.[234] (Oak Ridge National Laboratory)

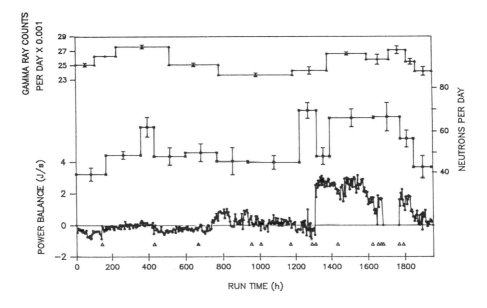

Figure 58. Heat, neutron, and gamma emission produced by a F&P-type cell. (ORNL)

Prompt gamma radiation also results when neutrons interact with normal hydrogen in the environment by the $^1H + n = {}^2D + \gamma$(2.22 MeV) reaction. This method has been used on occasion to detect possible neutron emission.[2,3,17,587,588] and resulted in grief[589] for Fleischmann and Pons when their evidence for neutron emission was found to be faulty because of a calibration error. This error was corrected by Petrasso et al.[589] and resulted in no peak being present at 2.22 MeV. Failure to detect neutrons using this method is understandable because only a small fraction of the neutron flux will result in gamma-ray production. In addition, neutron emission may not be associated with heat production at all, as is argued in Sections 5.2 and 6.2.9.

On the other hand, gamma rays near this energy might be produced and need to be distinguished from the possible 2.22 MeV peak. For example, Fleischmann *et al.*[3] detected a gamma energy at 2.8 MeV[1] while heat was being produced. Rather than this observation being used to discredit Fleishmann and Pons, as was done at the time, the corrected value provides useful information about the process. Other gamma energies have also been reported. For example, gamma emission is detected when an active nickel sample is held in hydrogen between 350 and 750 K.[342] The reported value of 661.5 keV, shown in Figure 59, was corrected by later measurements to 744 keV according to Takahashi.[536]

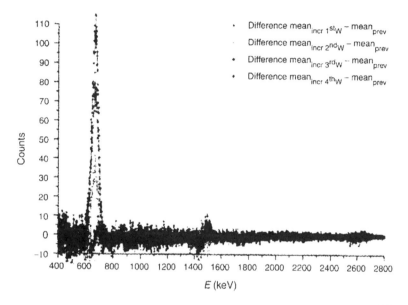

Figure 59. Gamma emission from nickel held in H$_2$ at 350-750 K.[342]

Radiation is frequently reported and expected when sufficient voltage is applied to produce a gas discharge.[55,290,293,296,590] However, in some cases the energy of this radiation is greater than would be expected to result from the applied voltage[492,591] alone. Gamma-ray energy between 80 and 230 keV, with an occasional pronounced energy at 228 keV, has been detected when no more than 6 keV was applied to palladium in D$_2$. After applying several hundred volts to low pressure gas, Karabut *et al.*[288] observed gamma emission from the cathode after the discharge was terminated, with an intensity *vs.* ion dose shown in Figure 60. Once the dose of H+ or D+ had reached a high value, significant radioactive decay by gamma emission was observed when the cathode was palladium, zirconium or silver. Shinya Narita *et al.*[591] detected gamma emission during glow discharge of palladium in D$_2$ using 4000-6000 V. Energy of peaks

---

[1] Based on the reported value after it was corrected by Petrasso et al.[589].

are shown for each sample in Figure 61. A variety of energies were produced depending on previous treatment of the sample. Hiroshi Yamada *et al.*[493] observed a single strong line at 106 keV when 660 V was applied to low pressure D_2 (Figure 62) using a palladium cathode. No neutrons were detected. Afterwards, a considerable increase in ^{63}Cu and ^{56}Fe was found on the palladium surface. These elements were not detected when the discharge did not produce this gamma radiation. The sample was also found to be radioactive, based on exposure of X-ray film, with two types of radiation being proposed.

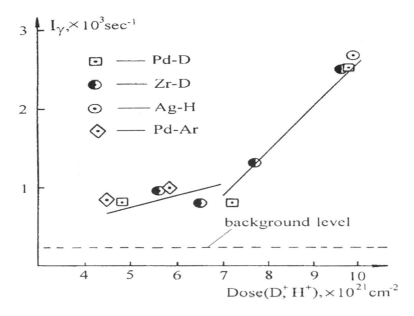

Figure 60. Intensity of delayed gamma emission from various cathodes as a function of ion dose. [288]

These observations are consistent with energy being emitted as a nucleus loses energy after a nuclear reaction has taken place. Some energy loss is prompt and some is delayed. In addition, a process able to raise electrons to a higher energy and generate energetic X-rays must be present. Such behavior is known to result from conventional nuclear reactions, but is not expected under these conditions.

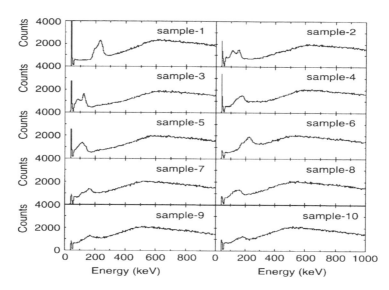

Figure 61. Gamma energy emitted during glow discharge at 4000-6000V in D_2.[591]

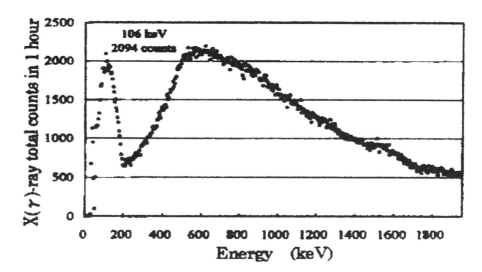

Figure 62. Gamma energy produced during glow discharge in D_2, after which Fe and Cu were found on the Pd cathode. The broad peak above 200 keV is background. (Yamada et al.)

4.6.3 Prompt Particle Emission

None of the possible charged particles can exit a cold fusion cell because they are easily stopped. As a result, such emissions must be detected within the cell.

A widely used method for detecting such particles involves a plastic called CR-39. When an energetic particle passes into this plastic, a change in chemical properties is produced where the particle traveled. This changed region can be dissolved by a solution of hot sodium hydroxide (NaOH) to produce a hole where the particle passed. The more energetic the particle, the more plastic will be dissolved, hence a larger hole is visible under a microscope. This method is described by Wang et al.[570] and by Roussetski[549,592] and has now been applied with success to detect energetic particle emission within cold fusion cells. Of course, some efforts have failed to find particles using this method, presumably because the NAE was not present. These failed efforts can be considered as "blanks" because they prove that the plastic does not always form particle tracks when exposed to the same chemical environment used when tracks are seen.

Early use of CR-39 produced no results.[593-597] One of the first successful observations was provided by Li et al.[572,573] after they had loaded and deloaded palladium in D_2 gas. They found that particle emission could be stopped by cleaning the sample with Aqua Regia or by exposing it to chlorine. Chlorine was discovered to have penetrated into the palladium surface, where it is known to prevent reaction between D_2 and Pd. In other words, conditions that prevented loading also did not generate energetic particles. Jin et al.[571,598] also found tracks in CR-39 after loading and deloading palladium in deuterium. For some reason, Russian palladium[569] was especially active in producing tracks, showing once again the importance of material properties. A similar result using a different source of palladium was obtained by Manduchi et al.[559] Roussetski[550] identified some of the particles emitted during deloading as being alpha and proton along with some neutrons. Lipson et al.,[547] using a PdO/Pd/Au heterostructure loaded with deuterium, concluded 3 MeV proton and 1.0 MeV triton tracks were produced during deloading. All emissions detected during these loading and deloading studies are at very low levels and some of the energetic particles, if not all, might result from crack formation in the metal rather than from normal cold fusion.

In addition to using the loading-deloading method, energetic particles have been sought in electrolytic cells under steady-state conditions. Lipson et al. (University of Illinois)[534] studied titanium and palladium coated with PdO and/or gold, using either D_2O or H_2O in the electrolyte after implanting helium into the surface of some samples. Tracks corresponding to 9-14 MeV alphas were found regardless of the hydrogen isotope used in the electrolyte. The presence of implanted helium increased track density. Other kinds of samples were also used. Palladium[541,545] was deposited as a thin layer on one side of a Al_2O_3 plate with two pieces of CR-39 clamped tight against each side. The piece clamped to the Al_2O_3 side served as a blank. This assembly was electrolyzed as the cathode using $Li_2SO_4 + H_2O$ as the electrolyte. Because the CR-39 plastic covered the Pd film from which emission was measured, no direct loading could occur in this region. This required any nuclear activity to be fueled by hydrogen diffusing from the edge of the shielded region, thereby providing a relatively low concentration of hydrogen to the active region. Nevertheless, 1.5-1.7 MeV protons and 11-16 MeV alphas were detected. In contrast, Taniguchi and co-workers (Osaka Prefectural Radiation Research Institute, Osaka)[575] found emission only when D_2O was used. They detected what appeared to be protons about 7 days after electrolyzing palladium deposited on copper, using a gold anode. Emission was detected using a $LiOD+D_2O$ electrolyte, but not when $LiOH+H_2O$ was used. In this case, gold would have been transferred to the surface of the palladium from the anode, with unknown consequences.

Researchers at SPAWAR Systems Center San Diego have recently produced tracks on CR-39 when it was placed in a cell containing various metal cathodes on which palladium was being deposited. The emission was observed only when an external electric or magnetic field was applied.[m] These fields are claimed to modify the morphology of the deposited palladium.

Besides various solids being a source of nuclear products, Richard Oriani and John Fisher[537,538,599] observe tracks in CR-39 when it is suspended in the vapor of an electrolytic cell, well out of the range of any particles produced at the cathode. Tracks were observed when the cathode was either palladium or nickel and when the electrolyte was based on either H_2O or D_2O. The track pattern on occasion indicated a point source near the CR-39 surface, which suggests a suspended small solid particle as the source. To make the observations even more challenging to understand, anomalous tracks were detected when the CR-39 was located within the cell, but well isolated from the electrodes or even outside of the cell.[600] An explanation offered by Fisher is discussed in Chapter 8.

Energetic particles are observed when the gas discharge method is used. When Lipson et al.[290] subjected titanium to high-current pulsed glow discharge in deuterium at voltages between 0.8 to 2.45 kV, the 3.0 MeV proton yield[n] was found to be higher than conventional cross-section would predict. This behavior is attributed to a screening potential of 620±140 eV. Other studies (Table 12) have reported this effect as well. Even though the rate of tritium-proton production is greater than expected, the absolute rate of this reaction decreases rapidly as applied energy is reduced. Consequently, simple shielding alone cannot explain the rarely observed high rates of tritium production in Fleischmann-Pons cells where the applied energy is much lower than used during these measurements.

---

[m] Details of the work can be found at http://www.newenergytimes.com/news/2006/NET19.htm#ee.

[n] Protons are produced when tritium is made by the normal fusion process.

Particle emission has been further confirmed by other researchers. Cecil and co-workers,[542] using a Si-SSB detector, found 6.8 MeV alpha during glow discharge of TiD. Deloading PdD at 20° C[565] and glow discharge of PdD[214] resulted in alpha emission at somewhat lower energy. Proton energy near 3 MeV has been detected by several studies.[549,574] "Strange" particles[483,518,563,576,590] have also been seen as tracks in photographic film. These emissions do not behave like normal particles because the paths are not linear and the particle frequently shows what appears to be sudden annihilation. Whether these particles can affect CR-39 is unknown.

Very intense alpha particle emission is expected when helium is detected and energetic protons and tritons should accompany the generation of tritium. Previous failure to detect these particles encouraged a search for mechanisms able to communicate the nuclear energy directly to the lattice rather than to individual particles, as is normally the case. As yet, particle emission has not been sought while detectable helium or tritium is produced. Instead, energetic particle emission has been found to occur under a variety of unexpected conditions having no clear relationship to other aspects of the cold fusion phenomenon and at very low levels. Nevertheless, these observations raise several questions. How much of the energy resulting from cold fusion is communicated directly to the entire lattice and how much, if any, first accelerates individual particles before being detected as heat? Do the detected energetic particles result from heat and helium production, but at a very low rate, or do they result from minor, easily initiated, secondary reactions—although ones that are unexpected?

In summary, energetic particles are found regardless of how much energy is applied or which hydrogen isotope is used. Apparently these nuclear reactions are easy to produce and most of the resulting nuclear energy is carried away by the emitted particles and not coupled directly to the lattice as some models have proposed. In other words, although the reactions are unexpected, they behave just like normal nuclear reactions. At the other end of the energy spectrum, such as bombardment by energetic particles or electrons, conventional nuclear reactions are produced, as listed in Table 12. However, the rates are sometimes much greater than conventional theory would predict. Unknown at the present time is whether a relationship exists between the low-energy and the high-energy conditions. Also unknown is the relationship between energetic particle emission and the heat-helium reaction. Is helium generated only as energetic alpha particles or is some of the energy coupled directly to the lattice? Is it possible for some emissions to remain undetected using present methods and these carry away part of the energy, especially when single-product nuclei are formed? Does "strange" radiation play a role in transferring energy?

Table 12. Selected studies of metals bombarded with energetic particles.

Source	Ion	Energy	Target	Result
Minari et al.[601]	D+	10-24 keV	Li(liquid)	no enhancement
Huke et al.[602-604] Takahashi et al.[62,77,605-607]	D+	5-60 keV 50-300 keV	Al,Zr,Pd,Ta,Li,Sr	branching ratio change
	D+	keV	TiD	3d fusion
Kitamura et al.[608]	D_2+	15-25 keV	Pd-Au	Enhancement increase
Miyamoto et al.[609]	D_2+	15 keV	Pd-Au, Pd	Enhancement Pd-Au>Pd
Goryachev et al.[490]	e-	27 MeV	$_{28}Ni^{58}$	$_{45}Rh^{102}$ produced
Wang et al.[494,610]	H+	330 keV	TiD	2.8 MeV alpha
Kubbota et al.[611]	D_2+	300 keV	TiD	5-8 MeV alpha
Kamada et al.[612-615]	e-	175 keV	AlD	melting, particle emission
Wang et al.[616]	D+	96 keV	TiD	3.3 MeV alpha, <30 MeV gamma
Wang et al.[617]	D_3+,D+	20-100 keV	Pd	unknown between 1-3 MeV
Wang et al.[618]	D_3+,D+	10-100 keV	Pd	D_3+ is more effective
Wang et al.[616]	D+	96 keV 240 keV	TiD PdD	3.2 MeV alpha, tritium? transmutation?
Takahashi et al.[84,619]	D+	20-300 keV	PdD,TiD	4.75 MeV triton and ^{3}He
Ochiai et al.[425,620]	D+	150-300 keV	TiD	2.8 MeV proton, , 4.75 MeV triton
Kasagi et al.[621]	D+	150 keV	TiD	2.75 MeV protons, 3d fusion
Kasagi et al.[622]	D+	2.5-10 keV	Pd, PdO-Au	enhancement increase
Wang et al.[623]	D+,H+	1-18 keV	Ti,Pd	40-60 KeV X-ray, 3-4 MeV protons
Shinojima et al.[624]	D+, D_2+	2.5-20 keV	Pd-Au	branching ratio 1:1
Kitamura et al.[625]	D_2+	30 keV	Pd-Al	expected p, n, and ^{3}He
Chindarkar et al.[626]	H+	5 keV	Pd, Ti	radioactivity
Baranov et al.[627]	e-	2.6 MeV	LaSm,LiSn,LaNd	radioactivity, transmutation
Kasagi et al.[628]	D+	150 keV	TiD	17.5 MeV proton
Iida et al.[629]	D+	240 keV	PdD, TiD	3d fusion
Rout et al.[630]	D+	12keV	Ti	tritium
Cecil and Hale[631]	D+	5-100 KeV	Li^6	branching ratio 1:1
Beuhler et al.[632,633]	D_2O cluster	200-325 keV	TiD, ZrD	proton, triton, ^{3}He
Chambers et al.[634,635]	D+	0.35-1 keV	Pd	5 MeV, unknown emission

4.6.4 Radioactive Decay

Radioactive decay resulting in gamma emission can be detected through the walls of the apparatus during an experiment or, with much greater sensitivity, after the sample is removed from the apparatus. Decay resulting in beta or alpha emission can only be detected after the sample is removed from the apparatus because such emission is easily absorbed by the walls. Although radioactive products, other than tritium, are seldom reported, clear examples of radioactive decay are occasionally observed.

A rather famous gamma spectrum produced by radioactive decay was obtained by Wolf (Texas A & M) in 1992 after electrolyzing a cell containing a palladium cathode with lithium, boron, and aluminum in the D_2O electrolyte. Part of the complex spectra is shown in Figure 63. At the time, the result was not widely publicized, but was well known nevertheless. Neither Wolf nor anyone else could reproduce the results and careful analysis revealed the spectrum could only result from high-energy deuteron bombardment.[636] To this day, the meaning of this data is unknown and generally discounted.

Other examples are not as ambiguous. Film exposure was observed by Bazhutov et al.[637] along with tritium production when titanium was used as the cathode. Autoradiographs have been made during numerous studies[286,404,507,551,555,556,638,639] after loading, which showed radiation having the low energy normally associated with X-rays, although some fogging might have resulted from tritium decay. In addition, radioactivity other than tritium[215,281,282,330,469,474,493,514,551,591] has been reported. Bush and Eagleton[163,474,519] produced a collection of gamma emitters with an average half-life of 3.8 days by electrolyzing rubidium carbonate in H_2O using a nickel cathode. This radioactive isotope might have resulted from the addition of two protons to common ^{85}Rb to produce ^{87}Y, a positron and gamma emitter having a 3.4 day half-life. Titanium electrolyzed in $K_2CO_3 + D_2O$ was found by Bernardini et al.[78] to become radioactive, which they explained by the proposed reactions $^{48}Ti + D = {}^{46}Sc$(83.8 day, γ+β) + ^4He and $^{50}Ti + D = {}^{48}Sc$(43.7 hour, γ−β) + ^4He. The sample also made excess energy. A gamma energy (76.8 keV), consistent with the presence of ^{197}Pt, was found by Mizuno[514] (Hokkaido University, Japan) to be emitted from a proton conductor (mixed metal oxide) after it was heated from 400° to 700°C in D_2 gas while a small alternating current was applied. This isotope is a beta emitter having a 18.3 hour half-life with an accompanying gamma ray. Presumably, adding a neutron to stable ^{196}Pt, which was present in the sample, could create this radioactive isotope. Gamma emission from what is claimed to be ^{24}Na was observed by Notoya[330] (Hokkaido University, Japan) after electrolyzing sodium carbonate in H_2O. This isotope is a beta emitter with a 14.95 hour half-life and accompanying gamma, which can be created by addition of a neutron to stable ^{23}Na. When an arc is created between nickel and carbon electrodes in an aqueous solution of ammonium molybdenate, a long-lived gamma emitter was produced by Nakamura and co-workers.[508] Gas discharge using palladium and deuterium produced radioactivity that lasted for more than three days after the discharge was turned off, as reported by Karabut *et al.*[146,214,215] In this case, the measured gamma energies correspond to emissions from [104m]Rh, [85m]Sr, [109m]Pd, and [105]Rh. The location of radioactive elements on their sample has been revealed by Savvatimova *et al.*[280,551] using autoradiography. They followed the decay over a 2-month period.

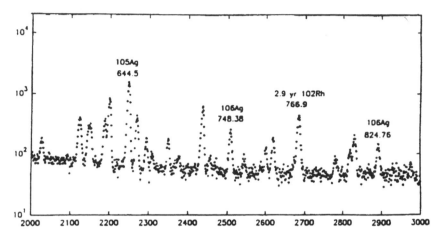

Figure 63. A partial gamma emission spectrum from 574 to 855 keV obtained by Wolf after electrolysis.

These examples of induced radioactivity are very rare and involve very small amounts of material. The isotopes proposed to be the source of detected radiation and the reactions from which they are created are not well established and may be incorrectly identified. Nevertheless, all of these observations are completely unexpected and can not be produced by normal processes. The results also show, once again, that energetic nuclear activity is located at the cathode of a cell.

An especially interesting study of emissions from palladium exposed to H_2 or D_2 was made by Rout *et al.* at the Bhabha Atomic Research Centre, India.[556,638] They found, by using autoradiography, weak radiation from palladium after it had been loaded with either H_2 or D_2 in a plasma focus device. This radiation continued as long as the sample contained some H_2 or D_2, but was not produced when other gases were used in the plasma device. Fogging of the X-ray film was not caused by chemical effects or emitted light. Later studies[555,640] revealed oxygen was required for fogging to take place, suggesting that recombination plays a role. The emissions caused exposure of films (>2 eV required) and of thermoluminescent detectors (>3eV required), but did not affect ionization detectors (>10 eV required), thereby establishing a range for the emitted energy. Fogging was enhanced by an electric field of either polarity, but reduced by crossed magnetic fields, which suggests that a mixture of particles having opposite charges was present. The emissions were able to pass through paper of 10 mg/cm^2 thickness, but were stopped by polyester film of the same thickness, ruling out normal emissions. As the authors conclude, this is a "new, strange, and unknown phenomenon". Regardless of the explanation, this phenomenon needs to be taken into account when claims for radioactivity in metals loaded with isotopes of hydrogen are based on autoradiography and perhaps when CR-39 is used.

4.7 Patterns of Behavior

The veracity of the observations can be evaluated in several ways. As noted earlier, many people must witness the same behavior. In addition, when the behavior occurs, it must be modified in the same way by changes in conditions. Therefore, the next question is, "Are the results affected by changes in conditions in the same way in all studies?" Again, the answer is yes as summarized in some recent reviews.[641,642] For example, when electrolysis is used, everyone finds the effect to be sensitive to the average deuterium content of the palladium cathode (Figure 32). Although the deuterium content has been explored in detail only when electrolysis is used, all methods are expected to be influenced by this variable. The increase in heat production as current density is increased is another universal behavior (Figure 44). Increasingly, people are detecting radiation and helium (Figure 47) when energy is produced. Unexpected elements with abnormal isotopic ratios are found on the cathode surface (Figure 53) and these frequently correlate with heat production and/or radiation. These consistent patterns are impossible to explain as error unless a variety of different errors happen to occur in each detector and in just the right combination in each independent study, which is very unlikely.

4.8 General Replication

Replication occurs when other people observe the same effects using essentially the same conditions. Unfortunately, in the case of cold fusion, the required conditions are not known. Occasionally, when a lucky combination of conditions has been created, the effects are observed. These effects have been seen many times, as the results listed throughout the book demonstrate, but not always on command. This failure of the effects to occur every time they are sought has become a major issue for the field and needs to be examined in detail because some confusion exists about what replication actually means.

Deciding whether sufficient replication has been achieved to justify accepting the effects as real phenomena is a complex problem. The issue can be examined in two opposing ways. On the one hand, when an anomalous effect is seen, it can be assumed to be neither novel nor important. On the other hand, when an expected anomalous effect is not seen, its absence can be assumed to be caused by failure to duplicate the required conditions. In other words, the approach taken to evaluate any observation is influenced by whether the observations are initially rejected or accepted. Let's discuss the former approach first.

Initially, many people assumed the behavior claimed by Fleischmann and Pons was not real. Instead, they thought that normal, but unexpected processes and errors were the cause. If so, the same behavior would be expected when light-water is used instead of heavy-water or when platinum is used as the cathode instead of palladium—so called blanks. Considerable criticism was directed at Fleischmann and Pons for not providing blanks, although they reported such information in a later paper.[10] Presumably, if enough blanks were run, the so-called anomalous effect would show up as often as it did using palladium and D_2O, thereby proving the effect was just error. As later work shows, a serious flaw exists in this demand. Light-water,[25,94,97,176,177,191,202,206,207,220,227,341,487,512,643] when used as an electrolyte, has been found to produce anomalous energy and nuclear products. In addition, platinum[67] will produce anomalous energy after extended electrolysis in D_2O because active material can deposit on its surface. As a result, without knowing this behavior might occur, the suggested blanks would have given a false conclusion to the Fleischmann and Pons work. Some of the best blanks turn out to be "bad" palladium used as the cathode and heavy-water containing a few percent light-water used as the electrolyte.[644,645] Ironically, this duplicates conditions used by some people who failed to make extra energy—work on which rejection of the claims was based. In other words, many attempted replications have actually used "blank" conditions without knowing and some "blanks" may have been active. This experience shows that much care and knowledge is required to properly evaluate "strange" effects. This is why good scientists keep an open mind until all knowledge required to make a proper judgment has been obtained.

On the other hand, if the anomalous effect is assumed to be real, replication requires that all of the important variables be known and controlled. As a trivial example, suppose a process is being explored that will not occur at temperatures below 20° C. This requirement must be known and care must be taken to keep the temperature above 20° for the process to be observed. If this requirement is not known and the temperature is allowed to change at random times, every variable may be exactly as it was before, but the sought for behavior will not be observed if the temperature should accidentally and without warning drop below 20°. In other words, replication is not possible unless the important variables are kept within their required range. Suppose the important variables are not known, what then? When this situation occurs, the sought after behavior will be infrequent and will appear to be random. Truly random behavior can be evaluated using statistical equations if enough values are accumulated. However, infrequent and nonrandom effects cannot be evaluated this way. Suppose, as in our example, observations of our mythical reaction were made over a period of time without providing any temperature control. Nothing would be observed in winter because the temperature would, we assume, always be below 20°, spring would produce sporadic behavior, and great success would be observed during the hot days of summer. Obviously, this is not random behavior, hence cannot be evaluated simply by increasing the number of measurements made in winter. Suppose we then explore the effect of temperature. It is winter so we start at 10° and duplicate the procedure while increasing temperature by 1°. As a result, we will see 10 failures before we get one success. Does this mean the effect is not real? No, because if we had started in summer, and cooled by one degree, we would have had many successes before we had one failure. Obviously, once the critical temperature is identified, we would have success every time. Conditions that control reactions in the real world are not as trivial and easy to identify as in this example, yet ignorance of the required conditions can have the same effect.

In summary, real but infrequent behaviors simply cannot be rejected based on reproducibility and statistical analysis. Such analysis only shows how often the behavior is likely to be seen when the experiment is repeated using knowledge available at the time. Nothing is learned about the reality of the proposed effect. Of course, if the behavior can not be reproduced at will, it can not be studied and it can not be made useful. This situation, by itself, is a good reason to ignore such an effect without judging whether it is real or not. By keeping the possibility open of the effect being real, new understanding might allow the reluctant behavior to be created on demand in the future. In contrast, rejecting an effect just because it conflicts with current theory is not a good reason and is a betrayal of the scientific method.

4.9 Questions About Individual Success Rate

The fraction of successful experiments a person might have produced is frequently requested, but often confused with general reproducibility as described above. Personal success is important because without frequent success the phenomenon simply can not be investigated by the particular researcher. A different set of methods and tools need to be used. Of course, personal failure generally leads the researcher to believe the effect is not real.

Repeated failure generally means the parameter space in which critical conditions fall is too small to be entered very often by accident using the available methods. Using our example, this would be like having success only when the temperature was between 20° C and 21° C, a range that would be entered very seldom by chance. In contrast, a high number means either the parameter space is very large, hence easy to enter by chance, or, better yet, the experimenter has identified the required values and can maintain the required conditions. In the case of cold fusion, the correct situation is not always made clear in the published description.

Few people can afford to spend time and money investigating something that occurs only when nature is in a good mood. How does this apply to cold fusion? Success is known to require very special conditions on a submicron scale. Examination of such regions cannot be made unless very expensive and specialized equipment is available. Without this equipment, the important structural condition and elemental composition remain unknowable when anomalous heat is occasionally produced. Without this knowledge, the special conditions can not be duplicated. One example of an invisible composition variable is the concentration of H_2O in the D_2O. Heavy-water absorbs normal water out of the air, which stops the cold fusion reaction when the H_2O concentration rises above about 1%.[645] Once this happens, success will not result no matter how many additional attempts are made. Because this is a known variable, it can be kept under control, but analysis for H_2O must be done with sufficient regularity to eliminate this as a cause of failure. Another potentially damaging condition occurs during electrolysis. A variety of elements react with the surface, some helping the nuclear process and some acting like poisons. Determination of the surface composition is expensive and the tools are rarely available. Without an ability to analyze for these elements, the required surface conditions can never be duplicated. Proper tools and tests are now being used more often as a few laboratories obtain the necessary funds and gain the knowledge to make cold fusion occur more often.

4.10 Duplication of Results (the Bottom Line)

How should a person judge whether these "impossible" observations are real or not? The gold standard in science is the answer to the question, "Do many independent observers witness the same behavior"? The answer in this case is an overwhelming YES. Many independent measurements of anomalous heat (Table 2), tritium (Table 6), helium (Table 7), transmutation products (Table 8), and radiation (Table 11) have been reported. These results are not easily explained by present theory. In addition, some studies reveal a high correlation between several different nuclear signatures. Failure to produce the effects every time should be taken as an example of ignorance rather than a reason for rejection.

Once this barrier to acceptance has been crossed, the question becomes, "Which of the various observations should be accepted and used to understand the mechanism of the process?" A detailed evaluation of individual studies now becomes an effort to extract just what nature is trying to tell us about a real but complex phenomenon, while using results that are sometimes confusing or completely wrong. In other words, cold fusion should be treated just like any other field of conventional science.

4.11 Explanation

Has the anomalous behavior been explained? No, a model has yet to account for all observations, although many attempts have been made and several theories look promising. Some of the theories are examined in Chapter 8. For a theory to be useful, the logical consequences of the model must be consistent with observation, including what is not observed, and with well established physical laws. Unfortunately, the absence of detailed knowledge about the NAE makes development of a useful model next to impossible. Without this knowledge, the model can not be applied to the unique conditions required to make the model work. As a result, most theories are based on a few assumptions that are justified by the author because a collection of equations is consistent with these assumptions. This approach has been taken largely out of ignorance about what has been observed and about the conditions actually present in the NAE, a problem this book may help clarify.

4.12 What Next?

Many essential questions still lack answers because the necessary measurements have not been made. In addition, current studies generally examine only one small anomalous behavior. Experimental design and description seldom allow the work to be related to other studies or to the general phenomenon. This approach is understandable in view of the need to prove that something strange was really happening. However, now the field needs to leave the provincial and defensive attitude of the past behind. A new and large window into the workings of nuclear chemistry has been opened and needs to be explored with gusto and imagination.

How should the phenomenon be explored? Heat and helium production provide the largest and most important indication that cold fusion has occurred. In addition, generation of these products at high rates is required for commercial application. Consequently, attention needs to be focused on the heat-helium reaction. Of course, this reaction can only be understood by measuring the associated radiation and particle emission. The other reactions have secondary interest, although they are important. For example, the rich assortment of transmutation products, while not presently produced at a high rate and are not of commercial interest, do provide a window into the basic nuclear mechanism. Although this branch of the nuclear process might someday be commercially important, it is too far from the necessary rate to be more than an example of a unusual secondary reaction. Neutron and tritium production fall into the same category. The problem is to learn why helium+heat is produced along with these minor additional reactions. What rules govern how the deuterons or protons are proportioned between the various reactions? What universal process is operating?

Why should we invest time and money to investigate this complex process? Fusion is the only known source of energy based on an unlimited supply of primary material. At the present time, attempts to harness this energy using hot fusion have not been successful. Even if the hot fusion reactors were able to make more energy than the complex machine needs to operate, the process might be too expensive and the generator too big to be accepted as a practical energy source. For this reason alone, many people would expect a rational society to explore a Plan B. Along came two chemists who claimed to make the fusion reaction work at useful power levels using a very simple device, a claim some people thought would make an excellent Plan B. As energy prices soar, the initial rejection is being overwhelmed by growing interest. Young students, who are not yet committed to the certainty of conventional thinking, are seeking information. A few companies have brought together the necessary money, skill, and tools. Mindless rejection of the idea is heard less frequently. Visits to the www.LENR.org website are increasing as curious people learn the facts for themselves. In other words, progress is being made toward having a Plan B. As Hurtak and Bailey[646] note,

> "The world's oceans contain a large amount of readily extractable heavy water, sufficient to meet the global energy needs for hundreds and perhaps thousands, of years. Heavy water production facilities will be needed. About one gallon out of every 7,000 gallons of ordinary water is heavy water (deuterium oxide or D_2O). The energy equivalent of a gallon of heavy water is about equal to 300,000 gallons of fuel oil. The cost of production of one gallon of heavy water is estimated at less than $1,000 or less than one cent per gallon of oil (energy equivalent)."

CHAPTER 5

Where Does Cold Fusion Occur and What Influences its Behavior?

5.1 Introduction

As noted previously, when nuclear products are formed by cold fusion, the reaction products are always found within small regions of the sample. Only the nuclear-active environment or NAE,[a] becomes active even though the entire sample is exposed to the same general conditions.[2] What distinguishes these unique regions from the surrounding inert material? Success in this field depends on finding an answer to this question.

A variety of environments have been suggested. Some are clearly present and some are imagined in order to make a theory plausible. Most theories completely ignore the real environments known to chemists and materials scientists. For example, the NAE is not pure, ideal PdD and it does not have a deuterium content equal to the measured average value, as is frequently assumed. As a guide to understanding what the NAE might look like, several candidates are worth examining even though none has been demonstrated to be a NAE. Future work is required to narrow the search.

5.2 Cracks

Cracks and small particles are the Yin and Yang of the cold fusion environment. Small particles are created between cracks and crack-like gaps are formed in the near-contact regions between small particles. The greater the numbers of cracks, the smaller are the isolated regions (particles) between cracks. The greater the number of small particles, the greater the number of small void spaces (cracks) where particles make contact. Consequently, these two kinds of environments cannot be separated when evaluating their possible effects, although they are discussed here as two distinct features.

Cracks and voids are the only environments obviously common to all successful experiments.[3-15] This realization caused many people[16-32] to suggest cracks as the location of nuclear activity. However, a distinction must be made between nuclear events when they occur as the crack forms (fractofusion) and when a process continues to operate well after a crack has formed. Generation of mild nuclear activity while cracks are forming has been frequently suggested,[16,33-53] although an occasional claim dismissing such a process[54] has been heard. While these observations are unique and important, the proposed

---

[a] A concept similar to NAE was suggested by Peter Gluck.[1]

fractofusion mechanism does not apply to the main cold fusion process, as has been occasionally suggested.[19,23,55,56] Conditions for crack formation are fleeting, the expected rate of formation is small, and the resulting nuclear products are typical of hot fusion rather than cold fusion. If cracks actually provide a location for a nuclear reaction, their final dimension and concentration must be major variables. What conditions influence these variables?

Electrolytic loading provides a steady supply of cracks because unavoidable concentration gradients produce stress.[57] For example, palladium expands by 10% (volume) upon reaching the beta phase boundary at $PdH_{0.7}$ and an additional 6% when the beta phase is increased to $PdH_{0.8}$. These cracks have with a wide range of dimensions. They range from being too small to be seen by a scanning electron microscope to being large enough to be visible to the unaided eye. Most cracks are invisible unless the sample is examined while fully loaded, hence fully expanded. Some cracks are only revealed as a line of bubbles as the material evolves gas while covered by a liquid, as shown in Figure 64. This process is described in more detail in Appendix C.

Figure 64. Bubbles rising from a palladium surface during deloading under acetone. (from Storms, unpublished)

Cracks are also present in other configurations as well. For example, palladium deposited on a surface by co-deposition[58-61] or bulk palladium black[62] will have space between individual particles that will behave like a crack. A sandwich made from CaO and Pd layers, as used by Iwamura and coworkers,[63] is expected to produce cracks as the palladium expands upon forming β-PdD and the CaO does not. Titanium is even more susceptible to crack formation than is palladium.[64-67] How might these cracks be involved in the cold fusion process? For the various observations to be explained, clusters are required and a mechanism must operate to allow them to penetrate the Coulomb barrier. Do cracks create these conditions?

Cracks and voids are the only demonstrated location of D_2 dimers (molecules) and to a lesser extent D_3 molecules that are proposed to be involved in a variety of observed nuclear reactions.[63,68-72] Such clusters are expected to be present on the surface of cracks were various processes might increase their energy and initiate nuclear reactions. Although similar clusters are proposed by Akito Takahashi and co-workers[73,74] (Osaka University) and Hagelstein[75] (MIT) to exist within the β-PdD lattice, this proposal is still based on an assumption rather than observed fact. Attempts to find D_2 dimers within a lattice at room temperature using neutron and X-ray diffraction have been unsuccessful[76-78] and calculations of whether such cluster can form give conflicting answers.[79,80,81]

Structures other than β-PdD might form on the surface of a sufficiently small particle, especially when it is exposed to a high deuterium activity. For example, a new phase[2] having the limiting composition of γ-PdD$_2$ might form when applied deuterium activity is sufficiently large, such as on the surface of a cathode when high-current is applied. This phase might also form when palladium-black is exposed to high-pressure deuterium gas. If the structure of this new compound contains closely spaced sites occupied by deuterons, as is the case for ZrH_2, dimers might easily form within the lattice. This phase is expected to be stabilized by lithium, which is known to slowly dissolve in the surface during electrolysis. Going from two atoms in a dimer to 6 atoms in a Iwamura cluster is a problem too difficult to discuss here.

Once clusters form, how might they be caused to pass through the Coulomb barrier? Cracks can have dimensions that are small enough to support a resonance process, as suggested by several people,[82-88] This might cause local accumulation of sufficient energy to overcome the Coulomb barrier. The Casimir effect might even operate within these small dimensions with unknown consequences.[89] If these processes are important, initiation of a nuclear reaction should be sensitive to the dimension of the crack. Even though a wide range of dimensions are available, the correct dimension might be present only in rare and isolated locations. Successful samples would have more cracks of the required dimension than would other samples. Some D_2 molecules in some cracks might even be subjected to conditions similar to those in a high-pressure gas. As suggested by Bockris[4,90,91] (Texas A & M), deuterium experiencing such conditions might contribute to the process.

The presence of cracks on the surface would allow laser light to be more easily absorbed, which could raise the energy level within the crack structure.[66,92-102] A method using a low-powered laser, pioneered by Letts and Cravens,[92,93] required application of a gold layer containing a high concentration of cracks and small particles. On the other hand, Nassisi[100](University of Lecce, Italy) observed the highest concentration of transmutation products to be located in ~1 μm diameter pits that had formed in a clean Pd surface after intense XeCl excimer laser irradiation. Apparently, lasers of sufficient power can make their own small crack-like regions, hence work without the need to apply a special layer of any kind.

What happens to radiation originating from within a crack? An answer might be provided by the observed very tightly focused laser-like X-ray emissions reported by Karabut.[103] ("LUTCH", Russia) X-rays originating in and emitted from cracks might be tightly focused by reflection from the walls of the crack. Such emissions would be frequently missed by detectors located at the wrong position, hence be more common than present experience would indicate. This observation might provide a method to explore the role of cracks.

As the reader can appreciate, cracks have some characteristics that make them good candidates to be a NAE. If so, the size of the crack must be very critical because otherwise the large range of values known to occur by chance should make the effect easy to produce, which it is not. In addition, the concentration of hydrogen or deuterium in an active crack would be important but hard to control and measure. Fortunately, the technology is now available to make repeating structures having small gaps of known dimensions. This technology, rather than relying on the chance production of gaps, should be applied.

5.3 Nanosize Particles

Arata and Zhang[62,104-107] at Osaka University demonstrated in a series of important studies, unusual results when certain batches of nanosized palladium (palladium-black) were exposed to pressurized deuterium gas. Significant energy, helium, and a little tritium[108] resulted. This behavior has been replicated by McKubre and co-workers[109] (SRI) with the help of Arata. In addition, Arata has made several of these active samples and finds energy can again be produced when samples are repressurized with deuterium after having been stored for more than eight years. This experience demonstrates that behavior based on nanometer-sized particles can be reproduced within a laboratory and between two laboratories.

Case,[110,111] with help from people at SRI,[109,112,113] demonstrated energy and helium production when palladium was deposited on charcoal and exposed to hot, pressurized D_2 gas. This method has been difficult to replicate[114] because too many variables are unknown about how to deposit palladium of the correct form on a carbon substrate. Some basic information about such materials is provided by Nag (Engelhard Corp.).[115] Because the active particles are too far apart to sinter into larger inactive particles, energy can be produced at higher temperatures than by the Arata method.

Arata[107,116] made a particularly small particle size by oxidizing $Pd_{35}Zr_{65}$, which is a mixture of the two phases $PdZr_2$ and $PdZr$. This process will result in ZrO_2 and a more dilute but very finely divided Pd-Zr alloy. Not all of the zirconium will be removed from the palladium and converted to ZrO_2. This dilute alloy will reach a higher D/Pd ratio than pure Pd of the same particle size because dissolved zirconium reacts more strongly with deuterium than does pure palladium. In addition, a large amount of deuterium will be absorbed on the particle surface rather than being dissolved.

When deposits are applied from a starved[b] electrolyte, as is the case in a normal cold fusion cell, small islands of deposited material grow at chemically active sites. These growths often have nanometer dimensions and produce the observed black color. Such structures are visible in many published pictures of surfaces obtained using an SEM.[117-120] However, most of these visible structures are not nuclear-active. In addition to small islands of material, the surface is frequently covered by a very thin and generally invisible layer containing a rich assortment of elements. The thickness of this layer needs to be considered as one of the important variables in forming the NAE at certain locations. This very complex surface structure resulting from electrolysis makes identifying and replicating the NAE very difficult using this method.

Deuterium is not the only active hydrogen isotope. Nickel exposed to hot H_2 gas was discovered by Piantelli to produce heat and nuclear activity. This behavior has been studied by Focardi[121,122] and co-workers at University of Bologna and University of Siena (Italy) for many years and has been replicated by Cammarota and co-workers.[123] The loading-deloading treatment they used to form nickel hydride is expected to result in many cracks and produce a large number of isolated small particles within the surface. Also, formation of black nickel oxide and its subsequent reduction by hydrogen, as has been used, is expected to result in many small particles of nickel hydride on the surface. Finely divided nickel (Fibrex) also has been found to host nuclear reactions when it is used as the cathode in an electrolytic cell containing a H_2O-based electrolyte.[124-127]

The smaller the particle size, the greater is the total D/Pd ratio. According to Cox et al.[128], the D/metal ratio for some elements can be as high as 8 when the cluster of metal atoms contains less than 5 metal atoms. When the number of metal atoms is sufficiently large to form a clearly defined crystal lattice, the amount of deuterium dissolved in the lattice is not affected by particle size.[129] The total D/Pd ratio becomes larger as a particle becomes smaller only because surface adsorption[130] overwhelms the fixed amount of deuterium dissolved within the lattice as the amount of surface area increases. With two different environments being occupied by deuterium, it is difficult to discover which environment is nuclear-active. Are small particles nuclear-active because their small physical size supports a resonance process within their

---

[b] A starved electrolyte is one in which the ion being deposited on the cathode is at very low concentration. This condition produces a much different deposit than does a larger concentration.

interior or are they active because they have a high deuterium concentration on their exterior surface? Because not all small particles are nuclear-active, other factors must play a role as well.

During glow discharge, but especially during electrolysis, the deuterium/hydrogen concentration and activity present at each particle is different and the concentration continuously changes. The deuterium content is controlled by competition between two processes. Deuterium is added by ion formation when current from the electrolyte or plasma flows into a particle. Some ions dissolve in the particle, but most either dissolve in the substrate or combine on the surface. In either case, gas leaves the surface. This gas forms bubbles during electrolysis, which shield some particles from further loading as the bubble grows. As a result, particles experience a fluctuating deuterium content as bubbles come and go. If the bulk composition is sufficiently high, it can supply deuterium to the surface fast enough to compensate for these interruptions, thereby keeping the average composition high. This rapid change in local composition and deuterium flux may play a role in initiating local nuclear-activity. This process also provides an explanation for so-called life-after-death.[131] This process describes heat generation that continues after current to the cell is stopped. Based on the mechanism described here, once cell current is stopped, deuterium continues to be supplied to the active particles from the surrounding material, just as occurred while bubbles covered the surface. This source of deuterium is proposed to support nuclear-activity until the average composition of surrounding material drops below a critical value. If this were how the process operates, a larger sample would be expected to have a longer life-after-death than a small one. The one-cm cube of palladium used by Fleischmann and Pons during their early studies, which melted through the beaker and bench after an explosion stopped the current, would seem to be the ultimate example of this process. Xing Zhong Li[85] (Tsinghua University, China) explains this effect by assuming that the energy release process has a lifetime of the order of 10^4 seconds. However, this lifetime is not consistent with many observations. Except for a few rare occasions, energy production stops just as soon as current stops.

Several questions need answers. Must the NAE have a small dimension, either as a particle or as a thin layer? Is size the only requirement or must the small particles also have a definite physical structure and chemical composition? Is the deuterium located within the lattice structure involved in the nuclear reactions or is adsorbed deuterium required?

5.4 Dendrites

Dendrites are small wire-like structures that grow from a surface. In many respects, they are small particles, but with some additional features. Because the tips have a very small radius, large voltage gradients may form during electrolysis or when they are present during glow discharge. Bockris and his students[132-134] propose that tritium production occurs at the tips because the tritium content in their cell increased when this structure was present and stopped when they were removed by shaking the cell. In this experiment, the dendrites contained Ni, Fe and Cr, which formed on palladium[135] either from cell components or as a result of transmutation reactions. Some dendrites, as explained above, would experience repeated loading and deloading cycles as bubbles formed. It is unclear which of these processes, voltage gradient or a flux caused by loading and deloading, is more important. This uncertainty applies to any mechanism proposed to operate on a surface exposed to electrolysis.

A high voltage gradient (V/cm) alone does not impart energy to ions. Energy is imparted by the total voltage through which an ion passes, not the gradient. The calculated gradient is large only because the dielectric barrier is very thin or because the tip radius is very small, not because high voltages are actually present. The total voltage at an electrode in an electrolytic cell is fixed at no more than a few volts by the chemical processes, hence it is not able to impart significant energy. In addition, to the extent that ions are created and attracted by the gradient, the voltage would be neutralized and its effective value reduced, thus making the effective magnitude of the gradient much smaller than calculations would predict. This uncertainty makes the loading-deloading process more likely as the important variable operating on dendrites. Nevertheless, the question remains, "Why would the loading-deloading process combined with a dendritic structure favor formation of tritium over other nuclear products or, indeed, produce any nuclear reaction at all?"

5.5 Role of Lithium and Other Alloys

Lithium salts consisting of Li_2SO_4 or LiOD are commonly used in electrolytic studies when palladium is used as the cathode. Although lithium will not deposit on the cathode as the metal, it can slowly react with palladium to form Li-Pd compounds,[136-143] of which a variety are known (Figure 65). Thermodynamic values and phase relationships have been calculated by Howald (Montana State University).[144] The palladium-rich alloys are inert, but when lithium-rich, they can dissolve in the electrolyte, thereby adding Pd^{++} ions to the liquid.[145] These ions gradually plate palladium on the cathode to form structures that were not initially present. These deposits will also gradually acquire enough lithium to dissolve and repeat the cycle. The process slowly changes the surface until steady-state is achieved between dissolution and redeposition. Long electrolysis might be required before enough NAE is produced to make detectable heat. In this case, the NAE would be the complex deposited structure. This process can also explain the sometimes limited lifetime for excess heat production and the burst production of neutrons and tritium. Presumably, the various nuclear reactions occur only while the surface has the correct lattice structure, thickness, and composition needed to initiate the particular nuclear reaction, conditions that are constantly changing during electrolysis.

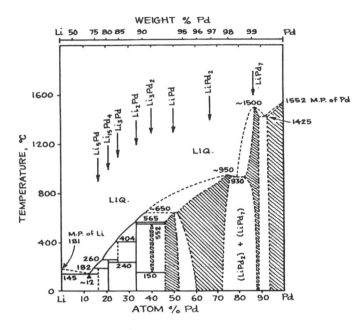

Figure 65. Phase diagram of the Li-Pd system. (from: W. G. Moffatt, Handbook of Binary Phase Diagrams, General Electric Co. 1978, based on the work of Loebich and Raub[152])

Extended electrolysis covers the surface with many other elements besides palladium and lithium. Platinum, silicon, and carbon are also deposited along with unexpected impurities in the heavy-water and cell components that dissolve in the electrolyte. As a result, the surface is a complex alloy, the composition of which changes with location on the surface and over a period of time. This behavior makes finding and identifying the NAE very difficult and is one reason the field has been slow to solve the reproducibility problem. Theory has been ineffective in explaining the process because only ideal, pure palladium is considered rather than what is actually present.

When gas discharge and ambient gas loading methods are used, lithium and the other impurity elements are present in very small amounts, if at all, yet heat and nuclear products are produced. Consequently, they are apparently not an essential part of the NAE. They might accelerate formation of the NAE during electrolysis and may be involved in a nuclear process like any other "seed" once the environment makes this possible.[146] Nevertheless, some impurity elements do seem to be important to success. Which ones are important and the role they play remain unknown.

5.6 Deuterium Flux

Iwamura and co-workers[147](Mitsubishi Heavy Industries Research Center, Japan) find that the flux of deuterium passing through their Pd-CaO sample to be important. X. Z. Li and his group[148-150] at Tsinghua University (China) find evidence for nuclear activity when deuterium is caused to diffuse through a thin tube made of only palladium. In addition, McKubre and co-workers[151] (SRI) extracted a flux effect from their observations of excess energy resulting from the electrolysis (see Figure 50). These studies are based on an overall average diffusion rate. However, only the local diffusion flux within the NAE can influence the energy-producing process. This local flux has a wide range of values depending on the presence of cracks, the size and orientation of grain structure, and the local deuterium concentration. These variables will have different values in different samples, hence will produce different behaviors. Energy will only result after a NAE forms because simple diffusion by itself is not expected to initiate energy-producing reactions.

5.7 Role of Hydrogen Isotope Content

From the beginning of this research, many people expected the nuclear reaction rates to be greater for samples containing a larger concentration of reactants. This means a large D/Pd ratio would be necessary, which was demonstrated by McKubre and co-workers at SRI (see Figure 33).[153] Since then, many workers have measured the deuterium content of their samples using various methods, as described in Appendix F. Most methods measure only the average deuterium content of the entire sample, not that in the NAE. The results give a lower limit to the composition of any NAE present on the surface when the electrolytic and glow discharge methods are used. Ambient gas loading will show less difference between the average and the surface composition if sufficient time is allowed for the sample to reach equilibrium. Even so, the amount of local concentration caused by absorption on the surface can be high, but it is frequently unknown and difficult to measure. For this reason, the actual deuterium or hydrogen concentration in the NAE is unknown.

In addition to the unknown difference between the average and the composition in the NAE, another serious uncertainty may influence understanding. All heavy-water (D_2O) contains some H_2O, which increases with time as exposure to the atmosphere allows water to be easily absorbed. All D_2 gas also contains some H_2 gas. In both cases, the hydrogen content can be greater than that specified by the supplier. Normal hydrogen enters palladium much more easily than does deuterium. Even if a small amount of H_2O is present during electrolysis or H_2 during gas loading, the H/D ratio in the palladium will be much greater than expected. Donohue and Petek[154] found 25% H in palladium after it was electrolyzed in 99.9% D_2O. Guilinger and co-workers[155]

(Sandia National Laboratory, NM) measured a separation factor (S) of 9.2-9.3 as defined by the relationship:

$$S = (H/D)metal/(H/D)liquid.$$

Dandapani and Fleischmann[156] obtained similar values. Other researchers[154,157-160] have observed this behavior as well. Using the measured separation factor, the hydrogen content of palladium can be calculated for various conditions and the result is shown in Figure 66. When 10% of the a D_2O-based electrolyte is hydrogen, 50% of the sites in the resulting palladium hydride lattice are occupied by hydrogen. Even when commercial D_2O containing 99.0% D_2O is used, about 10% of the positions in β-PdD are calculated to be occupied by hydrogen. This problem might be worse when lithium is dissolved in the lattice and if γ-PdD_2 should form on the surface. This means that a sample thought to be high in deuterium could actually be very dilute in this essential isotope, which would cause loss of nuclear activity based on deuterium.

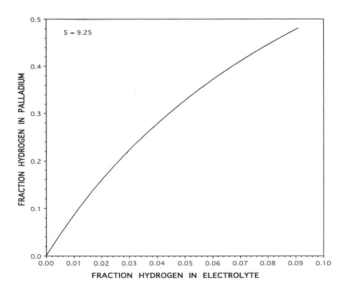

Figure 66. Effect of hydrogen in the electrolyte on the hydrogen content of a palladium cathode during electrolysis.[155]

Excess power production is stopped by addition of 1.6 atom % H_2O to D_2O in an electrolytic cell when palladium is used as the cathode, as shown in Figure 67. If enough deuterium were present in a sample loaded to D/Pd = 0.7 with pure deuterium to make observable heat, the same sample electrolyzed in 98.0% D_2O would have to be loaded to (D+H)/Pd = 0.83 to contain the same concentration of deuterium. This explains why very high D/Pd ratios are sometimes needed to make the electrolytic method work. To make matters worse, the separation factor is sensitive to conditions on the cathode surface so that the actual situation in a particular experiment might be much worse than these calculations would indicate. In addition, when PdD is analyzed and only deuterium is assumed to be present, the unexpected presence of hydrogen will produce a negative bias in the calculated D/Pd ratio. In other words, less deuterium will be present than the measurement indicates. Even the resistance method for measuring the D/Pd ratio can be affected by this error because deuterium produces a different resistance change than does hydrogen.

If deuterium clusters containing up to 6 deuterons are involved in the nuclear process, as the work of Iwamura (Table 10) indicates, the nuclear reactions will be very sensitive to the deuterium content of the NAE. Therefore, failure to produce heat or transmutation may be caused partly by the expected deuterium content being diluted by hydrogen even though the lattice sites are mostly filled and analysis indicates a high deuterium content. This problem, in addition to the absence of a NAE, might be one more reason why replication has been so difficult.

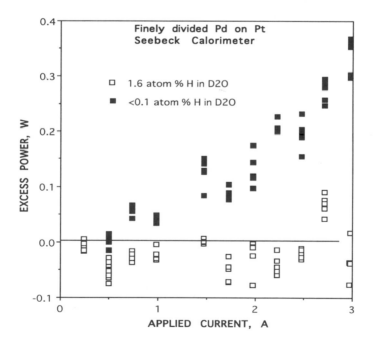

Figure 67. Effect of H$_2$O being present in D$_2$O on the production of excess power.(Storms[161])

5.8 Role of the Hydrino and Hydrex

Mills[c] [162-166] theorized and provided evidence for the electron in hydrogen or deuterium to occupy fractional quantum levels below the conventional Bohr level. This process is proposed to release more energy than is generated by a normal chemical process, but less than a nuclear reaction produces. Mills calls such atoms Hydrino or Deutrino and proposes that they form if a suitable catalyst is present to absorb the transition energy by a nonradiative process. During this process, an electron in the catalyst atom is raised to higher energy while the electron in the hydrogen drops to a lower energy, without a photon being involved. This energy eventually appears in the apparatus when the catalyst reverts back to its initial condition. Meanwhile, the hydrino or deutrino can form very stable compounds with various atoms, including other hydrogen or deuterium. When the orbiting electron has lost sufficient energy by moving into increasingly smaller fractional quantum levels, it may be close enough to the nucleus for the combination to have little apparent charge, allowing it to enter another nucleus as a shielded proton or deuteron. Dihydrino or dideutrino formation would allow two protons or deuterons to enter the target nucleus at the same time, as discussed in Section 8.4.3. The NAE would exist where a suitable catalyst is present to assist in the required orbital collapse and suitable elements are present that can participate in nuclear reactions. Energy released during hydrino and deutrino formation would add to subsequent nuclear energy generated within the NAE.

Rather than form a stable configuration between an electron and proton, Dufour and co-workers,[167-173] at the Shell/CNAM Laboratories (France), proposed an interaction that makes the assembly look like a neutron long enough for it to enter into a nuclear reaction, *i.e.* a short-lived shielded proton or deuteron. They call this fleeting assembly a Hydrex when a proton is involved or Deutrex when it contains a deuteron. The concept is similar to the one suggested by Mayer and Reitz[174] (Mayer Applied Research) and Moon.[175] Vigier and co-workers,[176-178] at Unversité Paris (France), proposed that a new "tight" Bohr orbit might allow these short-lived assemblies to be created. As proof for such assemblies, Dufour and co-workers[179] report the disappearance of hydrogen or deuterium gas from the apparatus when a high frequency discharge is generated with palladium used as an electrode. The hydrogen is not absorbed by the palladium, but is permanently changed so that it no longer appears as a normal gas. The rate of loss compared to the measured energy is linear and gives 7.1 ± 0.8 keV/atom for H_2 and 25.5 ± 1.1 keV/atom for D_2. This amount of energy is too small to result from a plausible nuclear reaction, yet it seems too large to have a chemical source. Obviously, the lifetime of this new form of hydrogen is not short, but instead acts like the stable form proposed by Mills.

---

[c] See: www.blacklightpower.com

These models provide a mechanism to lower the Coulomb barrier between hydrogen isotopes and another atom by forming a shielded nucleus. This would make transmutation possible without neutron adsorption and the required by beta decay. Such a mechanism would produce radioactive products less frequently than neutron addition, with a new element produced immediately without the need for beta emission. Also, the shrunken atom may have the potential to form clusters of shielded deuterons as required to explain the Iwamura (Table 10) observations. This mechanism is discussed in more detail in Section 8.4.3.

5.9 Role of Neutrons

Individual neutrons are unstable in nature, decaying with a half-life of 10.25 minutes into a proton, an electron, and a neutrino. Creation of a neutron requires these components be reassembled, to which energy of at least 0.78 MeV must be added. Because neutrons have no charge they can be absorbed into other nuclei with an ease that depends on the nature of the absorbing nuclei and the energy of the neutron. Energy from neutron capture is frequently released by gamma radiation, which is not observed during most cold fusion experiments. In the absence of prompt gamma emission, the resulting energy is not observed as heat until stored energy is released to the surrounding material by radioactive decay, which frequently involves beta and gamma emission. For example, if the stable isotope ^{110}Pd is the seed, addition of one neutron gives ^{111}Pd, which is a beta and gamma emitter with a 23.4 minute half-life accompanied by prompt gamma emission. Addition of two neutrons produces ^{112}Pd, a beta and gamma emitter with a 21.0 hour half-life. Additional neutrons produce beta + gamma emitters with increasingly shorter half-lives. Although such radioisotopes are easy to detect and identify, they are not found after excess heat production or transmutation reactions. A model involving neutrons must explain why these and similar easily detected radioactive products are absent. In addition, the lighter isotopes of palladium or any other seed would be shifted to heavier stable isotopes accompanied by gamma emission. Neither consequence is observed.

Kozima[180-183] at Shizuoka University (Japan) and Portland State University (US) proposes that solids contain clusters of neutrons. According to his TNCF (Trapped Neutron Catalyzed Fusion) theory, these are stabilized by a unique configuration and are released to cause cold fusion reactions under certain conditions. Two of the many questions not answered are: "Why is the extra mass added by these neutron structures not observed in normal material and why are neutrons not released in a detectable form when the stabilizing solid structure is vaporized?"

Fisher[184-186] suggests clusters (droplets) of neutrons are attached to the nuclei of certain atoms and these stable clusters can, under certain conditions, be released as polyneutrons to react with surrounding atoms. Conditions causing release are not obvious and are not present during normal chemical processes. Only the special conditions associated with cold fusion are assumed to awaken these polyneutrons. A complex reaction and regeneration process is proposed to feed a continuous and stable production of energy and nuclear product formation. On the negative side of the ledger, neither experimental nor general theoretical support exists for polyneutrons being real. In addition, the resulting reactions have to somehow avoid gamma emission and subsequent beta emission to be consistent with observations. Nevertheless, Oriani[187-189] (University of Minnesota) detected what appears to be extra heavy CO_2, which can be explained using the Fisher model.

Various people have suggested that neutrons or dineutrons can form within a sample by various mechanisms.[190-199] Support for this idea has been provided by Conte and Pierallice[200] who were able to detect neutron emission when they bombarded protons with very energetic electrons (2.28 MeV) from a $^{90}Sr+^{90}Y$ beta emitting source. Using the same beta source to bombard a La-Sm alloy, Baranov et al.[201] obtained evidence for neutron activation of the lanthanum. Apparently neutrons can form from protons and electrons if the electrons have sufficient energy, although this process is very inefficient. Being able to concentrate sufficient energy at the location of proposed neutron formation is a critical requirement for such neutrons to be involved in cold fusion.

Widom and Larsen[202] (Northeastern University and Lattice Energy, LLC) attempt to explain neutron induced transmutation by proposing a series of events, starting with formation of super-heavy electrons on an electrolyzing surface. These electrons make "cold" neutrons by combining with protons or deuterons. Next, the very low energy neutrons or dineutrons are proposed to react with elements (seeds) that are present and generate a range of transmutation products.[203] The authors[204] propose that the expected gamma radiation is absorbed while super-heavy electrons are present, thereby accounting for the absence of radiation from the expected (n,γ) reactions. They do not explain why gamma radiation is not detected once the super-heavy electrons stop forming. Yet previously made neutrons would continue to react and produce a decay chain of beta-gamma emitting isotopes. Observed behavior can only be explained if the half-life for super-heavy electron loss after production stops exactly matches the half-life for beta-gamma decay of the resulting radioactive isotopes, a very unlikely coincidence. They claim a match exists between a calculated cross-section for low-energy neutron capture and the distribution of elements reported by Miley (Figure 51). Based on the model, addition of neutrons to the seed and to all resulting isotopes would have to be extremely rapid so that only radioactive beta emitters of very short half-life are present in the sample. Presumably, these isotopes decayed away to produce the measured element distribution without their radiation being detected. Absence of detectable radioactivity after such a process is very unlikely. The NAE for this model would be the environment required to create the super-heavy electrons.

Neutron creation models have several very serious hurdles to overcome. They have to explain how the required energy can be localized in an electron. Once localized, this energy must appear either as kinetic energy of the electron or as an increase in mass. If the wave nature of the electron is used to describe the process, the frequency of the wave-packet must increase. Of course, mass, kinetic energy, and frequency of a wave are all related, but the behavior of the electron will be different if it is moving rapidly through a lattice compared to an extra heavy electron or wave-packet being fixed in a local region. An obvious question is, "Why would this energetic electron seek out a proton or a deuteron with which to make a neutron"? After all, 'hot' or heavy electrons have many other paths they can follow, the easiest of which is to generate X-rays.[d] Once formed, the neutrons have to be absorbed very rapidly before they decay or leave the apparatus. In addition, this absorption process normally produces easily detected gamma emission. Neither neutron nor gamma radiation resulting from such a process has been found even when transmutation occurs in the absence of the proposed heavy-electron shield. A change in atomic number after neutron absorption also requires beta emission, which is not observed. Also, any mechanism proposed to form heavy electrons has to address why the limitations discussed in Section 8.2.1 do not apply.

5.10 Role of Super-Heavy Electrons as a Shield of Nuclear Charge

The possible involvement of super-heavy electrons as a shield has been suggested by several people.[205-210] These electrons are proposed to enter an orbit so close to the nucleus that the proton or deuteron acts like a neutral particle. This shields the nuclear charge, similar to the known effect of a muon.[211,212] Szalewicz et al.[213] calculate that a five-fold increase in mass would be sufficient to explain the Fleischmann and Pons results. As noted in the previous section, for an electron to gain mass, it must acquire energy. This requirement does not apply to a muon, which has significant mass as a result of its basic nature. As a result, the roles played by an electron and a muon will be different. The manner by which this energy is accumulated and stored by the electron is critical to predicting its behavior. Failure to address this problem is the weakest part of the proposed models.

---

[d] When discussing heavy electrons, a distinction needs to be made between electrons to which energy is actually added, resulting in greater mass, and electrons that appear to be heavy just because this concept makes a theory more consistent.

5.11 Role of Superconductivity

Superconductivity is proposed to be one of the conditions that leads to cold fusion. This is an attractive idea because superconductivity results when electrons form Cooper pairs. These are able to pass through the sea of normal electrons without interaction and are proposed to shield the charge on certain nuclei, which allows transmutation and/or fusion to take place.[209,214-216] However, the transition temperature to the superconducting state for material known to be present in the cold fusion environment is much lower than the temperature at which nuclear reactions are found to occur. Consequently, for these electron clusters to be available, the NAE must be where a very high transition temperature exists. As yet, a localized, very high transition temperature has not been demonstrated in materials experiencing cold fusion, although efforts have been made to find such behavior.

The transition temperatures for β-PdD and β-PdH have been measured as a function of composition,[217-219] which, when extrapolated, gives 10 K for $PdD_{1.0}$ and 8 K for $PdH_{1.0}$. Lipson and co-workers[220,221] (University of Illinois) observed magnetic and resistance effects starting near 50 K, which they attribute to superconductivity. Tripodi and co-workers[222] (ENEA, HERA, and INFN, Italy) report resistance changes in PdH near the same temperature. Apparently, both workers were unaware of a structural transition in β–PdD(H) near this temperature,[223-225] which might cause the observed effect instead of superconductivity. No one has measured the transition temperature for the proposed PdD_2 phase or when cold fusion is actually occurring. If superconductivity were important, the NAE would be located where superconductivity occurs.

5.12 Role of Electron Cluster

Shoulders has explored[226-229] (http://www.svn.net/krscfs/) the properties of charge clusters, which he calls "EV" (Electrum Validum) or now "EVO". Similar clusters have been observed by Mesyats[230] (Institute of Electrophysics, Russia). These EVs are proposed to be 1-20 micron in size with an electron density of 6.6×10^{29} electrons/m^3,[231,232] and a charge that is at least a factor of 10^4 less than that expected based on the number of electrons present. Some of charge is proposed to be offset by positive ions within the cluster. These clusters are observed to form when spark discharge of any kind occurs—from the small unexpected spark originating at the finger to lightning. When an EV passes through a thin foil, the electronic structure of the solid is disrupted in its vicinity causing a 'cold' gas to be ejected and a hole to form. Shoulders and others[233,234] propose these clusters also have the ability to catalyze nuclear reactions. However, their formation and steady availability during electrolysis or when ambient gas is used seems unlikely. Nevertheless, they might play a role in initiating nuclear reactions during crack formation. Bhadkamkar and Fox[235]

(Fusion Information Center, Utah) suggest the Casimir force can provide enough energy during bubble formation to create EVs and Shoulders[227] suggests they can be created by charge separation in collapsing bubbles. This explanation might apply to the sonic method, but bubbles do not collapse during electrolysis—they only grow until they break at the surface. A demonstration that EVs can be generated without a spark in a benign environment is important to the acceptance of this structure as an universal mechanism to produce cold fusion.

Lochak and Urutskove[236] (Foundation Louis Broglie, France and RECOM, Russia) observed formation of similar 'strange' particles when they fused a titanium wire under water by discharging a capacitor through the metal wire. They observed the behavior of particles they call magnetic monopole as tracks on photographic film. Evidence for transmutation was also obtained. Whether the EV and the magnetic monopole are the same particle is unknown, but worth exploring. However, some caution should be taken in such a study because these particles seem to have an effect on the biology of living organisms.[237,238] Lewis[239-242] has called such particles plasmoids, as described by Bostick,[243] and related their behavior to ball lightning. Evidence for similar 'strange' particles also has been reported by Matsumoto[244,245] (Hokkaido University, Japan).

Other people have speculated about the possible existence of other unusual particles,[246-251] any one of which might play a role in creating a NAE. Although too little is known about the properties of these particles to permit a conclusion about their involvement in cold fusion, enough is known to make an investigation worthwhile.

5.13 Role of High-Energy Environment

High-energy environments are produced when a spark or plasma is formed between electrodes within a liquid, when a metal wire is rapidly melted under water, or when acoustic waves are generated in a liquid, causing bubbles to inject their contents into a solid. These three methods generate high local energy primarily by creating a very high temperature for a brief time rather than by producing a local high voltage. This condition generates energetic particles, but apparently not energetic enough to overcome the Coulomb barrier by brute force. Hence, a NAE is required to account for what has been observed. Generation of EV might play a role in each method as well as the presence of ions and electrons having a local density that is able to shield a nucleus. As discussed in Section 5.8, hydrino formation many play a role.

Several examples of these methods being used to make unexpected nuclear products have been published. Matsumoto[252] sparked palladium wires in K_2CO_3 + H_2O and produced transmutation products and evidence for a 'strange' particle similar to an EV.[253] Iron was produced by several studies when an arc was formed between carbon rods held under water,[254-256] with an increase in ^{58}Fe being observed.[257] Roger Stringham[258-263] pioneered the use of ultrasound to create bubbles in D_2O and cause them to collapse on a metal surface where they drive a bit of hot plasma into the metal. This method is entirely different from efforts to generate hot fusion within collapsing bubbles in a liquid, called sonofusion. When a wire made from titanium was melted under water, various transmutation products were made along with evidence for a 'strange' particle.[236] Each of these methods approaches the energy boundary between cold and hot fusion where the subtle nature of the NAE becomes less important.

Energetic deuterons can also be used to bombard a surface (Table 12) during which fusion is found to occur at a higher rate than expected. The fusion rate is sensitive to the nature of the bombarded surface because the chemical environment can influence the amount of electron shielding the dissolved deuterons experience. However, this behavior is a variation of hot fusion, not cold fusion, because the nuclear products are typical of hot fusion and are produced at a much lower rate than result from cold fusion.

5.14 Role of Wave-Like Behavior

Deuterons are proposed by Chubb and Chubb[264-266] (NRL) to exist as a wave structure, which they call Bose Bloch Condensate Matter. This in-phase wave structure is assumed to add energy by a resonance process until the Coulomb barrier is overcome. The resulting energy is slowly release into the lattice as the combined wave reverts to a normal particle structure of the product nucleus—in this case helium. The basic Quantum Mechanical (QM) model assumes such waves are a normal and universal aspect of matter. So far, no experimental evidence exists showing clear wave-like behavior of nuclei near room temperature, although atoms are observed to act wave-like at much lower temperatures. An easy to understand description of this concept is provided by Talbot Chubb.[267] A NAE would exist where waves can form and fuse.

Phonons, *i.e.* packets of wave-energy present in a lattice, are proposed by Hagelstein,[75,268] Swartz,[269] and F. S. Liu[270] to move energy between nuclei, thereby creating enough localized energy to overcome the Coulomb barrier. Double occupancy of sites by deuterons and quantum mechanical modes of interaction are assumed. Phonons are then able to couple the released energy to the surrounding structure. If this process were able to move large amounts of energy within the lattice after a nuclear reaction had occurred, any proposed process that depends on achieving a large localized concentration of energy to initiate nuclear reactions would not work. As energy accumulates, the phonons would quickly move this energy elsewhere before it had a chance to reach levels required to initiate a nuclear reaction. In other words, the process appears to be its own worst enemy for achieving a high local concentration of energy. This issue is further discussed in Section 8.2.2.

Proposed models need to show how a high local concentration of energy can occur, how this can happen often enough to make detectable energy, and why this happens only in the special and rare cold fusion environments. If the cold fusion effect is to be made reproducible based on these models, the nature of the special environment needs to be identified so that it can be created in large amount on purpose.

5.15 Living Organisms

Since 1967, Komaki[271,272] (The Biological and Agriculture Research Institute, Japan) and later Kervran[273,274] made the generally rejected claim for living organisms being able to initiate nuclear reactions as a way to obtain elements needed but not available in their environment. Because these claims were based largely on plants and animals, they were given very little attention. Komaki[275,276] improved the evidence by using various single-cell organisms to which modern analytical techniques were applied. Recently, Vysotskii and co-workers[277,278] at Kiev Shevchenko University (Ukraine) and Moscow State University (Russia) dissolved $MnSO_4$ in D_2O containing a bacteria (Deinococcus Radiodurans or Saccharomyces Cerevisiae T-8)) and found increased amounts of ^{57}Fe, a rare isotope of iron. This amazing result would be easy to dismiss except for the unusual method used to prove the presence of the iron isotope. The ^{57}Fe isotope is unique in being detectable using the Mössbauer method. This method uses the gamma ray emitted from ^{57}Co, which is passed through the culture while the source is moved with a changing velocity parallel to the gamma ray direction. At a certain velocity, the gamma ray is absorbed if ^{57}Fe is present and the amount of absorption is directly related to the amount. A detector located on the other side of the culture records the surviving gamma flux. No other element will cause such a change in absorption of the gamma ray. They measured in one experiment $8.7 \pm 2.4 \times 10^{15}$ atoms of ^{57}Fe, which would have generated 22 kJ at a rate of about 80 mW. Production of ^{57}Fe occurred only when $MnSO_4$ and D_2O were both present, as shown in Figure 68. In addition, some elements enhance and others inhibit the reaction.[279] The only logical process involves a deuteron entering the nucleus of ^{55}Mn, which is 100% of natural manganese, to produce the single isotope of iron. Later work produced ^{54}Fe when ^{23}Na and ^{31}P were claimed to fuse within a culture of Bacillus Subtillis.[280-282] If these results are accepted, we are forced to conclude that living organisms are able to create a NAE, perhaps within a large protein molecule.[283,284] Presence of only one observed nuclear product means the process is more complex than it first appears, with some kind of missed radiation being emitted. Direct coupling of such a large amount of energy directly to a molecule seems unlikely because this would surely result in its destruction, resulting in a steady loss of the unique environment and eventual termination of the reaction.

This raises the possibility of detecting energetic nuclear products at locations where bacteria are present, which might not be at the cathode of an electrolytic cell. Oriani[285] found charged particles to originate above the electrolyte at some distance from the cathode. Fisher attributed these to reactions to polyneutron clusters, but bacteria might also play a role.

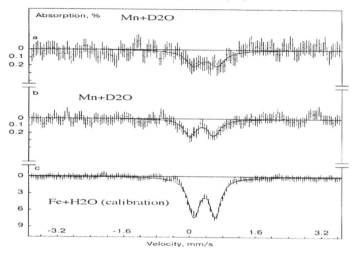

Figure 68. Mössbauer spectrum produced in MnSO₄+D₂O as a result of bacteria growth. The amount of gamma ray flux absorbed while passing through the cell is plotted as a function of relative velocity between the source and detector.

Researchers are attempting to reduce the level of radioactive contamination in soil using bacteria.[286] To the extent this process can occur, normal elements would also be changed into other elements, resulting in a change in their concentration and isotopic ratios over time. This possibility raises a very important question, "Has life created elements it needs out of a different collection present on the Earth initially?" Even more amazing are the over 400 recorded spontaneous human combustion events[287] that might be explained by sudden creation of an excessive amount of NAE in a body. These observations encourage a person to open the floodgates of imagination.

5.16 Conclusion

A variety of plausible and implausible NAE models have been explored, with little evidence that an actual one has been identified. Nevertheless, several conditions appear to be important. As expected, the greater the hydrogen (deuterium) content, the faster the reaction. Indeed, the reaction rate is probably related to some power of concentration. Also, the more rapidly protons or deuterons diffuse through the NAE, the faster the reaction occurs. Application of extra energy by using increased temperature or laser light accelerates the process. Nanoparticles, either deliberately created or as an accidental consequence of the treatment seem to be important, but they are not the only essential condition. The role of impurity elements is unknown, although they will have an effect on the general environment and may be essential if hydrino formation is important. Impurities are apparently important as 'seeds' for transmutation reactions. Because nuclear reactions are found to occur in living systems, the possibility exists for the NAE to be a complex molecule. Studies designed to determine which, if any, of these proposed NAE are present would be worth undertaking. Finally, radiation from what appear to be single-particle producing reactions is missing and needs to be sought before more exotic explanations are accepted.

A distinction needs to be made between reactions resulting in measurable heat and those producing very little nuclear product, such as transmutation, tritium, and neutron production. The difference in rates is so huge that a variety of NAE may be operating by different mechanisms. A more detailed discussion of some proposed mechanisms is provided in Chapter 8.

How does a person go about increasing the amount of anomalous energy? That will be the subject discussed in the following chapters. However, a few generalities are worth noting here. Past efforts have focused on acquiring the 'right' kind of palladium. A special alloy (Type A) made by Johnson-Matthey is claimed by Fleischmann to be active. On other occasions, a Pd-Ag alloy[288] and Pd containing boron are said to be active.[289-293] In general, palladium obtained from certain sources has been found to produce success more often than from others. Reduced crack formation seems to correlate with success.[294] Other metals besides palladium are also found to be active. Dash[295] (Portland State University) produced energy using titanium (Ti), Bockris et al. deposited copper on palladium to produce tritium.[134] Letts and Cravens[92] plated Au on Pd to produce anomalous energy while using laser stimulation. I was able to obtain anomalous energy using various layers of palladium deposited on platinum,[60,296] even though clean platinum is completely inert. In other words, the basic material used as the cathode is not active initially even when it is made of palladium—activation is required. Nevertheless, the base material does affect the morphology and subsequent activity of the deposited layer. The challenge is to learn how to create a layer that has the required properties.

CHAPTER 6

What Conditions Initiate Cold Fusion?

6.1 Introduction

A study of cold fusion or LENR is difficult because the active regions are tiny and occur at random locations. In addition, a useful theory is not available to guide a search for these sites. As a result, success is achieved largely by chance. A successful effort requires a background in chemistry as well as in physics, a rare combination in a researcher. As a result, many mistakes are made. Although the apparatus used by most people looks simple, it is not. Many of the techniques, such as calorimetry and surface analysis, require significant skill. The field has become too complex for casual efforts to make much difference.

If you intend to study cold fusion, you should expect to spend at least a year mastering the experimental techniques and the literature, with very little reward for your efforts from nature. If you are lucky, you may see an anomalous event, which you are advised to treat with caution. On the other hand, if you have a major laboratory available, adequate funding, and extreme patience, you might contribute to one of the most important discoveries of this century.

The equipment needed depends on which of the many anomalous behaviors you want to use to see the effect and which method you want to use to initiate it. Many possibilities are available. Also, do you want to explore the nuclear processes or try to understand the environment in which these processes operate? In other words, do you want to practice physics or chemistry?

6.2 Initiation Methods

The methods known to initiate an anomalous effect are discussed below in approximate order of increasing energy applied to the NAE. This ranking starts with the claim hardest to accept and ends with a behavior currently accepted by conventional science. The amount of applied energy affects which nuclear reactions are initiated and their reproducibility. Each method provides a different window for examining the process. Care must be taken when conclusions obtained from one method are applied to the other methods because different amounts of energy and different environments are involved. Even when a fixed average energy is applied, this average will not be uniform. Greater energy well be applied to some regions compared to others resulting in a range of behavior. In addition, a sample is not uniform in its physical characteristics. As a result, within the same sample neutrons might be emitted from cracks as they form; tritium could be generated at the tips of dendrites; transmutation might occur where susceptible "seeds" are located; and helium and heat could result from various reactions taking place where the deuterium content was very large. The challenge is to separate the observations and place them into a proper role within a theory without assuming the same process produces every effect just because they are observed to occur at the same time from the same sample.

6.2.1 Living Organisms

As noted in Section 5.15, living organisms appear to be able to initiate nuclear reactions to acquire essential elements denied by their environment. This process is studied by culturing single cell organisms in the absence of essential elements. Growth of the missing element is measured using various techniques. Transmutation has been detected in cultures based on either D_2O[1-3] or H_2O.[4-7] Evidence for heat, radiation, and energetic particles has been neither sought nor obtained. So many unanswered questions, unresolved logical conflicts, and incredible consequences haunt these claims that a rational interpretation is very difficult. Additional replication is essential because the implications of this research are so important.

6.2.2 Ambient Gas

Exposure of certain materials to either D_2 or H_2 produces extra energy and nuclear products on occasion. Increased temperature and increased pressure enhance the effect. Evidence for nuclear reactions has been observed when finely divided palladium[8-12] is exposed to D_2 gas (see Section 5.3). Exposure of bulk titanium[13-16] and bulk palladium[17,18] to D_2 gas also produces anomalous behavior.

Specially treated nickel[19-22] produces energy and evidence for nuclear activity when it is exposed to hot H_2 gas. However, available information is not sufficient to allow easy replication by other scientists (see Section 4.3.2).

Of the various methods, exposure of specially treated metal powder to ambient gas is the closest to being developed into a practical energy source.[23] In principle, very little energy needs to be supplied to the apparatus to cause energy production, making this source of energy very efficient.

6.2.3 Proton Conductors

A modest voltage applied to a hydrogen-containing material will cause the dissolved hydrogen to move. This process is an example of well-known electromigration, by which a current passing through a material causes an increase in the transport rate of any dissolved ion. When electromigration of hydrogen occurs in a solid, the material is said to be a proton conductor and when this process occurs in palladium, the result is called the Coehn effect.[24] Attempts to obtain heat from the Coehn effect have been explored by several people using PdD.[25,26] Other metal hydrides, such as TiD_2 and ZrD_2, should also be explored to determine whether the greater deuterium content they contain would be helpful.

Various proton conductors, based on mixtures of metal oxides, have been examined by heating them in the presence of D_2 gas. Extra energy is produced on a few occasions by electromigration when a small current is passed through the material.[27-33] The method has a large energy amplification factor because very little power is applied, but energy density has been rather low so far.

6.2.4 Electrolysis Under Faraday Conditions

Faraday conditions involve chemical reactions initiated by electron transfer when voltage is applied to an electrolytic cell without formation of a plasma or spark. This is the most studied and perhaps the most challenging method used to explore cold fusion. The challenge is tolerated because the method is simple to use and applies deuterium to the palladium with a higher hydrogen (deuterium) activity than can be easily achieved any other way. In addition, the method has become convention, following Fleischmann and Pons (see Section 4.3.1).

The most commonly studied cathode is various forms of palladium. This metal is used because it is unique in dissolving a large amount of deuterium without serious structural damage. More information about palladium is provided in Appendix C. Bulk metal as plates or wires, or thin deposits applied to various inert substrates have been used as the cathode. These materials will have a wide range of characteristics even when they have been formed by what appears to be an identical process.

The D_2O used in the electrolyte should be purified by distillation in contact with $KMnO_4$ before use to remove dissolved organic material,[a] metals,[34] and bacteria.[35] LiOD, Li_2SO_4, D_2SO_4, or K_2CO_3 are then added to make the solution conductive. Dissolved nitrates and chlorides should be avoided because they rapidly attack the anode. Ethyl alcohol[36] has also been used as an electrolyte solvent. Even pure D_2O without dissolved salts has produced success[37], although high voltages are required to achieve useful current.

Platinum is commonly used as the anode. Although platinum slowly transfers to the cathode, its presence is apparently not harmful. Nevertheless, transfer changes the behavior of the sample and may influence creation of a NAE. Nickel has also been used as the anode on occasion. This metal transfers very slowly once black NiO forms on its surface. Gold, silver, and copper rapidly transfer to the cathode and should not be used.

---

[a] The analytical report provided with a recent purchase of 99.96% D_2O from Cambridge Isotope Laboratories gives a total organic carbon content of 3.2 ppm, pH = 6.01, and conductivity = 1.5 µS/cm. Past measurements by the author have found up to 9 ppm solids in commercial D_2O.

Current is applied as DC or in various waveforms,[38] conditions that have an important effect on the D/Pd ratio and on deposited material. Application of an initial small current is frequently used to reduce the tendency of the metal to crack as it loads with deuterium. Increased loading can be encouraged by applying alternating high and low currents. However, transfer of material from the anode and reaction with lithium in the electrolyte is increased when high current is used.

Temperatures up to the boiling point have been explored, with greater heat production being achieved at higher temperatures.[39-41] A temperature of 450°C was explored[42-46] using a fused salt electrolyte consisting of LiCl+KCl+LiD. Glow discharge might result in still higher temperatures at the active surface, but such temperature measurements have not been reported.

When nickel is electrolyzed as the cathode in H_2O containing various salts, excess energy is occasionally reported. In addition, a wide range of transmutation products are found,[47] but the results are only partially consistent with other work and difficult to explain. Some of the extra energy is proposed to result from hydrino formation[48]. Evidence for induced radioactivity has been reported (see Section 4.3.2).

The electrolytic method has a lot of supporting experience and can be used to explore a variety of conditions at temperatures below 100° C. However, the results are not reproducible, energy production can be very unstable, and the nuclear products are difficult to measure and sometimes ambiguous. In addition, energy amplification is generally low.

6.2.5 Electrolysis Under Plasma Conditions

When voltage in excess of about 100 V is applied to an electrolytic cell, a plasma forms. This creates a high-energy environment, broad spectrum light emission, destruction of the cathode, chemical changes in the electrolyte, and heating of the solution.[49-55] Occasionally, extra energy and transmutation products are observed. Study of this method has been difficult because conditions are very chaotic and the local temperature is very high. Voltage and current spikes occur, making applied power measurement difficult. Decomposition of the electrolyte makes a chemical balance hard to achieve. More attention needs to be applied to improving the design of the apparatus so that the many corrections presently required need not be made.

Of the various methods, this is the most spectacular and the one for which the highest power production has been reported.

6.2.6 Plasma Discharge

When sufficient voltage is applied to a low-pressure gas consisting of deuterium or hydrogen, a plasma forms[56-63,64-69] by a process called glow discharge. This fourth form of matter[b] is created by ions and electrons being accelerated in opposite directions and colliding with the ambient gas to make more ions and electrons. Light is emitted when the ions and electrons recombine. The cathode surface is bombarded by ions having energy considerably greater than can be obtained during electrolysis creating a high activity of deuterium in the cathode surface. However, a disadvantage is discovered when attempts are made to increase ion flux (current) to values easily obtained using electrolysis. The required higher voltage adds energy by causing an increase in the product of current and voltage. This causes excessive heating of the cathode surface, which requires cooling to prevent loss of deuterium. As a result, calorimetry is made more difficult and a lot of energy is wasted. Nevertheless, excess power at high levels has been reported.

A somewhat different kind of plasma can be created by applying a high AC voltage through a dielectric barrier to gas at atmospheric pressure.[72,73] This creates plasma having higher energy and lower ion flux than does glow discharge and, as a result, the method shows different characteristics (see Section 5.8). Nevertheless, excess energy and transmutation products are reported.

The gas-plasma method is attractive because well documented materials can be applied to the surface and they can be studied without fear of change. A major advantage is provided because real time measurement of tritium[70,71] (Section 4.4.1) and other nuclear products is relatively easy because the cathode can be viewed directly using a variety of diagnostic tools.

6.2.7 Laser Light

Laser light of various wavelengths and intensity has been applied to surfaces, but energy transfer can be sensitive to the nature of the surface.[74] Low-intensity radiation (9 W/cm^2) near 660 nm requires a special surface to produce extra energy.[75] This surface was created on occasion by using deposited gold.[76-78] Application of a similar wave length was found to be effective in producing extra energy provided the electrolyzed palladium cathode was already making extra energy.[79] A XeCl laser (308 nm) generated transmutation products in a D_2 gas-loaded thin film of palladium[80] and in gas-loaded palladium wire[81-83] without previous activation. A wave length of 1054 nm and pulsed power of 1-2 x 10^{18} W/cm^2 caused TiD and TiH to emit alpha particles and protons.[84] Pulsed power of 5 x 10^{19} W/cm^2 was found to generate protons with energy up to 30 MeV and highly ionized Pb^{46+} having energy up to 430 MeV[85]. These energetic particles alone will produce measurable nuclear reactions of various kinds. An ideal

---

[b] The other clearly identified forms are gas, liquid, and solid. Nanosized material might eventually be given its own classification. Ambiguous material exists in the transition regions between these forms and provides fuel for debate.

wavelength probably exists to initiate the effect without significant application of energy, as calculated by Bass.[86]

The method opens a window through which to explore how energy couples to the nuclear process. Once the ideal wavelength is demonstrated, the method can be combined with other methods to increase energy amplification.

6.2.8 Sonic Implantation

Bubbles can be generated in a liquid in various ways and these bubbles, if they collapse on a metal surface, inject their contents into the metal. As a result, both deuterium and oxygen are injected into the metal. Stringham[87] generated such bubbles using an ultrasonic generator located in the liquid and claims production of significant extra power and helium. Sonic frequencies up to several megahertz have been used.

Griggs[88] used mechanical rotation within ordinary water to generate cavitation bubbles and claimed to produce significant extra energy. Jorné[89] applied ultrasound to a slurry of palladium particles. Application of ultrasound during electrolysis does not improve general loading[90] or generate extra energy.

This method has no relationship to bubble collapse within a liquid,[91] so-called sonofusion. This method relies on achieving sufficient temperature within the bubble to initiate a hot fusion reaction and neutron emission.

6.2.9 Crack Formation

As discussed in Section 5.2, cracks form as a consequence of reacting a metal with hydrogen or by application of mechanical stress. Efforts to produce large numbers of cracks were rewarded by production of nuclear activity. Crack formation can even be heard as acoustic emissions using a sensitive microphone.[92,93] Because the effects are fleeting and small, the method offers little information about the more robust and useful mechanisms. In addition, radiation and nuclear products resulting from crack formation may superimpose on the results obtained from other methods, making interpretation more difficult.

6.2.10 Ion Bombardment

Table 12 lists various studies during which targets were bombarded by very energetic ions or electrons. Although the results are not directly related to cold fusion, many nuclear reactions are initiated at unexpectedly high rates, as discussed in Section 5.13. Apparently, the Coulomb barrier is easier to overcome than expected when nuclear reactions take place in a solid, rather than in a plasma. The reaction rate is sensitive to the nature of the solid, which is a surprise to conventional theory. As applied energy is reduced, the difference between the observed and predicted rates becomes even greater. One has to wonder how large the screening would be if the NAE that operates during cold fusion were bombarded by high-energy ions.

The glow discharge and ion bombardment methods allow the transition between conventional hot fusion and cold fusion to be explored within a solid. Presumably, the NAE is very important at very low energy. As applied energy is increased, the NAE becomes less important and the resulting nuclear products change. At high energy, the process is independent of the environment, is energy dominated, and behaves like typical hot fusion. Maximum energy production will probably occur when the right amount of energy is applied to the NAE, either as bombarding ions or as laser light. This energy level has yet to be determined.

6.3 Summary

Many methods, besides electrolysis, are available to initiate the effects. The phenomenon is not limited to one method. This is a very important discovery because most normal phenomenon in nature can be initiated many ways using many combinations of materials and processes. Rather than going away, as 'pathological' observations are said to do, the evidence is only growing stronger. The observations involving cold fusion have the characteristics we normally expect find in nature.

In addition to excess energy, the phenomenon produces many types of radiation. How this radiation relates to heat production has yet to be discovered. Nevertheless, energy is being generated and released at levels far in excess of any chemical process. In addition, helium is clearly produced when heat is generated. Many transmutation products as well as tritium are occasionally found in amounts far in excess of what can be explained by conventional models or prosaic processes. The nuclear products demanded by skeptics have been found, although neither in the amounts nor in the kinds expected.

CHAPTER 7

What Is Detected and How Is It Measured?

7.1 Introduction

Detected radiation emitted from the NAE includes X-rays, γ–rays, alpha particles (^{4}He), protons (hydrogen), beta particles (electrons), neutrons, and abnormal radiation of unknown types. If emissions occur immediately, they are called "prompt" and when they occur over a period of time, the result is called radioactive decay or "delayed". Both types have been produced, but radioactive products are rare. These emissions occur over a range of energy that is determined by the originating nuclear reaction or unstable nucleus. In general, each type of emission requires a different method for its detection, some of which are described in the literature[1] and below.

7.2 Neutron

Neutron emission has been detected at very low levels using several different kinds of detectors. However, as noted previously, this emission will not be discussed here because the amount is too small and found too infrequently to provide useful information about the main nuclear processes.

7.3 Tritium

As discussed in Section 4.4.1, tritium is a radioactive isotope[a] of hydrogen. It decays by beta emission with an energy of 18.58 keV and with a half-life of 12.346±0.002 years.[2] Heavy-water (D_2O) always contains some tritium at concentrations near 10^8 atoms/ml, depending on its source. This amount can be highly variable and needs to be checked before beginning a study of tritium production.

When tritium originates from a nuclear reaction, it normally starts as an energetic particle (triton), the presence of which can be detected using any of the tools designed to detect energetic particles. Once the ejected particle has lost translational energy, it can be detected as ambient tritium. If the resulting nuclear energy should be communicated directly to the lattice immediately upon formation of tritium, energetic particle emission would not be detected.

Four methods of analysis are commonly used.

---

[a] A sample of tritium having an activity of 1 μCi or $3.7x10^4$ Bq has a disintegration rate of $2.2 x10^6$ dpm and contains $2.1x10^{13}$ atoms of tritium.

1. Tritium can be converted to its oxide (water) and mixed with an organic fluid (liquid scintillator). This liquid emits light when a beta particle passes through it and photomultiplier tubes are used to detect the light as individual pulses. This is a very accurate and sensitive method and commercial devices are available. A major error can occur if chemicals are present in the sample to cause light production called chemical luminescence. This error is avoided by purifying the sample by distillation before examination[3] or by waiting for sufficient time (usually 1 hour) after the sample is mixed with the scintillator.

2. The electron+ion current generated as the beta passes through a gas can be measured using a sensitive electrometer. The presence of certain impurity molecules in the gas can affect the signal in unexpected ways.[4]

3. The sample can be placed on film (autoradiograph) and viewed as regions of exposure. Because exposure might be caused by emitters other than tritium, radiation absorbers are needed to determine whether the radiation has the energy expected from tritium.[5,6]

4. The elements in a sample can be separated using a mass spectrometer. This method requires a very good instrument to distinguish between ions generated by a mixture of tritium (T), deuterium (D), and hydrogen (H), which is the usual situation. Tritium frequently exchanges with the other isotopes of hydrogen, causing a variety of ions containing tritium to be present. These include T^+(3.018), $\mathbf{D^+(2.0141)}$, $\mathbf{DD^+(4.0282)}$, DT^+(5.033), $\mathbf{DDD^+(6.0423)}$, TT^+(6.037), and DDT^+(7.047), with the ions in bold having the greater intensity. Only DT at mass 5.033 can be used to measure the tritium content, but with poor accuracy because DDH^+(5.036)[7] is frequently present in small amounts. Helium-3 (3.016), the decay product, can be used to calculated the amount of tritium in a sample if the time since tritium was added[8,9] is known or can be determined by extrapolation of the decay rate.

Because tritium is easy to identify, its presence is usually not in doubt. The challenge is to determine how the tritium got into the experimental apparatus. Several possible answers have been explored. Deuterium gas and heavy-water always contains tritium. This initial amount needs to be subtracted from the amount measured after the study. Tritium will concentrate in the electrolyte if the cell does not contain a recombiner[10-14] and corrections can not always be trusted. This process is discussed in more detail in Appendix E. Most problems in interpreting the results can be reduced if the cells are sealed, preventing any change in initial tritium content;[15] a total inventory is made;[16,17] or if the amount detected is too great to be explained by any plausible error.[18-21]

7.4 Gamma and X-ray Radiation

Gamma and X-rays are electromagnetic radiation (photons) and are detected using similar devices, with γ-rays being much more energetic than X-rays, making them easier to detect outside of a cell. X-rays result when electrons return to a lower energy by emitting electromagnetic radiation and gamma radiation results when particles in the nucleus return to a lower energy.

Energetic particles moving through a solid will generate X-radiation as the particles are slowed by collisions. When produced this way, the resulting X-rays are called bremsstrahlung[24,25] and have a wide range of energy. Voltage alone in a Fleischmann and Pons electrolytic cell is not sufficient to generate detectable X-rays, nor can they be produced by chemical reactions during gas loading. However, glow discharge and similar energetic processes are expected to generate low-energy X-ray even in the absence of nuclear reactions. Most X-radiation will be absorbed by the apparatus, thereby making its detection unlikely.

Gamma radiation frequently accompanies other types of emission as part of radioactive decay. In addition, this kind of radiation is normally produced when two nuclei fuse into one, when a neutron is absorbed by a nucleus, and by positron annihilation (511 keV). Gamma radiation normally has enough energy to pass through the walls of a typical cold fusion apparatus and be detected if it is present.

Many instruments are now available to detect X- and γ-ray. A good description is located at www.canberra.com/literature/931.asp. Common detectors use a variety of methods. Gas-filled Geiger-Müeller counter is one frequently used method to show the presence of radiation. Light pulses generated by a sodium iodide crystal (NaI) or voltage pulses generated by germanium (Ge) can be used to determine the energy of the initiating X-ray. Exposure of specially designed film can be used to locate the source within a sample. Knowledge of the energy gives essential information about the source. Common errors result if background radiation, for example from radon, potassium, and other natural sources, is not taken into account. Modern instruments provide sufficient energy resolution to identify these very weak natural emitters.

7.5 Charged Particle Radiation

$^4He^{++}$(alpha), $^1H^+$(proton), $^3H^+$(triton) or $^3He^{++}$ (helium 3) are possible energetic particles. Alpha radiation results when $^4He^{++}$ is ejected from a nucleus. Unstable isotopes, generally of the heavier elements, are known to emit such particles with energies as high as 10 MeV. Protons, tritons, and ^3He are not emitted from a normal nucleus and, if they are detected in a cold fusion experiment, would result only from one of the fusion paths listed in Table 5.

Charged particles can be detected a number of different ways. CR-39 is a plastic used to make eye glasses that also is sensitive to the effect of charged particle and neutron bombardment. These particles modify the plastic where they pass, making it soluble in a concentrated solution of hot NaOH. After "development" small pits (visible at 500x) are formed, the size and shape of which reveals the energy and type of radiation. The method is useful because the plastic can be placed directly in an electrolyte next to an active surface without significant damage and particles can be detected before they are stropped by intervening liquid. In addition, the plastic stores information, much like an autoradiograph, so that a very small flux can be detected just by using longer exposure. Details about the use of this material can be obtained at http://newenergytimes.com/news/2007/NET21.htm#apsreport.

Individual particles can be counted by gluing thin sheets of plastic- and glass-scintillator material to the active surface of a photomultiplier tube.[26] Thickness of the plastic scintillator is chosen to stop 5 MeV tritons but to allow 5 MeV protons to pass through, which are stopped in the glass scintillator. The different shapes of the light pulses produced in the different scintillators allow the two types of particles to be separately measured. Of course, many variations of this method can be used to measure the energy of other types of particles having different energies.

Proportional counters detect the transit of a particle through a gas subjected to an electric field. Ions are produced within the gas and these are detected either as a small current or a voltage pulse across a large resistor between the chamber and power supply. The gas is a mixture of 90% argon and 10% methane. Because the pressure is one atmosphere, a very thin window can be used to allow passage of alpha particles.

A cloud chamber, although seldom used these days, can reveal the area from which emission is occurring. The length of the track gives information about the energy of the particle, and introduction of a magnetic field can give additional information about the type of radiation. Although primitive by modern standards, the method can clearly reveal delayed energetic particle emission and locate its source without the ambiguity of possible electronic noise.

A silicon barrier detector measures particles when they cause a resistance change in a solid-state diode. A reverse voltage is placed across a diode, located on a thin layer of silicon. A particle passing through this diode causes a momentary current to flow, resulting in a voltage pulse that is proportional to the energy of the particle. These devices are very sensitive and have a low background.

7.6 Beta Radiation

Beta radiation consists of energetic electrons, the emission of which causes the atomic number of the emitting nucleus to increase by one without having a significant effect on atomic weight. Beta-emitting isotopes can have a half-life as long as 10^{16} years or as short as milliseconds. The energy of the radiation combined with the half-life allows the emitting isotope to be identified.

When beta radiation has been detected, it results from radioactive emission after the sample has been removed from the cell. Only a few delayed beta emitters, other than tritium, have been detected. However, many beta emitters have an accompanying gamma that might be detected without removing the sample from the apparatus.

Several models, as discussed in Chapter 8 and Sections 5.8 and 5.10, involve shielding of a deuteron or proton by a special electron. This electron might be emitted after its job has been done. Such electron emission may be the missing radiation required when the nuclear products appear to consist of only one atom. A search for this radiation is important.

7.7 Transmutation

Transmutation products are elements resulting from deuterons, protons or other elements fusing with the nucleus of an element previously present in the NAE. Transmutation can also result if an element splits into two smaller parts, *i.e.* fission. Transmutation products can only be determined by examining the material in which they form after the experiment. Because transmutation always involves overcoming a very large Coulomb barrier, the process is impossible to explain by conventional processes or by most mechanisms used to explain d-d fusion. Consequently, acceptance of the idea is difficult for some people. As a result, great pains are taken to demonstrate that the claimed nuclear products are really novel and not contamination—a proof difficult to provide. The best indication that contamination is not the source of an unexpected element is based on finding an abnormal isotopic ratio (see Table 8). Significant changes in isotopic ratio of the heavier elements do not naturally occur and would not be expected to involve a single-stage process operating in a cold fusion cell. The second best indication is the presence of an unusually large amount of the transmuted element, greatly exceeding the analyzed amount initially present in the cell. Recent work used a "seed" enriched in one of its isotopes.[27] This method provides especially compelling evidence when the nuclear product shows the same abnormal enrichment pattern in its isotopic ratio as does the seed.

Methods are now available to analyze even very small amounts of some elements within very small regions. This includes Secondary Ion Mass Spectrometry (SIMS), Electron Dispersive X-ray (EDX), Auger Electron Spectroscopy (AES), and X-Ray Fluorescence Spectrometry (XFS). SIMS is used to measure the isotopic composition within very small regions of the surface. A small amount of bulk sample can be analyzed to determine the elements it contains by Neutron Activation Analysis (NAA), Optical Emission Spectroscopy (OES), or by Inductively Coupled Mass Spectrometry(ICP-MS).

During NAA, the sample is exposed to neutrons, which make certain isotopes of some elements radioactive. The characteristic gamma emissions are detected and used to identify and determine the amounts of the activated isotopes present throughout the sample. However, only a few isotopes can be determined this way. A more complete assay of isotope content can be made by ionizing the entire sample and separating the ions using a mass spectrometer (ICP-MS). In addition, the entire sample can be ionized at high temperature and analyzed for element content using the optical spectrum (OES). However, this method cannot provide the isotopic composition.

For SIMS analysis, the surface is bombarded by energetic ions, causing a very small region to leave as ions, which are separated and identified using a mass spectrometer. Errors can be introduced when complex ions are formed from normal sources having the same mass-to-charge ratio as the unknown isotope.

Several methods use X-rays to analyze the surface. EDX involves bombarding the surface with electrons and analyzing the energy of the resulting X-rays. This method is frequently used with an electron microscope that provides the electrons. When XFS is used, the sample is bombarded by a tightly focused X-ray beam of a single high energy. The energy of the resulting secondary X-rays is used to identify the presence and amounts of certain elements. AES involves bombarding the surface with energetic electrons and then measuring the energy spectrum of the emitted secondary electrons. Each of these methods has different limitations, accuracy, and a different set of elements it can detect. An increasing number of studies use a combination of methods, adding to the reliability of the results. As tools for analysis gain sensitivity, normal material and surfaces are found to contain a large and unexpected collection of impurities. Perhaps it is a minor exaggeration to suggest that, at sufficient sensitivity, all surfaces are found to contain some of almost all common elements before the sample is properly cleaned. Even casual cleaning, such as is done without a clean-room, is not always successful in removing every detectable impurity. As a result, a change in element concentration becomes more important than the absolute amount.

Contamination has many sources. If calcium or magnesium are present on the surface, oxygen will also be present even after the cathode has been exposed to hydrogen. Carbonate ions in the electrolyte, resulting from CO_2 being absorbed from air, can deposit carbon. Any metal used as the anode, including platinum, will eventually migrate to the cathode. Metal wires used to connect the electrodes to the outside will contribute their elements to the cathode even when they have been protected by Teflon shrink tubing or epoxy. All of these impurities will concentrate at certain sites on the cathode surface where their concentration can be very high. Polishing abrasive is frequently seen in small spots as particles of Al_2O_3, MgO, or SiO_2. Chlorine may be trapped in cracks if Aqua Regia cleaning has been used without subsequent heating to high temperature. In addition to the material that might come from the outside, Afonichev[28] (Inst. for Metals Superplasticity Problems, Russia) found the small amount of zinc alloyed with titanium to migrate to the surface when hydrogen is removed at 600° C. This process might also occur in other metals when impurity diffusion is enhanced by diffusion of hydrogen, causing any dissolved element to migrate to the surface.

Surface analysis alone is not definitive in identifying the source of some elements. A very complex mixture of elements having very uneven topography covers the surface of a cathode after extended electrolysis. The amount and kind of element depends on where examination is made. Frequently, impurity atoms will deposit at certain active sites and cause a high local concentration. Use of samples containing enriched isotopes would eliminate many of these uncertainties. Methods using glow discharge, laser stimulation, or ambient gas are less affected by contamination and provide a more reliable indication of transmutation than does electrolysis.

7.8 Helium

As alpha particles are slowed by the surroundings, helium gas accumulates. This gas can diffuse rapidly through solids as single atoms, but its diffusion rate is greatly reduced when it forms gas bubbles (clusters) in grain boundaries.[29,30] When helium is made by the electrolytic process near the surface, and perhaps by other methods as well, no more than 50% can be retained by the cathode,[30] because some will diffuse outward to the surface, where it will be released into the surrounding gas. The rest will diffuse inward and be captured by the metal. Of course, the deeper within the structure helium is formed, the more will be retained. The observed retention of less than 50% shows that helium is produced very near the surface. Once helium has been trapped in palladium, it is only released by heating the metal to a temperature near its melting point.[29,31,32] Some generated helium is missed when measurements are made only of the ambient gas. By not considering the helium trapped in the metal, a positive bias for the energy/helium atom value results.

Helium is usually measured using a mass spectrometer. The mass of helium is 4.002603, which is very close to that of D_2 (4.028202). As a result, the resolution of the mass spectrometer must be exceptional if clean separation is required. Such high resolution is available for a price, which eliminates the need to remove deuterium. However, a cheaper residual gas analyzer (RGA) can be used after deuterium has been chemically removed. Fortunately, the two elements can be easily separated because deuterium is chemically reactive and helium is not. Two methods of separation have been described. The D_2 is reacted with extra oxygen and the resulting D_2O is removed from the remaining gas by freezing in liquid nitrogen. The other method pumps the gases using a titanium getter pump. This type of pump traps deuterium and all other reactive gases while largely ignoring helium.

Because air contains helium, an air leak can introduce significant helium into the apparatus. Detection of argon at mass 39.962 is used to indicate the presence of air. Of course, this heavy atom will diffuse less rapidly through a leak than would the lighter helium atom. Nevertheless, absence of argon in the apparatus is a good indication that an air-leak is not present. An addition, when detected helium rises above its concentration in air (~6 ppm), the source can not be a leak. Even a small air-leak will add much less helium than 6 ppm, an amount that can be calculated based on the measured leak rate, from which a corrected value for anomalous helium can be obtained.[35,36]

A less used and less sensitive, but more convenient method of analysis is available. A gas containing helium can be ionized using a high-voltage discharge to generate an optical emission spectrum containing helium lines, shown in Figure 69. The strongest line for helium is at 587.6 nm, which is well removed from the nearest interfering lines produced by deuterium at 486 nm and 660 nm. As a result D_2 does not have to be removed. The intensity of a chosen line will be proportional to the amount of helium present. Elements producing broad-spectrum emission must be excluded from the plasma to gain maximum sensitivity to helium.

587.6 nm

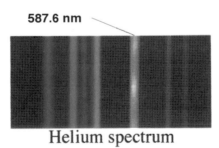

Helium spectrum

Figure 69. Helium spectrum produced by a 5000 V discharge. (from
http://hyperphysics.phy-astr.gsu.edu.)

Although cold fusion normally produces ^4He, ^3He might be produced by a proposed condensation reaction,[37] if a deuteron should fuse with a proton,[38] or when tritium decays[2,8] by emitting a beta particle. These two isotopes of helium can only be separately analyzed using a mass spectrometer. So far, all detected ^3He appears to result from tritium decay.

7.9 Heat Energy

Heat measurement requires a calorimeter, as described by a number of researchers.[39-50] Several basic types are available, based on different physical properties. The adiabatic method and flow calorimetry use the heat capacity, isoperibolic and Seebeck types use the thermal conductivity of a thermal barrier, and the ice and boiling calorimeters use a transition from one phase to another. Unfortunately, description of the method is frequently clouded in complexity when the method is applied to cold fusion. To make a useful contribution, the method must be simple to explain and understand. It may seem extreme, but in my opinion, if a 10-year-old can't understand how the measurement is being made, it has failed.

A suitable calorimeter costs between $5,000 and $15,000, depending on the type and the choice of components. Additional funds are needed to buy D_2O and the expensive metals. After the apparatus is assembled, a person can count on spending a year of their life getting the apparatus to work properly and learning all of the ways it can give false information. For those intrepid researchers choosing this journey, seeing anomalous heat for the first time will be like the birth of your first child, with many joys and disappointments to follow.

Having built and used most of the different types, I have learned to appreciate some of the criticisms. In general, heat is not difficult to measure accurately. The problem comes when a small amount of abnormal heat is superimposed on a rather large background heat, in this case, that which is added by chaotic electrolytic action when the Fleischmann-Pons method is used. Applied power is assumed equal to the product of current through the cell times the voltage at the calorimeter boundary. Because this power can contain low- and high-frequency fluctuations,[51] many values need to be averaged using high-speed data acquisition. Use of analog meters, even if they have digital display, is not recommended because they will not capture values produced by voltage spikes. Good agreement between the behavior of an internal electrical heater and a "dead" cathode reduces concern for many of these errors.

This electrolytic action also causes D_2 and O_2 gases to form, which if allowed to leave the cell would remove energy[b]. Early studies allowed this loss and were frustrated by not knowing just how much of these gases left the cell and how much remained after some unknown amount had recombined back to D_2O. Now most studies place a catalyst in the cell, which converts all D_2 to D_2O by reacting it with oxygen. In addition, use of a gas-tight cell allows the D/Pd ratio of the cathode to be determined by measuring orphaned oxygen[c] in the cell[d]. A

---

[b] See Appendix A for instructions about correcting for this energy loss.
[c] See Appendix F for a description of the orphaned oxygen method.
[d] Other methods to determine the D/Pd ratio are described in Appendix F.

number of such catalysts are available from Alfa Aesar, Inc.[e] The cell must also be gas-tight to avoid picking up H_2O from the atmosphere, which will stop the LENR reaction when it dissolves in the electrolyte, as discussed previously (see Figure 67).

The flow-, isoperibolic- and Seebeck-type calorimeters measure steady-state power. Consequently, the temperature of the cell needs to be constant before the values have any meaning. When power being generated within the calorimeter is changed, as was done to obtain Figure 70, the calorimeter can no longer provide an accurate value until it has again stabilized, after about 50 minutes in this case. This requirement is more clearly shown in Figure 71 where the apparent excess power shows a sudden change when the applied power is changed and returns to a stable and accurate value only after about 50 minutes. In addition to this requirement, stability can be achieved only when the temperature of the cooling water and power applied to the cell are held constant. Brief bursts of power can be incorrectly measured, depending on their rate and magnitude, and can be overwhelmed by random fluctuations if all conditions are not held constant.

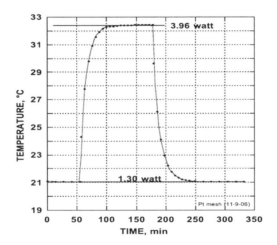

Figure 70. Change in cell temperature caused by increasing the amount of applied power.

Power is frequently expressed several different ways, which makes data comparison ambiguous. Watt/cm^2 or watt/cm^3 values are occasionally reported as a way to eliminate differences in sample size when a comparison of power is made. These quantities have neither practical nor theoretical meaning at this stage, except to reveal production of a large amount of power in a small space. Contrary to intention, use of such values does not even allow samples to be accurately compared. The amount of active material is highly variable between

---

[e] Alfa Aesar, (800)343-0660, www.alfa.com

samples, regardless of the area or volume and most of the sample is completely inert. As a result, the area and volume has no simple relationship to the amount of energy producing material (NAE) that might be present. Being able to produce a large power density is an important discovery, but the reported overall values are actually much smaller than the true power density of the active regions. When means are discovered to make the active sites in greater amount, power density will be greatly increased and these reported values will be considered insignificant.

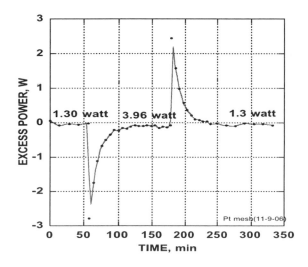

Figure 71. Behavior of apparent excess power when applied power is changed.

Another value occasionally reported is the heating efficiency or heat balance, which is Power Out/Power In. Sometimes the Excess Power/Applied Power is also reported. These values are also not useful in comparing measurements because experimental systems are not designed to maximize these ratios, hence a large variation in values can be expected. Such values are only important when a practical device is evaluated, in contrast to a research tool. Swartz[52] finds a maximum value to be produced when the excess power is plotted as a function of applied power, which he calls the OOP (optimum operating point). This behavior is expected because anomalous power does not increase as rapidly as applied power above a critical value. In other words, the process making anomalous heat eventually saturates, which naturally produces a maximum in a value based on this ratio. At this stage, the goal is to produce the maximum amount of excess energy, not the most efficient system. Conditions producing the maximum amount of excess power are more important than those producing a maximum OOP.

The next problem involves the choice of calorimeter type. All types can be used to study any of the methods used to initiate the effect when properly modified. The following discussion is designed to help choose a calorimeter and to avoid unexpected errors.

7.9.1 Adiabadic Type

All types can be used as adiabatic calorimeters. Sudden heat production is measured by combining the rate of temperature change and the effective heat capacity of the cell, while taking into account loss of energy through the cell wall while this temperature change takes place. In principle, the total temperature changed combined with the heat capacity of material experiencing the temperature change gives the total energy involved. However, because an unknown and variable fraction of the electrolyte and cell components participate in this temperature increase, the amount of generated heat will be uncertain unless the cell is well stirred. The calorimeter is calibrated by applying a pulse of power, as shown in Figure 70, and noting the shape and size of the rising and falling temperature curves. A complex mathematical analysis allows the basic characteristics of the calorimeter to be extracted.

7.9.2 Isoperibolic Type

The simplest type of calorimeter is called isoperibolic. Heat passing through a thermal barrier, usually the glass wall of the cell, is determined by measuring the temperature difference between the inside and outside of the cell, once this difference has become constant. The main problem involves measurement of the inside temperature. Thermal gradients and convection currents can cause non-uniform temperature even when the cell is stirred. For a calorimeter to be useful, these gradients must not change during a study and must be the same as those existing during calibration, which is seldom the case. Calibration using an internal Joule heater is not recommended unless it is used while electrolytic current is flowing between the cathode and anode. The resulting bubble action greatly reduces thermal gradients. A calorimeter can also be calibrated using a "dead" cathode while applying electrolytic current. Unfortunately, when this dead cathode is removed and replaced by the electrode of interest, gradients within the cell can change. This problem can be avoided because many cathodes take considerable time to become active, which allows the calorimeter to be calibrated during this time without having to change samples.

The basic descriptive equation is $W = C*\Delta T$ for a cell losing heat by conduction, where W is the sought for generated power, ΔT is the average temperature across the cell wall, and C is a constant obtained from calibration. As an example, a value for C can be obtained from Figure 70 by dividing the change in applied power by the change in temperature. However, such a single value is not much use. Instead, the calibration must be expressed as a linear equation, $W = A + C*\Delta T$, because ΔT at zero power occasionally does not correspond to a heating power of zero. This offset error is unimportant as long as it does not change after calibration. Anomalous power is claimed when $W - I*V$ is greater than zero, where I is the current through the electrolyte and V is voltage measured at the calorimeter boundary. Of course, a catalyst is assumed to be present that combines all extra D_2 and O_2 back to D_2O. (See: Appendix A) This simplified equation has to be modified if other factors affect the calibration constant, such as vaporization of the electrolyte, change in electrolyte level, or change in electrolyte concentration as LiOD deposits on the components above the liquid. Rather than trying to correct for these changes, the calorimeter should be designed to avoid the need for such corrections.

A calorimeter using heat transport by radiation through a Dewar wall, rather than by thermal conductivity through a solid,[53-55] is more complicated to evaluate. Heat transport by radiation through a vacuum Dewar is determined by $(T^4in - T^4out)$ rather than being a simple function of ΔT. This design offers few advantages to compensate for the added complexity.

Maintaining a constant and uniform temperature around the cell, usually in the form of a water-cooled jacket or a water bath into which the entire cell is placed, is an additional problem. Both methods of cooling have their own set of limitations, the main one being the difficulty in completely surrounding the cell by a constant temperature environment. The lid usually represents an uncontrolled heat loss. However, this heat loss is not important as long as it is constant. This requirement can be met by placing the entire apparatus in a constant temperature environment in addition to using water cooling.

Other changes in conditions can also produce false information. For example, Figure 72 shows an apparent production of excess energy when the resistance of the cell suddenly decreased, causing power to be reduced faster than the thermal inertia of the cell could match. This behavior could be mistaken for a heat burst if the cell voltage were not examined closely and the reason discovered. When all of these potential problems are considered, this method is hard to make accurate to better than ±250 mW, although much smaller changes in anomalous power can be detected and measured when they occur over a brief time. Fleischmann and Pons obtained better accuracy than this by careful design and many calibrations.

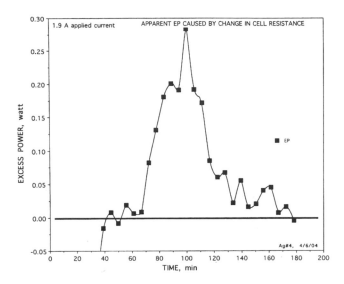

Figure 72. False excess power caused by a change in cell resistance resulting in a change in applied power.

A simple and serviceable design is shown in Figure 73. This isoperibolic calorimeter was used by Storms and it is similar to the design used by Bush and Eagleton.[56] A magnetic stirrer is covered by Teflon and the thermistors are enclosed in Pyrex tubes, the ends of which are made thin and are filled with oil to provide good thermal contact. Two thermistors are located at two different levels, which when combined, provide an average temperature. The cell also contains a recombiner, it is gas-tight, and it is connected to the atmosphere through an oil seal. A Teflon screw clamp allows the cathode to be easily removed. All Pt leads are fused through Pyrex glass tubes[f], which pass through O-ring compression seals in the Teflon lid. A jacket through which constant-temperature water flows surrounds the cell. Equipment able to pump constant-temperature water can be purchased from various companies.[g] In addition, the cell sits on a motor that causes the internal magnetic stirring bar to turn at a constant rate. Use of constant stirring rate is essential because the calibration constant is changed by changes in stirring rate, as described previously in Figure 22. Because some heat is lost through the Teflon lid, the cell is somewhat sensitive to changes in room

---

[f] A Pt-Pyrex seal is stable, but it is not gas-tight. It can be made gas-tight by applying a few drops of superglue to the inside.
[g] Cole-Parmer, (800) 323-4340, www.coleparmer.com
Fisher Scientific, (800) 766-7000, www.fishersci.com
Omega , (800) 826-6342, www.omega.com

temperature. Nevertheless, the calorimeter works well and is simple to construct, although it is not as sensitive as other designs, as described below.

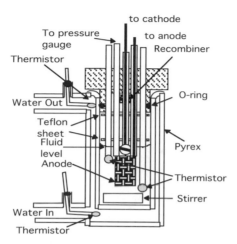

Figure 73. Drawing of a simple isoperibolic calorimeter contained in a water-cooled jacket.

Fleischmann and Pons solved the temperature gradient problem without mechanical stirring. Their calorimeter, shown in Figure 74, consists of a cell having a much smaller diameter compared to its height, with a partially silvered vacuum jacket used as the thermal wall. The lower part is surrounded by an unsilvered region, which is immersed in a constant-temperature bath. Use of a long-thin design encourages good thermal mixing with heat loss caused mostly by radiation through the walls rather than through the lid. Unfortunately, although the calorimeter worked as well as they claimed,[57-59] their description is too complex for most people to understand, which contributed to the rejection. Also, absence of a recombiner raised additional questions.

7.9.3 Double-Wall Isoperibolic Calorimeter

The gradient problem can be avoided altogether by building what is called a double-walled isoperibolic calorimeter.[43,60] In this design, the cell is placed in a container and the temperature drop across the wall of this container is measured rather than across the wall of the electrolytic cell. This method can be calibrated using an internal Joule heater and the electrolyte does not need to be stirred. As a result, error can be reduced to below 10 mW. However, the highest sensitivity requires applied power to be relatively small to prevent overheating, which requires use of a small sample. Figure 75 shows a calorimeter built by Huggins and his students.[61] The two blocks are cylinders made of aluminum between which is placed powdered Al_2O_3. The cell is Pyrex glass, which is sealed except for a gas vent to allow release of orphaned oxygen to the atmosphere through an oil bubbler. Just above the coin-shaped cathode is located a platinum coil used as a Joule heater or as an inert cathode during calibration. Like all calorimeters, this one needs to be kept in a constant-temperature environment in order to keep the reference temperature (T2) constant.

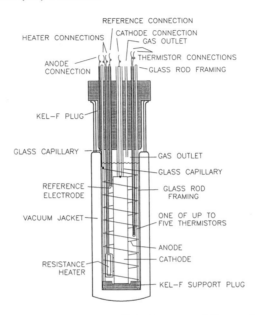

Figure 74. Drawing of an early Fleischmann and Pons calorimeter.

The descriptive equation is the same as the one applied to the single-wall isoperibolic calorimeter as discussed above. Because variable temperature gradients are not present, the calibration equation generally shows much less scatter and is generally more reproducible than the one obtained from the single-wall type. A detailed description of this type also has been published by Miles and Johnson.[62]

7.9.4 Flow Calorimeter

Further improvement in accuracy is gained by using a flow calorimeter. Heat is extracted from the cell by flowing fluid, usually water. The amount of heat can be calculated by knowing the flow rate, the increase in temperature of the fluid after passing through the cell, and its heat capacity. Capture of 98% of the heat can be achieved and the acceptable amount should be no less. The amount of captured energy can be determined by applying a known amount of power using a resistor within the electrolyte or by using electrolytic heating with a dead cathode. Maintaining a known and constant flow rate is the most serious challenge. Constant rate pumps can be obtained from Fluid Metering, Inc.[h] A picture of the calorimeter used by Storms is shown in Figure 76, which uses a cell similar to the one shown in Figure 73. When the apparatus is completely assembled during use, the cell is completely surrounded by the Dewar that is closed at the top by an insulating cover. This design allows 98% of the heat to be captured. During use, the external box is closed to allow the environment to be held at 20±0.01° C. All resistors used to measure current and temperature are contained within the constant temperature box.

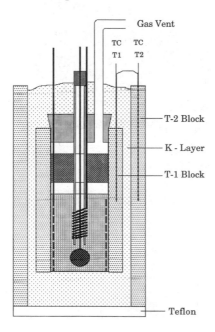

Figure 75. The double-wall calorimeter used at Stanford University.

A more complex version used by McKubre and co-workers (SRI)[49] can be seen in Figure 77. The entire cell is submerged in oil held at constant temperature and the entire apparatus is placed in a constant-temperature room. The constant-temperature oil is pumped past the cell, entering at the bottom and exiting at the top where its temperature is measured. The cell contains a recombiner and is gas tight except for the Gas Pipe that allows constant pressure to be maintained within the cell. A heater surrounds the cell, which is used for calibration and for holding the cell temperature constant. Unlike most calorimeters, this one operates at constant temperature with electric power applied to an internal heater to achieve this condition. The amount of power

[h] Fluid Metering, Inc., (800) 223-3388, www.fmipump.com

needed to maintain constant temperature is used to calculate generated power. Heat capture is 99.7%.

Figure 76. Flow calorimeter used by Storms.

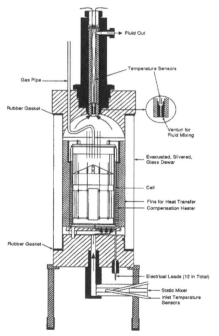

Figure 77. Calorimeter used at SRI, International. (McKubre *et al.*[64])

A clever variation requiring no external constant temperature bath is described by Närger et al.[46]

The flow calorimeter can be used as an absolute instrument, requiring no calibration, if the heat capacity and flow rate of the fluid are known and if all heat is captured by the fluid. However, total heat capture is almost impossible to achieve, hence this quantity must be determined. Consequently, the calorimeter must be calibrated by introducing a known amount of power. Once this calibration is undertaken, knowledge of the heat capacity becomes unnecessary because the defining equation becomes $W = C*F*\Delta T$, where W is measured power, C is a constant obtained from the calibration, ΔT is the increase in temperature experienced by the cooling fluid, and F is the flow rate of the cooling fluid. As a result, the heat-loss correction is automatically combined with the heat capacity to create a calibration constant, C. Once again, C should be described by a linear equation, which results in an equation of the form $W = A + B*F*\Delta T$, because ΔT might not be zero when no energy is produced in the calorimeter. This off-set error is not important as long as it is constant.

7.9.5 Dual-Cell Reference Calorimeter

This method is based on comparing two nearly identical isoperibolic cells, one containing an inert cathode in H_2O, and the other containing an active cathode in D_2O.[65-67] While this method is simple and cheap, it has several important disadvantages. First, the internal temperature of two cells must be determined, each of which may have different errors caused by different temperature gradients. Second, keeping equal and constant power applied to each cell is almost impossible to accomplish. Changes in internal resistance will always occur and this will cause changes in power between the two calorimeters even though the same current is passed through both cells. Consequently, both cells will not have identical behavior, hence each may be affected by a different error. If this method is used, a comparison should be made using calorimeter types not susceptible to temperature gradient problems, such as double wall isoperibolic, flow, or Seebeck.

7.9.6 Seebeck Calorimeter

The Seebeck[41,45,68-75] type calorimeter has many advantages, including sensitivity, stability, and simplicity. For its construction, an enclosure is made with walls in which many thermocouples are located. These walls can also be formed by gluing together several commercial thermoelectric converters, as described in Appendix B. When thermocouples are used, they are connected in series, with one junction on the inside and the other junction on the outside. As a result, a temperature gradient between the inside and outside will generate a voltage proportional to heat loss no matter where this loss occurs, provided the effective thermal conductivity of the wall is uniform. The challenge is to achieve a uniform thermal conductivity and a uniform sensitivity of the generated voltage to temperature. This requirement calls for many thermocouples be used with uniform spacing. A fan is generally required to distribute heat within the Seebeck enclosure because variations can occur in both of these variables. This method can achieve accuracy better than ±20 mW over a wide range of generated power and can be calibrated using a resistor within the cell. Some designs have even achieved sensitivity in the µW range when the chaotic behavior of an electrolytic cell is not involved. However, because heat is lost from the cell only through the surrounding air, the cell tends to run hotter for the same applied power compared to when water cooling is used. This effect can make a comparison between excess power obtained from different calorimeter types difficult to interpret because increased temperature can contribute to greater reaction rate or no reaction at all if the high temperature causes loading to be too small.

Seebeck calorimeters can be purchased from Thermonetics, Inc.[i] or they can be constructed[41] for about $150, but with a significant investment of time. Figure 78 shows a picture of the Thermonetics version, which is contained in a constant-temperature enclosure because it is sensitive to changes in room temperature. In addition, a fan is located within the Seebeck enclosure to better distribute heat. A handmade version is shown in Figure 79. This design has a sensitivity and accuracy similar to the commercial model, but with the advantage of being insensitive to changes in room temperature. The white material is PVC into which 1000 iron-constantan thermocouples have been placed. Water-proof paint is used to protect the water-cooled reference thermocouples contained within the cooling jacket. Only the upper fan is shown, with another fan located at the bottom. A better design using commercially available thermoelectric plates[j] is described in Appendix B.

Figure 78. Thermonetics Seebeck calorimeter with lid removed.

---

[i] Thermonetics Inc., (858) 453-5483, www.electrici.com/~thermo
[j] Hi-Z Technology, Inc., (858) 695-6660, www.hi-z.com

The descriptive equation is W = C*V, where W is produced power, C is the calibration constant and V is generated voltage. For highest accuracy, the calibration constant needs to be expressed by a quadratic equation, which results in an equation having the form $W = A + B*V + C*V^2$.

7.10 Accuracy of Calorimetry

How is the accuracy of a calorimeter evaluated? Calorimeters suffer from two kinds of error, random and systematic. Each measured value has a random variation that hides any excess power smaller than this variation. In addition, the calibration constant shows a random variation, with each constant in the calibration equation clustering within a range. These variations, when combined, create a limit to the smallest amount of anomalous power that can be considered real. A second, more serious potential error, results when the measured values drift with time. This drift can be caused by changes in room temperature, changes in flow rate when the flow-type is used, changes in the sensitivity of the devices used to measure temperature, drifts in applied power, and other changes that are sometimes hard to identify. The source can only be evaluated by observing how the calorimeter behaves over a period of time or when various changes are deliberately made. This uncertainty is always greater than the random error and is important because the drift can be mistaken for anomalous power, although frequent calibrations can reduce this error. Unfortunately, the occasions and frequency of calibration are seldom mentioned in publications. Even though the magnitude of both errors should be reported, only random error is commonly mentioned. Also, calorimeters of the types described here only measure power. Determination of total energy requires this power be integrated over time, which adds additional uncertainty.

Figure 79. A handmade calorimeter containing 1000 thermocouples. (Storms[41])

Methods for measuring heat are well understood and provide the basis for much information used in chemistry. The technique does not fail just because it is applied to cold fusion. Failure comes from poor use of the method, for which there are many examples in Table 2. The important question is, are the well-done studies sufficient to demonstrate production of anomalous energy? Of course, what you might consider to be "well-done" depends on which potential error you are willing to ignore. The type of potential error is limited only by imagination, many examples of which are summarized and then rejected by Storms[76,77] in an extensive review. Nevertheless, a few potential errors are real and need to be understood. Shanahan[78] has proposed that changes in locations where heat is produced within an electrolytic cell could introduce error when flow calorimetry is used. This error is shown by Storms to apply to neither flow[79] nor to Seebeck calorimetry,[42] although the isoperibolic method can be affected. Swartz[80] used a computer model based on hypothetical temperature errors to question the accuracy of flow calorimetry. No demonstration of the proposed mechanism has been reported. On the other hand, a potential error may occur when D_2 and O_2 gases are allowed to leave the cell. Jones and co-workers[81] (BYU), and Shkedi and co-workers[82,83] (Bose Corp, MA) observe the obvious, that an uncertain amount of recombination between D_2 and O_2 within a cell could introduce an uncertain error. Using this argument, Jones[84,85] criticized heat reported by Miles,[86] who replied in a series of exchanges.[87-89] Miles answered by pointing out that he, as do many people, measured the amount of internal recombination occurring in his open cell, for which corrections were made. Accurate corrections can be applied as described in Appendix A. This error does not occur in a closed cell,[90] which is now used by most people when anomalous heat is observed. In addition, open cells seldom have this error when applied current is increased above 0.1 A[79] because very little recombination occurs on the electrodes at currents above this value. Shelton and co-workers[91] at BYU provide a critique of general calorimetry and make some useful suggestions. In particular, they suggest the calorimeter should be tested by measuring a known chemical reaction. This can be done, as described in Appendix B, by measuring the energy associated with formation of PdD during the loading process. Surprisingly, no one has done this although Keesing et al.[92] saw the endothermic reaction and attributed it incorrectly to a Peltier effect. Of course, the importance of the error depends on how big it is compared to the signal. Some measurements totally overwhelm all error.

7.11 Summary

The anomalous products produced by the cold fusion process are being detected using conventional, well-understood methods. In addition, the methods are being applied with success in a growing number of laboratories. Of the detected nuclear products, only ^{4}He is seen in an amount consistent with heat production.

CHAPTER 8

Explanations, the Hopes and Dreams of Theoreticians

8.1 Introduction

When Fleischmann and Pons first made their claims, the observed behavior was very difficult to explain without introducing many untested and novel assumptions, many of which continue to provide the basic building blocks of theory in this field. A theory has to contend with several very difficult challenges. An explanation must account for a very high rate of nuclear interaction, which produces heat and helium without gamma emission. At the same time, it must show how the many complex nuclear reactions involving heavy isotopes with very high Coulomb barriers are possible. The theory must show why most attempts are failures, yet show how a proper combination of material properties can lead to success. Finally, the theory must be consistent with well accepted Laws of Nature and the well known behavior of nuclear reactions based on decades of experience. No published theory has met all of these requirements.

As with all science, discovery is a moving target for theoreticians. In this case, theoreticians must also ask themselves just how far from conventional thinking they are willing to go in order to explain the observations? A model too unique will be neither accepted nor understood. On the other hand, easy explanations do not account for the observations. As a result, study of the field still requires an open mind, creative thinking, and a willingness to follow where nature leads.

Most mechanisms either involve the deuteron being pushed over the Coulomb barrier by localized energy or electrons shielding the charge on the deuteron in various ways. Variations on these basic mechanisms have now generated over 300 published papers. A few models even propose neutrons to be involved or that the Coulomb barrier simply disappears under certain conditions. With so little information available initially, theory was based mostly on imagination and speculation, including the approach that encouraged Fleischmann and Pons.[1-3] While much more information is now available to define the limits of theory, this situation has not changed significantly.

In addition to describing the nuclear mechanism, a model must be able to suggest conditions that initiate the mechanism. In this field, so many conditions and variables are possible that success is very unlikely without a guiding model. The model must be related to the chemical and physical environment in a way that allows a wide range of conditions to be rejected. Failure of most theoreticians to address those variable over which we in the experimental world can have some control is one reason progress has been so slow.

With a few exceptions, the large collection of published theories will not be discussed here. Instead, attention is directed to the evaluations provided by several qualified theoreticians. Takahashi[4] examined some recent theories, Chechin and co-workers[5] provide an extensive review of theories published before 1994, Preparata[6] gives a similar review for ideas published before 1991, and Kirkinskii and Novikov[7] discuss a small part of the subject published before 1999. People who propose mechanisms are best suited to evaluate their own work, hopefully using information provided here. In the process, please remember that incorrect assumptions in theory are equivalent to experimental error in observations. Both lead to false conclusions and distract from general acceptance. The intention of this chapter is to focus thinking in directions more consistent with observation than most theories have achieved. In the process, many questions have been asked for which I hope answers can be provided in the future.

Several basic observed facts need to be acknowledged to keep from being too distracted by assumptions. As the tables in Chapter 4 make obvious, all types of nuclear products are observed, but with a huge range of production rates. The reaction rate for heat production is about 10^{12}-10^{14} events/sec, rare tritium production gives about 10^6-10^8 events/sec when it occurs at all, and neutrons are produced at a few tens event/sec on a good day. Tritium and neutron appear to be produced independent of any other nuclear product. Transmutation reactions involving clusters fall between the tritium and neutron generation rates, but are near the low end of this range. The temptation to explain a rapid reaction, such as heat production, using the behavior of a slow reaction, such as neutron production, needs to be resisted or at least justified.

The reaction rates respond to applied energy in different ways. Can evidence for screening obtained by bombarding a metal with ions having many keV of energy[9-12] be applied to conditions existing in nanoparticles of palladium exposed to ambient D_2 gas? Theoreticians need to decide whether one basic process operates over the entire energy range or whether several independent processes operate, each at its own independent rate and under it own unique conditions.

Additional confusion results when a variable is changed because several different effects are produced at the same time. For example, increasing applied current to a cathode will raise its temperature. Does the additional excess power result from the increased number of deuterons being delivered to the NAE or because additional energy is provided by the increased temperature? Does the additional excess power result because more NAE is created at the higher temperature or because the mechanism causing nuclear interaction becomes more efficient? Such multiple results need to be separated in theory and in practice.

The nuclear reactions, rather than occurring throughout the palladium, are observed only in certain small regions containing very little palladium along with many other elements. Clearly, these regions must have unexpected and unusual properties to be able to initiate a nuclear process—properties not found in ordinary materials. These properties do not include some that are commonly accepted. For example, a high deuterium content does not, by itself, initiate nuclear reactions. A large deuterium concentration becomes important only after the NAE forms. This element is one of the reactants; hence its concentration determines the reaction rate once a reaction is possible. Also, simply having small nano-sized particles present does not guarantee success. These particles must have unusual features to be active.

A listing of observations considered to be well established is provided in Table 13. These observations need to be made part of any theory. What is missing is also important and this is listed in Table 14. Because some missing behavior might eventually be observed or its absence understood, conclusions based on what is missing may change.

Table 13. Observations needed to be addressed by theory.

1. When ^{4}He and anomalous power are produced at the same time, the observed energy/^{4}He is 25±5 MeV.
2. At least 10^6 more tritium is found than is expected to accompany detected neutron emission and about 10^6 less than can account for heat production, assuming the hot fusion branching ratio.
3. Tritium can be produced using D_2O or D_2 and perhaps H_2O or H_2.
4. Neutron emission is less than 10 n/sec with occasional larger bursts.
5. Radioactive products, including tritium, are rare but occasionally observed.
6. Prompt energetic particle emission, including alpha and proton, is frequently observed but with a small rate.
7. Both protium and deuterium are involved in a variety of nuclear reactions resulting in different nuclear products.
8. Transmutation occurs during which 1, 2, 4, or 6 atoms of deuterium can be added to a target nucleus.
9. The most abundant transmutation elements are Fe, Zn, and Cu.
10. The nuclear reactions occur in very small and isolated locations.
11. Strange emissions are occasionally observed having behavior much different from normal particles and radiation.
12. Living organisms are able to host transmutation reactions.

Energetic particle emission, X-rays, and gamma rays are occasionally found, as listed in Table 11. This new and growing collection of energetic emissions require many of the theories to be re-evaluated. After all, if the released nuclear energy appears as conventional energetic particles, where does the proposed lattice coupling mechanisms fit in the process? Perhaps two mechanisms are operating. One occasionally makes energetic particles and neutrons, with an energy similar to that resulting from hot fusion, and the other dumps energy directly into the lattice. The rules that determine which of these processes operate under various conditions need to be made part of any lattice-coupling theory.

Table 14. Expected but missing behavior.

1. Gamma emission is rare.
2. Neutron emission is rare.
3. Alpha emission rate is not consistent with accumulated helium.
4. X-rays expected when a significant alpha flux is absorbed are missing.
5. The second nuclear product resulting from transmutation is frequently missing.

Finally, theoreticians need to agree on several general ground rules about what features are acceptable in a theory and what features are too inconsistent with general experience to be easily believed. In other words, a process known from general experience to be rare or impossible should be avoided unless it can be well defended. Each person who has given this problem much thought has their own list of forbidden behavior, which they apply to other people's models with enthusiasm. Here are four of my suggested Limitations. These are proposed as a means to start a debate about why these Limitations might be unnecessary. In this way, a challenge is created for theoreticians to prove the suggested Limitations wrong and, in the process, strengthen their own theories.

8.2 Limitations to Theory

8.2.1 Limitation #1:
Neutrons do not initiate cold fusion reactions.

This is based on the following considerations:

1. When a neutron is absorbed by a nucleus in the absence of a (n,α) reaction or similar prompt particle emission, a gamma ray is emitted. Additional energy can be released over time if a radioactive isotope is formed. Neither process is observed during cold fusion at a rate consistent with measured transmutation rates. However, a few rarely detected radioactive products are consistent with neutron addition and need to be explained.
2. A new element does not form after a neutron is absorbed until a beta or another particle is emitted. If the isotope formed is stable, which is frequently the case, a new element will not result. Therefore, a wide range of transmutation elements cannot be produced without a large number of neutrons being available to create many short-lived beta emitters. These beta emitters

have not been detected. In addition, this process would cause the concentration of stable isotopes of target elements to be shifted to the high end of the stable range, which is not observed.

3.   Helium can only result from alpha decay after a neutron is adsorbed by certain elements that are not present in a cold fusion cell. The energy resulting from this alpha decay is too small compared to the measured value. Therefore, neutron absorption cannot precede helium production.

4.   Neutrons are unstable and emit beta radiation as they decay. Neither the beta radiation nor the neutrons themselves have been detected at sufficient levels to be consistent with transmutation or helium production.

5.   Neutron formation requires about 0.78 MeV. Accumulation of such a large amount of energy in one location violates Limitation #2.

8.2.2 Limitation #2:

<u>Spontaneous</u> *local concentration of energy cannot be the cause of nuclear reactions.*

This is based on the following considerations:

1.   Spontaneous localization of energy violates the Second Law of Thermodynamics and the less well-known Le Chatelier's Theorem. Except for minor statistical fluctuations, energy always goes downhill. Within the range of statistical fluctuations, most atoms in a solid have energy near the average. Only a few atoms having energy within the Boltzmann tail are exceptions. The probability of an atom within this tail having energy sufficient to overcome the Coulomb barrier is extremely low. For the Second Law to apply, the additional energy must come from a source having an energy greater than is present in the local region where energy is accumulating. If a high-energy source does not exist within a system, the energy must be applied from a source outside of the system.

       Spontaneous absorption by an atom of many low-energy phonons originating from within the system is example of what is not permitted because energy is required to go uphill. In addition, such accumulation requires the existence of an energy storage mechanism. This mechanism has not been demonstrated to exist.

2.   If energy could be concentrated in an amount sufficient to initiate a nuclear reaction, chemistry as we know it would not exist. For example, chemical reactions that require a catalyst could be initiated simply by local energetic regions forming with sufficient energy to overcome the small chemical barrier

without a catalyst being required.[a] Chemical explosives would be set off spontaneously with unpleasant frequency. Why should we believe that local concentration of energy can occur spontaneously with sufficient intensity to initiate nuclear reactions without causing easily observed chemical reactions?

3.   When local energy exceeds chemical bond energy, bonds are broken and the energy changes form. This process prevents thermal or vibrational energy from being concentrated in a chemical system because it can be quickly converted to chemical energy as various chemical bonds are broken and reformed so as to absorb the energy. This is why a phase change occurs when the average energy exceeds a critical value. Consequently, the amount of energy able to concentrate in one atom by any mechanism has an upper limit that is too low to initiate nuclear reactions at the observed rate.

4.   Metal atoms in a crystal lattice are not fixed in position. They are vibrating and would vibrate even more in response to local increase in energy. Therefore, resonance cannot concentrate energy on a single atom because energy will be quickly communicated to the surrounding atoms. The concept of phonons and thermal conductivity rests on this process working very efficiently. In addition, the D nuclei are so far apart in PdD that excessive vibrational energy is required[8] to bring them close enough to fuse. If enough energy were applied to overcome the barrier, as is done in a plasma, behavior similar to that resulting from hot fusion would be expected.

8.2.3 Limitation #3:

Compact clusters of deuterons cannot form <u>spontaneously</u> simply by occupying sites in palladium that are too small to permit normal bond lengths.

This is based on the following considerations:

1.   The calculated size of locations between palladium atoms only applies to equilibrium conditions in the absence of atoms in these locations. When atoms are placed in these sites, the metal atoms move apart, while applying a force to move the larger atom out of the space. The force applied to the proposed large cluster of deuterons cannot exceed the chemical bond strength that holds the metal atoms in position. Stated another way, atoms will occupy positions within the lattice that result in a minimum energy for the entire lattice. Addition of two or more hydrogen to the same site within this structure does not produce an energy minimum. Forcing the metal atoms apart by multi-

---

[a] Of course, chemical reactions do occur spontaneously when the required energy to overcome the chemical barrier is almost present. This extra energy is available in the Boltzmann tail, unlike that required to overcome the Coulomb barrier.

particle occupancy requires energy be accumulated in the lattice, in violation of Limitation #2.

2. Unless energy is supplied to the cluster, the cluster cannot form because all such reactions between normal ions are expected to be endothermic. Limitation #2 prevents energy accumulation.

3. If a cluster should form, it must have a barrier to the loss of energy that prevents it from immediately decomposing before it can react with nearby nuclei. Proof that of such a barrier exists has not been provided in current theories.

8.2.4 Limitation #4:

For energy to be released from a nuclear reaction, at least two products must be produced.

This is based on the following considerations:

1. When energy is released during a nuclear reaction, momentum is conserved and energy is shared between emitted particles and/or photons, as they are emitted from the reaction site in opposite directions. In other words, the momentum vector sum of recoiling nucleus, emitted particles, and/or photons, and/or phonons, and/or scattered electrons, and/or recoiling host must be zero in center-of-mass coordinates.[b] As a result, two or more particles are normally observed as reaction products.

2. Occasionally, when a normal nuclear reaction appears to form a single product, a very weak photon is emitted but is undetected and most energy remains in an unstable nucleus. This energy will be released slowly over time by emitting radiation of various types. This delayed emission is not observed. Consequently, weak prompt radiation needs to be sought.

3. Direct coupling of nuclear energy to a lattice is observed during the Mössbauer process. The amount of energy coupled to the lattice by this process is very small compared to that being released by the cold fusion reactions. No evidence exists to support the belief that this process can couple high levels of nuclear energy. Consequently, a true absence of energetic particles resulting from the reaction of interest must be demonstrated before concluding that direct energy transfer to the host lattice can occur by a similar process.

---

[b] The description is provided by Talbot Chubb.

8.3 Plausible Models and Explanations

8.3.1 Proposed Sources of Heat and Helium

Using the collection of observations and limitations provided, we can now explore a few explanations for heat and helium production. The field was started based on the presumption of deuterium-deuterium fusion being the source of heat, tritium, and neutrons. When helium was found to be the main source of heat, people wondered why the expected gamma radiation was not observed. Models based on direct coupling of energy to the lattice were suggested to avoid the need for gamma emission. In addition, if a simple fusion reaction were possible, d-p fusion should be more common[13] than d-d fusion and d-t fusion should be detected when tritium is produced[14]. Neither p-d fusion nor d-t fusion is detected. These unmet expectations make a simple d-d fusion look less like the source of energy from the cold fusion process. What is the source of heat and helium if it is not d-d fusion? Several credible reactions have been suggested and can be evaluated by comparing their energy production to the measured value.

Kim and Passell[15] and Frodl et al.[16] propose the reaction d(^{6}Li, α)α is the source. This reaction produces 11.2 MeV/helium, with a value that is too small to be consistent with the observed value of 25±5 MeV/He (see Section 4.4.2). Kozima[17] suggests the reaction ^{7}Li(n, α+β)α, which gives 8 MeV/helium. Again the energy is less than the measured value. Taplin,[18] in a similar vein, proposed the reaction ^{7}Li(p, α)α, which generates a similar small value of 8.6 MeV/helium. Passell[19] analyzed palladium after energy had been made and found a consistent reduction in the ^{10}B content compared to virgin metal. He proposed the reaction ^{10}B+ ^{2}d = 3^{4}He as the source of helium. However, this reaction produces only 5.97 MeV/helium. Any or all of these reactions might operate, but it is unlikely they are a major source of energy associated with helium production. Takahashi[20] proposed that four deuterons condense to make ^{8}Be, which quickly decomposes into two alpha particles, each with 23.8 MeV. This energy is consistent with the measurements provided formation of ^{8}Be can be justified. Other sources of helium are discussed next when transmutation reactions are examined.

8.3.2 Proposed Sources of Transmutation Products

Observed transmutation products can be evaluated by examining the beginning and ending of the reaction, without proposing a mechanism, as was done above for helium production. This approach is possible because no matter which mechanism is accepted, the number of protons and neutrons involved in the reaction must be conserved and the reaction must be exothermic. For the sake of discussion, several assumptions are made. These are: the seed is palladium and fusion with deuterium is followed by fission. In addition, the most common transmutation products are ^{63}Cu, ^{57}Fe and $^{?}Zn$, along with non-radioactive isotopes of other elements.

What is the consequence of zinc (Zn) being made from fission of palladium after reacting with deuterium? An answer to this question can be explored by examining the consequences of various hypothetical reactions. Addition of one deuteron to palladium is forbidden because this results only in radioactive products. Addition of two deuterons gives argon (Ar), with six combinations giving stable isotopes of argon and zinc, as listed in Table 15. The sample would be enriched in ^{70}Zn, provided all isotopes are equally involved.

Suppose helium is made as one of the products when two deuterons are added to palladium isotopes. This combination gives isotopes of sulfur (S), some of which are stable, and stable argon. Neither argon nor sulfur would remain on the cathode in significant amounts to be detected by later analysis because argon is an inert gas and sulfur would form D_2S gas, which would dissolve in the electrolyte.

What happens when copper (Cu) is one of the nuclear products instead of zinc? This reaction results in potassium (K) with only one stable combination of products. The isotopes ^{65}Cu and ^{41}K should be enriched by this process, in contrast to ^{63}Cu being the observed enriched isotope. This lack of agreement with observation would seem to rule out this reaction.

In a similar manner, the consequences of forming iron (Fe) can be explored. Addition of two deuterons to palladium produces four stable combinations of $^{26}Fe + ^{22}Ti$. If all isotopes of palladium participate in the process equally, the process should enrich ^{58}Fe and ^{50}Ti. If iron is produced along with helium, the other element would be calcium. If the isotopes of palladium were equally involved in this process, ^{58}Fe would be the most enriched isotope. However, ^{57}Fe is observed to be the enriched isotope when this element is found in abnormal amounts on the cathode, again giving poor agreement with this proposed reaction.

Does this lack of agreement between predicted enrichment and the various measurements mean these proposed reactions do not occur or are only certain isotopes of palladium active? This question is impossible to answer at the present time. If helium is produced as part of the fusion-fission process, the absence of detectable radiation associated with helium formation can be explained. However, the energy/He resulting from such reactions is too great to be consistent with the measurements. Is it possible for helium to result from two simultaneous reactions, one having a low He/energy ratio and the other having a high ratio, with an average equal to the measured value? This possibility is worth exploring.

These suggested possibilities are only provided to guide thinking away from fusion and toward other possibilities. Clearly, more work is needed along with theoretical justification. For example, why would palladium experience fission after fusing with deuterons? Why would only non-radioactive isotopes result? In addition, light hydrogen is always present and this might react in the same manner as deuterium, thereby complicating interpretation of the final mixture of nuclear products.

Table 15. Examples of possible nuclear products that are nonradioactive.

	Atomic weight of each isotope shown		
	Pd+2d =	Zn+S+He	
%[a]	Pd$_{46}$	Zn$_{30}$	S$_{16}$	Q, MeV
1.02	102	66	36	35.5
1.02	102	68	34	35.9
11.14	104	68	36	35.2
1.02	102	70	32	31.5
11.34	104	70	34	34.0
27.33	106	70	36	34.2
	Pd+2d=	Fe+Ca+He	
	Pd$_{46}$	Fe$_{26}$	Ca$_{20}$	
1.02	102	58	44	39.6
1.02	102	56	46	39.7
1.02	102	54	48	36.4
11.14	104	58	46	39.8
11.14	104	56	48	39.3
22.33	105	57	48	39.9
27.33	106	58	48	40.3
	Pd+2d=	Zn+Ar	
	Pd$_{46}$	Zn$_{30}$	Ar$_{18}$	
1.02	102	66	40	41.2
1.02	102	68	38	42.0
1.02	102	70	36	37.1
11.14	104	68	40	40.9
11.14	104	70	38	40.1
27.33	106	70	40	39.9
	Pd+2d=	Cu+K	
	Pd$_{46}$	Cu$_{29}$	K$_{19}$	
1.02	102	65	41	40.1
	Pd+2d=	Fe+Ti	
	Pd$_{46}$	Fe$_{26}$	Ti$_{22}$	
1.02	102	56	50	49.4
1.02	102	57	49	46.1
1.02	102	58	48	48.0
11.14	104	58	50	49.4

a = % abundance of Pd isotopes

Takahashi et al.[21] have explored the fusion reaction in detail by assuming that energy is accumulated in palladium by multi-photon resonance absorption of X-rays. These X-rays are proposed to be made by other nuclear reactions occurring in the PdD lattice. The various fission channels are evaluated for their energy requirement and yield using a liquid drop model. This approach, generates a wide spectrum of transmutation products that are in general agreement with the observations of Mizuno et al.[22]

Elements much heavier than hydrogen are proposed to fuse. For example, iron (Fe) is claimed to form when an arc is struck between carbon rods (C) under water.[23-25] Vysotskii[26] gives evidence for the reaction $^{23}Na_{11}$ + $^{31}P_{15}$ = $^{54}Fe_{26}$ in biological cultures. All of these reactions, if they occur at all, involve a large Coulomb barrier and result in elements much heaver than can result from fusion with isotopes of hydrogen. These observations and proposed reaction paths strain all of the proposed models for an explanation.

Finally, Miley has reported a broad range of elements to result from electrolysis of electrodes containing Ni+Pd using a H_2O electrolyte. Many, but not all, of the elements lighter than nickel can be explained as fission products, as was done previously using palladium. Some elements might even be fission products from the palladium, which is known to be present in some samples or from platinum that might have transferred from the anode. However, the elements near lead (Pb) require a different explanation. Such an explanation might include addition of four protons to ^{198}Pt to give ^{202}Pb, which has a very long half-life. As noted earlier, the observations are not consistent with neutrons being involved, even though this has been suggested by Widom and Larsen.[27] Consequently, the transmutation results of Miley still lack a satisfactory explanation.

8.3.3 Proposed Mechanism to Initiate LENR

For the reported nuclear reactions to be explained when deuterium is used, at least two deuterons must come together under conditions permitting formation of helium and/or transmutation products. This requirement implies the formation of clusters. The simplest cluster is the D_2 molecule, but this has not been found in the β-PdD structure. Even if it should form in another phase, such as PdD_2, the d-d distance would be much too large to allow fusion to take place. As various calculations have shown, forcing this molecule to form within β-PdD or forcing a smaller bond length requires considerable. The problem becomes overwhelming when a cluster of up to 6 deuterons is required to explain the Iwamura results. How are such clusters formed and what process allows the large Coulomb barrier to be breached? No theory has provided answers to these questions without violating one or more of the Limitations noted previously. Can an answer be suggested that does not suffer from this problem?

The Mills model of hydrino (deutrino) formation provides a possible answer. Mills[28] proposes that the electron in a hydrogen isotope can occupy energy levels below those associated with normal Bohr orbits. A barrier exists between conventional Bohr levels and the unique Mills levels that prevents transition under normal conditions. For an electron to enter these fractional quantum states, a special catalyst atom is required. This atom must have a special energy level into which energy released by formation of a hydrino (deutrino) can go. Repeated operation of a suitable catalyst is proposed to cause the Mills electron to lose energy and drop to increasingly lower energy levels. This process is exothermic, thus satisfying Limitation #2. Once the electron has reached a sufficiently collapsed state, the nuclear charge would be shielded and the deuteron could fuse or enter another nucleus, thus providing a solution to the Coulomb barrier problem as first proposed by Mills[c]. During fusion, the Mills electron might be ejected as a prompt beta particle, thus providing a solution to the single-product problem (Limitation #4). In addition, this electron would have sufficient energy to form a neutron by combining with a proton on occasion,[29] thus accounting for the occasional detection of neutrons. In summary, this model provides a solution to several major problems plaguing cold fusion if the reality of hydrino formation can be accepted. Assuming a hydrino-like structure can form, how does this process work?

Normally, most deuterons are not close enough to a Mills catalyst to react. However, when deuterons are forced to diffuse through the lattice, they have a much greater probability of contacting a rare catalyst atom. The resulting deutrino structure continues to move with the flux and eventually finds another catalyst atom where the electron energy level can be reduced still further. The flux has an additional role to play because, as local deuterium is converted to deutrino, the reaction would slow down and eventually stop unless diffusion supplied more deuterium to the catalyst and removed the reaction products. In addition to generating the conditions required to initiate nuclear reactions, this process can also be used to explain the claimed production of some anomalous energy when deuterium is simply caused to diffuse through palladium.

This process continues until the electron reaches a level that can shield the deuteron and allow a nuclear reaction to take place. This process takes time for sufficient deutrino concentration to build up, hence the delay in heat production. The Mills level required to shield the nucleus will depend to some extent on other shielding mechanisms operating at the same time. In other words, a deutrino might be more effective in fusing with another nuclei in a solid because various additional shielding mechanisms are available that do not

---

[c] This idea has been widely discussed on the internet without consensus following Mills' initial proposal.

operate in a plasma. As a result, a highly shrunken Mills structure, as Mills opines, might not be required.

Once the Mills structure has formed, how are clusters created? When an electron enters the fractional quantum states, Bohr energy levels become available and these are occupied by another electron, thereby creating a negatively charged ion. These levels allow chemical bonds to form between deutrinos and other atoms. As a result, complex clusters of deutrino molecules are proposed to form by an exothermic process, making multiple deuterons available for nuclear reactions. Several problems becomes immediately apparent. Presumably, a cluster of two deutrinos could fuse to make one helium and two prompt beta particles before additional deutrinos could be added to make the cluster bigger. Even if this reaction were inhibited, a cluster of four deutrinos, would be unstable because the deutrinos could collapse into ^{8}Be, followed by decomposition into two alpha particles as Takahashi has suggested based on his proposed TSC structure. These processes would seem to make formation of clusters containing more than four deuterons impossible.

In addition, why do the deutrinos only interact with certain nuclei (seeds) within the palladium lattice and not very often with palladium itself? When deutrinos do react with palladium, why is fusion apparently followed by fission? Why do the other elements not experience this additional reaction? More study is clearly required. Consequently, to be successful, this and all similar models relying on cluster formation must be controlled by some unknown rules.

The difficulty in replicating the effect might be caused by two interacting variables; the presence of a catalyst having sufficient concentration and the concentration of deuterons. For example, if two clusters of deutrinos (four atoms) must come together to make helium, the reaction rate can be expected to be sensitive to the deuterium concentration raised to the fourth power, with the concentration near the catalyst being important. A sample having a low concentration of catalyst would require a very high concentration of deuterium, such as is typical of the Fleischmann-Pons process, for detectable power to be produced. On the other hand, a sample containing a high concentration of catalyst could be expected to produce measurable energy without having to acquire a high concentration of deuterium, such as is experienced using the gas-loading method. In other words the wide variation in behavior of individual samples and of the various methods depends on the amount of catalyst that happens to be in the sample and the deuterium concentration that can be achieved where the catalyst is located.

The question then becomes, "What catalytic elements must be present?" The catalysts suggested by Mills, based on the ionization energy of isolated atoms, will not apply to atoms within a solid alloy because the type of bond and its energy will modify the energy levels. Only certain elements in certain chemical relationships may have the required energy levels, thereby making the catalyst rarer than available ionization energies would indicate. Can a suitable material be suggested? Iwamura *et al.* (Table 10) observe that CaO must be present to make the observed transmutation occur in palladium. Suppose either element in CaO, when in contact with palladium, is a Mills catalyst. How would this process operate? Deuterium passing through layers of this material is converted to deutrino clusters and some of these diffuse back to the surface where they react with the deposited seed. Energy is communicated to the lattice by prompt emission of the Mills electrons or of unused deuterons present in the cluster. These emitted energetic electrons or deuterons would be the missing additional product required to carry away the energy and momentum of the transmutation reaction. Of course, some helium would be expected to result from fusion within the clusters, in addition to the observed transmutation products.

So far, we have focused on the behavior of deuterium. How does hydrogen behave? Hydrogen has not been observed to fuse. Instead, transmutation occurs when hydrogen adds to a heavier nucleus. Transmutation reactions apparently have a much lower reaction rate than does d-d fusion. Consequently, heat can be produced but at a lower level than when deuterium is used. Energy produced by hydrino formation alone can be expected to provide an important fraction of the measured energy when hydrogen is used.

8.4 Conclusions

Although this proposed process has, as yet, no generally accepted theoretical justification, the basic mechanism meets all of the proposed Limitations, provides a method to overcome the Coulomb barrier, and explains both transmutation and helium production. In addition, it shows why the diffusion flux is important, why certain samples are active and others are not, and why nuclear activity is frequently delayed. It even accounts for occasional neutron production. No other single model has all of these features. In addition, the mechanism does not preclude other mechanisms from operating to help lower the Coulomb barrier. Detection of prompt energetic electron emission and/or radiation of the proper energy would go a long way to support the proposal and its consequences. In addition, some question, as noted, need answers.

As anyone who reads theory papers will quickly realize, all theories have one or more major flaws, which other theoreticians are not shy to point out. The challenge is to agree on which assumptions are flawed and then find a way to prove whether they are flawed or not, without too much personal conflict. Hopefully, the summary of observations provided here combined with the Limitations listed above will help this effort.

CHAPTER 9

What Should Happen Next?

Even though some people may find total rejection of all anomalous observations easier than accepting a nuclear process as an explanation, a choice needs to be made. If nuclear reactions are not the cause of these anomalous observations, something else of importance is operating. If many trained scientists are deceived by such simple experiments, how can we trust anything claimed by science? If so many errors are made using conventional methods, how can similar measurements be trusted when applied to other behavior? The mountain of evidence causes one to wonder what more needs to be discovered before science will accept the reality of cold fusion? These are basic questions that go to the heart of the scientific method and the integrity of the scientific profession—issues that are not going away as some skeptics predicted.

Skepticism is a common and necessary approach to life. Science, especially, thrives on doubts and questions as a means to weed out crazy ideas created by rich imagination and to overcome the less noble aspects of human nature. When Fleischmann and Pons revealed their amazing claims, most good scientists put their doubts aside and sincerely tried to make cold fusion work. A few of us were successful, but most were not. Those who saw nothing concluded nothing of importance was happening. This example of normal human nature separated people into so-called "skeptics" and "believers". In science, a mechanism exists for these two viewpoints to be tested and resolved, without excessive conflict. Instead, the skeptics went to war—a war they have now lost. They went to war because the ideas were too great a threat to accepted knowledge and would cause too great an economic loss to some people. In other words, self-interest and arrogant certainty dominated the response of leaders in science and government rather than objectivity and concern for the common good. This war has been costly to everyone, especially now that the need for a different primary energy source has become so important. Mankind can no longer afford the delay. This is not to suggest that skeptics should simply give up. Results should be criticized, but people need the freedom and encouragement to explore cold fusion without risking rejection of the entire program or their careers. Naturally, a lot of nonsense and useless ideas will distract from this process, providing skeptics with a useful role. However, this role needs to be directed toward extracting what is real and useful rather than stopping progress all together.

In the beginning, available observation made novel assumptions necessary, many of which were hard to believe. Recent work shows nuclear reactions can occur in solids at low energy and they behave much like ordinary nuclear reactions, with a few important exceptions. Like "normal" reactions, energy is communicated to the nuclear products that are emitted as energetic particles. Like "normal reactions", energy can be measured and related to the amount of energy expected from the reaction. At the same time, several novel processes are at work and some novel emissions are detected. These unexpected behaviors open the door to many other discoveries, which hopefully will be better accepted. For example, the role of living systems in changing the elemental composition of our world needs to be appreciated and explored.

The cold fusion saga has revealed serious flaws in the way science and the media, both popular and scientific, handle new ideas. Most myths, when created to protect special interest, go largely unnoticed because the harm to society is not sufficiently obvious, hence it is difficult to counter. In contrast, the reality about cold fusion was and still is so distorted and the possible consequences of this distortion are so important, a serious examination of what went wrong is essential. Some influential scientists, followed by the press, created a false myth. The process permitting this to happen needs to be changed to prevent society from being harmed in the future by being denied important technologies, especially in the US where this rejection process was taken to extreme.

At present, the field is handicapped by problems with intellectual property constraints. The US Patent and Trademark Office will not issue patents on the subject, with a few exceptions. This rejection is not just based on the patent application being bad, which many are, but because the claims are believed to be based on fantasy. This unimaginative approach has distorted how patents should be used and has forced people to protect their ideas in unproductive ways. To avoid triggering rejection, an honest description involving cold fusion is avoided. Instead, attempts are made to patent vague theories and imagined applications, frequently without mentioning the dreaded words. If someone should later make a useful device and the described theory is found to apply, the people who developed the theory hope to make money. This approach interferes with open academic discussion about the basic processes and is unfair to people who will actually make cold fusion practical, perhaps without knowing anything about the proposed theory and its imagined, unproven claims. This approach, fostered by the failure of the Patent Office, will have a negative effect on the future of the field. Other shortcomings of the Patent Office are discussed by E. D. O'Brian.[1] The sooner the US Patent Office recognize the reality of what is called cold fusion, the sooner the patent mess can be cleaned up and the phenomenon can be developed in a normal way.

The world is at a crossroad. Energy is getting more expensive and the oil supply will continue to be under threat by political tension, the growth of China and India, exhaustion of oil fields, and impact of severe weather on oil fields and distribution. Global warming is real and being made worse by use of fossil fuels.

Surely these are sufficient reasons to explore a new source of energy no matter how difficult it is to understand. Future generations will have little sympathy for a society that allowed these conditions to become worse when a better alternative might have been available, but was ignored.

CHAPTER 10

Brief Summary of Cold Fusion

The phenomenon of cold fusion or low energy nuclear reaction occurs in an unusual solid or even within complex organic molecules. A variety of nuclear reactions are initiated, depending on the atoms present. Some of these reactions occur at a rate sufficient to make measurable heat. The most active reaction produces ^{4}He when deuterium is present. Other reactions occur at lesser rates, but rapidly enough to accumulate detectable nuclear products. These products can be tritium, and/or transmutation products, with all isotopes of hydrogen being potential reactants. Some nuclear products have also been detected as energetic particles along with gamma and occasional neutron emission. Many nuclear reactions can occur at the same time and originate from different environments and mechanisms within the same sample, with most of the material being inert. This behavior complicates interpretation.

Active environments appear to form when cracks are produced, as well as in very small particles of the special material. As yet, the unique nature of this special material, or so-called nuclear-active environment (NAE), has not been identified and it cannot be created very easily or in significant amount. This ignorance has been the main reason replication has been difficult, although some methods and samples work better than others.

Cold fusion is distinguished from hot fusion by needing relatively low energy to be initiated, by being very sensitive to the environment in which it occurs, and by making nuclear products that are different from those made by hot fusion. Many advantages are obtained by using cold fusion as a power source instead of hot fusion. Cold fusion does not generate significant hazardous radiation or make significant radioactivity while generating heat energy. In addition, energy is produced with greater density (watt/cm^3) than by hot fusion or even by terrestrial fission reactors. The nuclear-active material, deuterium, is in plentiful supply and its extraction from common water does not generate hazardous waste. Because the nuclear reactions can be initiated in small regions of material and do not produce a threat to safety, useful generators can be made consistent in size to the need. As a result, a single generator could be located in each home with the capability to generate all energy, heat and electricity, needed for the dwelling and for personal transportation. Absence of a large central source of power along with its transmission lines would increase efficiency, lower the cost of installation, and increase reliability.

The main limitation facing the method is the relatively low temperature at which heat energy is produced. While this property reduces the threat of fire or explosion, it requires the use of efficient methods to convert heat to electric power or hydrogen, methods that are now being developed. In addition, the lifetime of the nuclear-active environment is unknown, a property that will affect the cost of power. Of course, when this source of power becomes practical, the conventional sources of energy will be obsolete and could be phased out, resulting in potential economic disruption. This threat is and will continue to be a major handicap to rapid development of the energy source in some countries.

In addition to being a potentially ideal source of energy, a mechanism has been revealed to accelerate decay of radioactive waste and contamination. In other words, the phenomenon has the additional potential to undo some of the damage caused by the use of fission power. This mechanism, once understood, will no doubt lead to many other unanticipated benefits and insights into the behavior of nuclear interactions.

Appendix A

Calculation of the "Neutral Potential"

When electrolysis occurs in a cell containing H_2O or D_2O, H_2 or D_2 is produced at the cathode and O_2 is produced at the anode. If these gases are allowed to escape from the cell, corrections must be made for the energy they carry away. The energy lost from the cell is the energy that would have been gained by the cell if the gases had reacted within the cell to produce the compound from which they had been produced.

This correction can be determined as follows using decomposition of H_2O as an example. The rate at which water is decomposed by applied current (I) can be calculated using the following equation:

$$\text{Rate(moles of } H_2O \text{ decomposed/sec)} = I / (2* 96485). \qquad (1)$$

Based on this equation, current flow of 53.605 Ampere-hours will decompose 1 mole of H_2O (18 ml) to give one mole of H_2 (22414 ml STP[a]) and 1/2 mole of O_2 (11207 ml STP). The energy required to generate these gases is equal to the enthalpy of formation of water per mole (ΔH=68.300 kcal/mole, 285.83 kJ/mole at 25°C) multiplied by the number of moles of H_2O consumed. Insertion of the enthalpy of formation into Equation (1) results in Equation (2) for the total energy (Q) used.

$$Q = \Delta H * I * t / (96485 * 2) \qquad (2)$$

where time (t) is expressed in seconds.

If electrolysis is done in the absence of recombination, this amount of energy will be lost from an electrolytic cell as the gases leave.

The amount of lost energy can be calculated another way as follows: Equation (3) is the defined relationship between energy (Q) in calories and power in watts.

$$Q * 4.186 / t = \text{Power in watts being lost} \qquad (3)$$

In an electrolytic cell, the amount of applied power can be obtained from Equation (4)

---

[a] STP = standard temperature and pressure. Temperature = 273.16 K, Pressure = 1 atmosphere.

$$\text{Watts} = V * I \qquad\qquad (4)$$

where V is the voltage being applied to the cell and I is the current passing through it. Combining Equations 1, 2, 3 and 4 gives the following equations for what is called the neutral potential of H_2O and D_2O.

$V(np) = (68300 * 4.186)/(2*96485) = 1.482$ volts for H_2O at 25°C and 1 atm
$V(np) = (70394 * 4.186)/(2*96485) = 1.527$ volts for D_2O at 25°C and 1 atm

   This voltage provides a simple method to calculate how much energy is lost from a cell as the decomposition products of water. For better accuracy, the values need to be modified slightly if the pressure is not 1 atm and the temperature is not 25° C.
   When the neutral potential is multiplied by the amount of current passing through a cell, the power used to decompose water is obtained. This power can be added to applied power as a correction (Equation 5) for gasses lost from the cell. Of course, this correction is unnecessary if all of the gas reacts to produce water without leaving the cell.

$$W(\text{correct power}) = [V + V(np)] * I \qquad (5)$$

   Besides being used to calculate a correction for energy lost, the neutral potential is also the voltage required to produce a current flow through a cell that results in decomposition of water. Figure 80 shows an example of how this applies to a typical cell used for cold fusion. No current is seen to flow below 1.53 V, thus showing that decomposition of D_2O does not take place below this voltage as expected.
   A voltage can be calculated when the Gibbs energy of formation is used in the formula instead of the enthalpy of formation. This results in a voltage of 1.262 V for D_2O. This value cannot be used to predict the minimum voltage required to decompose water for the following reason.
   The relationship between enthalpy(H), entropy(S), temperature(T), and Gibbs energy (G) is given by the following defining equation.

$$\Delta G = \Delta H - T\Delta S \qquad (6)$$

Enthalpy of formation describes the total amount of energy that is used or generated by a reaction and the Gibbs energy is the amount of energy available to do work after the entropy requirement of the system has been satisfied. If, for example, D_2 and O_2 gases at 1 atm and 25° C are mixed and caused to combine to form D_2O in an electrolytic cell, the maximum voltage generated by this reaction is 1.262 V because this is the maximum voltage available to do work. On the other hand, this voltage is insufficient to reverse the reaction because the entropy requirements cannot be met until applied voltage exceeds 1.527 V.

An example of how a Fleischmann-Pons cell behaves is shown in Figures 80 and 81, where the voltage being applied to or generated by a sample of Pd loaded with D_2 is shown. In these figures, voltages applied to the cell are positive. Two voltages are noted in the figure, the voltage based on the Gibbs energy of formation and the voltage based on the Enthalpy of formation. The difference between these voltages is the energy (voltage) required to satisfy the entropy requirement of the reaction before any reaction can occur. The voltage of 0.9 volt, where the values become negative in the figure, results because the D_2 is not reacting with oxygen, but is simply forming D_2, which generates a lower voltage because the reaction is less energetic than formation of D_2O. This voltage is the open-circuit voltage (OCV), which provides a way to measure the deuterium activity in the surface where the $2D=D_2$ reaction occurs. Under the conditions existing when the measurements in Figure 81 were made, the OCV was typical of the alpha-beta two-phase region.

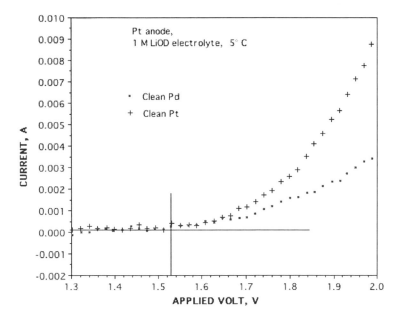

Figure 80. Voltage required to cause a current to flow in a cell containing D_2O. The cathode material is shown.

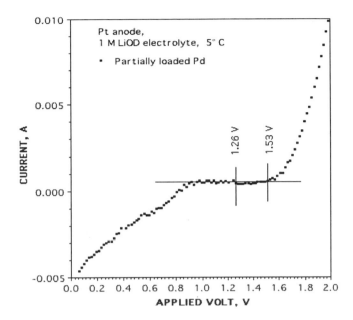

Figure 81. Example of the voltage required to decompose D₂O and the voltage generated when D⁺ combines to form D₂ gas. The slight current offset from zero has no meaning.

When a cell is used without a recombining catalyst, a problem arises because the amount of recombination is uncertain. As a result, the amount of correction is uncertain, resulting in apparent excess energy if a correction is made assuming no or little recombination when, in fact, a lot of recombination occurs. This problem is reduced if the behavior shown in Figure 82 is considered. As the figure shows, if applied current is sufficiently high, no recombination can take place within a cell, thereby allowing an accurate correction to be made. At 0.1 A, the recombination fraction is 0.05 and the correction would be 0.145 W, with no more than 0.008 W uncertainty. If the current were 0.05 A, the recombination fraction would about 0.25, resulting in a correction of only 0.058 W, but with a larger uncertainty. The values attributed to Jones *et al.* [1] in the figure are from a paper that gives a good example of biased reasoning. They measured the recombination fraction at very low currents, where it is high, and used these values to dismiss all measurements using open cells, without acknowledging that most successful studies used much higher currents or closed cells where this correction is unnecessary.

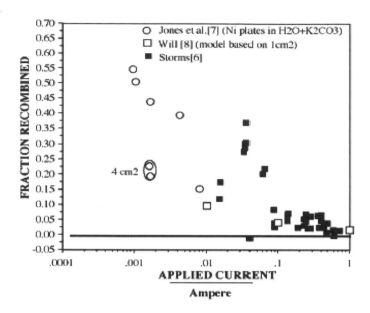

Figure 82. Effect of applied current on the fraction of recombination in an open cell.[2]

Appendix B

Construction and Evaluation of a Seebeck Calorimeter

B.1 Introduction

A Seebeck (Kalvin) calorimeter is very easy to use, is simple to construct, its function is easy to describe and evaluate, and it can achieve a very high sensitivity and accuracy. In addition, the active region can be made as big or as small as is necessary to hold any size apparatus from which heat is being released. This great adaptability and simplicity makes the method very attractive to use when studying cold fusion. Although the device can be applied to any method, its application to the electrolytic method is described here.

In this design, thermoelectric converters completely surround the source of heat. Temperature outside of the converters is held constant while temperature within the calorimeter is allowed to increase. The average temperature difference generates a voltage, which after calibration is used to determine the rate at which heat passes through the converters. Because the design is very simple, operation is easy to understand and potential errors are easy to determine. As described previously, calibration of an ideal device is based on a linear equation, watt = A + B*V. However, a real device benefits from using of a quadratic equation, watt = A + B*V + C*V², where V is the generated voltage.

When the calorimeter is used to study cold fusion, a gas-tight glass cell containing an electrolyte and electrodes is placed in the enclosure. Because the measured voltage represents an average heat loss through all parts of the barrier, the device is only slightly sensitive to where the cell is placed within the enclosure. Nevertheless, a fan is used to distribute heat more evenly and to reduce the cell temperature by removing heat from it more rapidly. The calorimeter is completely insensitive to where heat is being generated within the cell itself.

A calorimeter suitable for measuring the cold fusion effect must be sufficiently sensitive to detect a few tens of milliwatts superimposed on tens of watts. In addition, it must remain stable over long periods of time. Power production can be calibrated by generating heat using a resistor contained in the device. A dead cell or conditions expected to produce no anomalous energy can also be used. If the calorimeter is sufficiently sensitive, the total amount of energy given off by a known chemical reaction can also be measured. For example, the calorimeter can be used to measure the total amount of energy absorbed when D_2O is decomposed and β-PdD is formed at the cathode. Because all enthalpies of formation are well known[1,2], the method can give further evidence for the ability to detect small amounts of energy. Shelton et al.[3] used two other chemical reactions to test their calorimeter. Enthalpy of dilution of 1-propanol with H_2O and reaction between perchloric acid solution and tris[hydroxymethyl]aminomethane (47.24 kJ/mol) were used. Because such calorimeters measure power and not energy, the reaction used must be slow compared to the time constant of the calorimeter, requiring the reactants to be combined slowly so as to provide constant heating power.

Defining the accuracy of a calorimeter using only one criteria is not practical because several different and independent potential errors exist. Several examples are worth exploring. The amount of power being applied to a Fleischmann and Pons cell is noisy because bubble action generates a fluctuating load for the power supply. In addition, use of a fan adds additional electrical noise. If this fluctuation is too great, it can mask small changes in anomalous power, but without introducing an error that might be interpreted as being anomalous power. An additional error can occur of the calibration constant or the sensitivity of the calorimeter changes with time. These changes can be caused by changes in reference temperature, room temperature, the amount of recombination taking place in the cell, or physical parameters when new samples are placed in the cell. This potential drift is the main source of incorrect results.

B.2 Description of Construction

The calorimeter is made by gluing together commercially available thermoelectric converters using waterproof epoxy glue, as shown in Figure 83.

Figure 83. Glued panels assembled into two halves of a calorimeter. The length is 13.9 cm, the width is 6.9 cm and the total depth when assembled is 14.8 cm.

The panels are connected electrically in series. Once assembled, the outer surface is covered with electrically insulating, waterproof epoxy paint. The electrical resistance of this coating must be tested and found to be high (>1 Mohm) before final assembly. If the resistance is too low, unwanted voltages will be generated by chemical reaction between the cooling water and the metal plates. These assemblies are placed within watertight plastic boxes designed to direct water evenly over the outside surface. When assembled, the two boxes are stacked one on top the other as shown in Figure 84. Figure 85 shows a typical open calorimeter containing an electrolytic cell and small fan. Baffles are provided inside the Seebeck enclosure to pass air evenly over and around the cell. Wires and plastic tubes are passed into the cell through channels having good thermal contact with the cooling water. In the design shown in Figure 85, these channels are stainless steel tubes, which pass the length of the device within the cooling water. By passing the wires through these tubes, any heat loss or gain is held constant regardless of changes in room temperature.

Figure 84. Assembled calorimeter with water cooling jacket in place. In this design, the wires pass out of the cell through plastic water cooled channels.

B.3 Calibration

Circuits are arranged so that current and voltage used for calibration and for electrolysis are measured using the same resistor and data acquisition (DA) channels. In this way, measurement errors caused by errors in the DA channels are cancelled. Measurements are made using National Instruments data acquisition boards and Labview. Switching from electrolysis to calibration is accomplished by throwing one hardware and one software switch, which allows automatic calibration over the entire power range. A typical calibration is shown in Figure 86. Four points are taken going up in power and four are taken going down, in sequence. A random error is obtained from the standard deviation of points from the least-squares line drawn through the values. Data are taken after a delay of 90 min, which allows the calorimeter to reach steady-state. Table 16 lists calibration equations obtained over five months of examination.

Figure 85. Completed calorimeter with cell and fan in place.

Experience and analysis of this information indicate an uncertainty immediately after calibration of about ±16 mW. Drift caused by changes in room temperature and other factors can increase this uncertainty to ±30 mW during long runs. If the average coefficients were assumed to be constant during the time shown in the table, the maximum uncertainty at an applied power of 8.3 W would be ± 60 mW. In other words, the calorimeter is stable to within ±60 mW or 0.7% over 5 months if no effort is made to recalibrate. Because calibration is so easy, these small drifts can be easily identified as error.

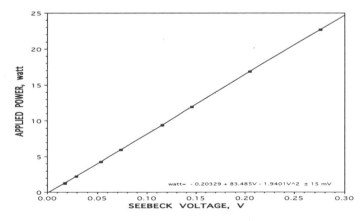

Figure 86. Typical calibration using an internal resistor.

Table 16. Calibration equations obtained over a period of 5 months.

Date	A	B	C	error, mW	W @ 0.1V
2/14/05	-0.020	83.02	0.12833	7	8.283
2/17/05	0.006	82.85	2.012	17	8.311
2/21/05	0.001	83.04	0.366	17	8.309
2/26/05	0.001	83.35	-1.220	11	8.324
3/10/05	-0.051	83.18	0.325	24	8.270
3/15/05	-0.022	83.86	-1.927	10	8.345
3/21/05	0.002	83.30	-0.764	16	8.324
3/24/05	-0.027	83.76	-1.737	28	8.332
3/29/05	0.000	83.40	-1.556	19	8.324
4/29/05	-0.013	83.82	-1.998	16	8.349
5/7/05	-0.052	83.40	-1.516	24	8.273
5/19/05	-0.076	83.25	-1.258	10	8.236
5/22/05	-0.116	83.33	-0.841	5	8.209
5/26/05	-0.093	82.94	1.053	17	8.212
5/29/05	-0.065	83.04	0.587	16	8.245
6/1/05	-0.078	83.46	-2.371	13	8.244
7/2/05	-0.203	83.49	-1.940	15	8.127
average=	-0.047	83.323	-0.745	16	8.277±0.060

The cell contains a recombiner so that no gas except orphaned oxygen leaves the cell. To determine if the recombiner is working and to measure the D/Pd ratio using the orphaned oxygen method, a small plastic tube carries gas from the cell to a reservoir of oil. Any change in gas pressure within the cell is detected as a weight change of oil being applied to a balance (±0.01g). This method allows the amount of orphaned oxygen resulting from D entering the Pd cathode to be determined and, from this, the D/Pd ratio, a method that is described in more detail in Appendix F.

B.4 Measurement of the Heat

When current is applied, the surrounding D_2O is decomposed into D_2 at the cathode and O_2 at the anode, with D_2 reacting with the Pd cathode to produce β-PdD_x. This reaction is endothermic, as can be seen in Figure 87 because more energy is used to decompose D_2O than is produced when forming PdD. Note that when heat is supplied from the internal resistor (heater), the calorimeter takes about 80 min to reach equilibrium. Consequently, the initial decrease in excess power shown during loading is caused mainly by the delay in reaching thermal equilibrium. When the applied current is increased, the reaction is more rapid, causing more power to be absorbed for a shorter time, as expected.

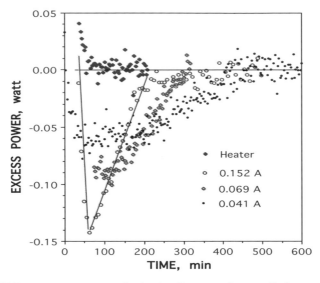

Figure 87. Power measurement during loading at various applied currents.

This reaction requires energy that can be calculated from the reaction $(x/2)D_2O + Pd = PdD_x + (x/4)O_2$, where x is equal to the average D/Pd ratio. This reaction passes through several stages on the way to the end-product. As the cathode takes up deuterium, palladium converts to α-PdD followed by formation of β-PdD, with increasing thickness and a range of composition. Eventually, reaction stops and conditions return to zero power. The final composition is in steady-state with a stable gradient and a constant rate of diffusion between the surface and random cracks. Deuterium continuously enters the cathode as deuterium ions and leaves as D_2 gas. No additional chemical power is produced in the cell by this processes.

During loading, deuterium is added to the cathode to produce the composition change shown in Figure 88. Starting at zero time, complete loading of the alpha phase becomes visible as a small break in slope. As loading continues, the combined composition begins to deviate from what would be expected if all deuterium reacted, as shown by the straight lines. Clearly, loading at low current is not 100% efficient. Nevertheless, net heat is only generated when deuterium actually reacts with palladium. The horizontal line shows the published composition at the β-PdD phase boundary for 1 atm D_2 and room temperature in the presence of α-PdD. As the deuterium content increases above this value, a smaller fraction of available D_2 is absorbed, with a sudden termination as a limit is reached. The unreacted D_2 and O_2 being generated by electrolytic action are recombined by a catalyst and remain in the cell as D_2O.

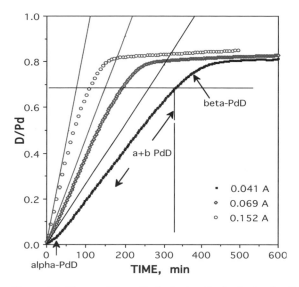

Figure 88. Composition of a typical cathode during loading.

B.5 Calculation of Enthalpy of Formation

Energy used during the loading reaction can be calculated by integrating data shown in Figure 89 between zero power and the curve drawn through the data points over the duration of the loading process. Note the first three points were taken while the calorimeter was approaching steady-state and must be ignored. This curve gives a value of -106 kJ/mol Pd. If termination of the reaction is assumed to occur at an average D/Pd of 0.8 at 460 min, the reaction equation becomes:

$$0.4D_2O + Pd = PdD_{0.8} + 0.2O_2.$$

Because Pd and O_2 are assumed to be at standard-state, their enthalpy values are zero. The enthalpy of formation of $PdD_{0.8}$ can be calculated by combining the published enthalpy of formation of D_2O (294.6 kJ/mol) and the value shown in Figure 89. The result gives $-\Delta H_f (PdD_{0.8}) = -106 + 0.4 * 294.6 = 11.8$ kJ/mol Pd. Scatter in the power measurements and a nonuniform composition, create a potential of error of about ± 2 kJ/mole.

Sakamoto and co-workers[2] determined the enthalpy of formation of β-PdD as a function of composition by measuring the heat given off when D_2 gas combined with palladium metal. This work gives the equation

$$-\Delta H_f \, (PdDx) = 44.99 - 41.89 * x = 11.5 \text{ kJ/mol Pd for } x=0.8.$$

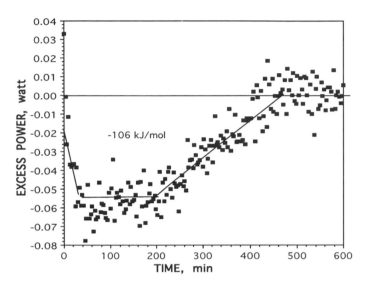

Figure 89. Data used to calculate enthalpy of formation of PdD$_{0.80}$ using 0.00974 mole of Pd.

Values obtained from this equation can be compared to -13.3 kJ/mole for PdD$_{0.77}$ reported by Flanagan and co-workers[4], using the same method as Sakamoto *et al.* Correction for dilution of deuterium by a possible protium impurity is not considered. Much more information is available for PdH, for which the enthalpy of formation is about 2 kJ more negative than PdD.

Agreement between calorimetric measurement reported here and those published earlier is well within the possible error of all measurements, thereby demonstrating the potential accuracy of this calorimeter.

B.6 Summary

A Seebeck calorimeter can be easily constructed with characteristics that eliminate most errors thought to cause false reporting of anomalous energy from cold fusion. In addition to being stable and accurate, the calorimeter is sufficiently sensitive to accurately measure the enthalpy of formation of PdD$_x$ to give 11.8 kJ/mole while using only 1 g of palladium.

Appendix C

What Makes Palladium Special?

C.1 Introduction

The LENR process first and foremost involves material properties. A nuclear reaction may be the ultimate goal, but this will not happen unless the necessary conditions have been achieved by a suitable arrangement of atoms and electrons. Fleischmann and Pons chose palladium because it is able to take up a large amount of deuterium and, according to their model, place the atoms in close proximity. In fact, palladium is not the best absorber of hydrogen and the deuterium atoms are not as close as those in the D_2 molecule.[1-3] Nevertheless, palladium worked for some unknown reason, but so does titanium,[4-7] nickel[8-16] and perhaps other materials as well.[17-19] Even platinum and copper, which do not react with hydrogen, can be made active when proper material is deposited on their surface. When pure palladium is used, the surface region where the nuclear reaction is proposed to occur, is actually a complex alloy containing a significant amount of lithium[20-30] as well as other impurities.[31] Consequently, properties of the base metal cannot be used to understand processes taking place in the surface when the electrolytic method is used. The other methods suffer from this problem to a lesser extent. For example, impurities are expected to be slowly deposited on the cathode during gas discharge because sputtering cannot be avoided. Even finely divided palladium-black is expected attract impurities to its surface during its formation, with unknown consequences.

C.2 Phase Relationship in the Pd-D System

Because β-PdD has been the major focus of the CMNS field, a brief description of its important properties is worthwhile. Much general information about the hydride can be found in publications by Fanagan and students[32-37] and Lewis.[38,39] Surface cleaning of palladium has been explored by Musket.[40]

The phase relationship of the Pd-D system is shown in Figure 90 at various pressures and temperatures. Pure palladium has a face-centered cubic (fcc) crystal structure and the alpha phase forms when a small amount of deuterium, located in random sites between the palladium atoms, creates what is called an interstitial solid-solution. This phase becomes saturated at about α-$PdD_{0.05}$ when the D_2 pressure exceeds 0.031 atm at 20° C. Increasing pressure to 1 atm at 20°C converts α-$PdD_{0.05}$ to a beta phase with a composition of β-$PdD_{0.675}$. Deuterium in this structure occupies a face-centered-cubic (fcc) sublattice within the palladium sublattice, called a defect compound. Sites in this new sublattice are randomly filled until all sites are occupied at $PdD_{1.0}$. Above

275°C and at a D_2 pressure of 35 atm, the two-phase region between α-PdD and β-PdD disappears.

Transition from the alpha phase to the beta phase occurs when deuterons shift from being randomly located in alpha-PdD to a more regular structure in beta-PdD. This shift can be viewed as a transition from a solid-solution of D^+ ions dissolved in palladium to the formation of a compound with a variable number of filled sites in the deuterium sublattice, without a change in the basic crystal structure of the metal atoms. Ordering of the filled and unfilled sites can occur under high pressure[41] or when the material is cooled. This transition occurs at 55 K for β-PdH, based on a heat capacity anomaly.[39,42] According to the neutron diffraction studies of Ferguson et al.,[43] this transition is caused by some hydrogen shifting to tetrahedral occupancy. No evidence exists for tetrahedral occupancy above this temperature. Lipson and co-workers[44] report observing magnetic anomalies below this temperature in PdH.

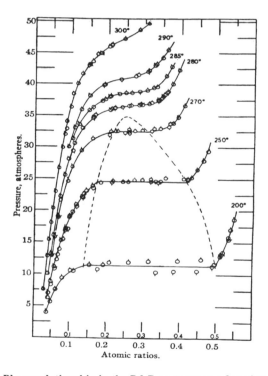

Figure 90. Phase relationship in the Pd-D systems as a function of pressure and temperature[45].

Other phases might form in the presence of lithium or when the applied deuterium activity is increased above that required to form $PdD_{1.0}$. Indeed, recent measurements of resistivity by McKubre and Tanzella,[46] and by Spallone[47] suggest another two-phase region above $PdD_{1.0}$ becomes accessible when using electrolytic loading. This additional phase might be PdD_2, although calculations show formation of this phase requires a great deal of energy, which could be supplied by the high activity created during electrolysis.[48] Presumably, the normal single-atom sites are each occupied by two atoms in this structure. This phase is not expected under normal conditions.[1,2,49-51] Formation of D_2 clusters within the β-PdD lattice is proposed to occur by a resonance process,[52] although calculations show that they are not stable under equilibrium conditions.[53]

C.3 What Happens When Palladium Reacts with Deuterium (Hydrogen)?

Achieving a high average D/Pd ratio is a necessary but insufficient requirement to initiate the cold fusion reactions when bulk palladium is used in an electrolytic cell. A great deal of effort has been devoted in many laboratories to achieve this goal. Various additives[54-56] and annealing procedures[57-61] have been tried, as well as application of applied current having complex waveforms.[62,63] Methods using complex wave forms, very low initial current, periodic current reversal, and/or alternating high and low applied current are found to improve loading, presumably because the methods reduce stress concentration and allow periodic strain relief in the growing PdD(H) layer, thereby avoiding crack formation. These methods will also modify the morphology and composition of material as it deposits.

Palladium having large grains loads better than when small grains are present.[64,65] Impure palladium is able to grow larger grains during annealing compared to very pure palladium, which perhaps explains why impure palladium has been generally more successful in making excess energy. Annealing at 900° C for 1 hour in a good vacuum or hydrogen appears to work best. Preloading at high D_2 pressure (>34 atm) above the α–β transition temperature (270°C) to avoid crack formation also helps.[58,66,67] If the palladium contains the proper impurities,[68,69] this step can be avoided. In general, thin palladium, including that studied as deposited films, has a higher average composition than does thick palladium. A larger limiting composition is also achieved at lower temperatures.

As PdD(H) forms, the surface changes from being smooth to containing many ridges caused by the slip of various atomic planes, resulting in a very complex topography. If stress has not been properly relieved, visible cracks will form. As electrolysis continues, impurities deposit on this complex surface, with the atomic arrangement of the deposit being influenced by the underlying structure. In addition, certain elements are more attracted to certain chemically active sites than to others. As a result, the surface becomes chemically and structurally very complex. Somewhere within this mixture, a structure may form occasionally that is able to initiate nuclear reactions. If the deuterium(hydrogen) concentration is sufficiently high at these locations, detectable nuclear products will result along with heat energy. This requirement applies to all methods used to initiate the effects.

C.4 Deuterium Pressure over the Pd-D System

Santandrea and Behrens[70] have reviewed the various published measurements of H_2, D_2 or T_2 in the Pd-gas system, from which the following equations are taken. Pressure of deuterium gas over the two-phase region as a function of temperature is given by Equation 1. As deuterium fills the unoccupied lattice sites in the beta phase, pressure of deuterium in equilibrium with the compound increases, as can be calculated using Equation 2. However, the mathematical form of this equation only applies to the behavior of chemical activity. When used to calculate pressure, which assumes ideal behavior, it predicts a pressure that approaches infinity as the D/Pd ratio approaches unity, which is not realistic. Such behavior will not occur in the real world because the system will become increasingly non-ideal as pressure is increased, thus causing a disconnect between pressure and activity called fugacity. In addition, the structure will probably change, forming another solid phase when the pressure exceeds a critical value, perhaps before all deuterium sites are completely filled in β-PdD. Nevertheless, the equations are useful to predict behavior within most available pressure ranges.

$$\ln P[D_2, atm] = -4469/T + 11.78, \text{ where } T = \text{temperature (K)} \tag{1}$$

$$\ln P[D_2, atm] = 12.8 + 2\ln [r/(1-r)] - [11490-10830r]/T, \text{ where } r = \text{D/Pd ratio} \tag{2}$$

C.5 Consequences of Reacting Palladium with Deuterium in an Electrolytic Cell

When palladium is loaded with deuterium in an electrolytic cell, two things happen. Once the beta phase starts to form, cracks are generated, some of which reach the surface where they allow deuterium to leave as D_2 gas. The number of cracks is sensitive to how the metal was previously treated.[71] As the limiting composition of the beta phase is reached, a concentration gradient forms between where deuterium enters the lattice and where it is lost as D_2 gas. Generally, this gradient occurs between a chemically active site where D enters the lattice and the nearest surface penetrating crack. On an atomic scale, these gradients result in a wide variation in surface composition, but with a much greater average than the bulk composition. The maximum average D/Pd is determined by the total rates of D entering the metal and D_2 leaving the metal, a typical leaky bucket effect. The loss rate can be measured by weighing the cathode as a function of time and plotting the result as weight *vs.* square root of time, as shown in Figure 91. A linear relationship reveals diffusion to be the rate determining step. A break in slope occurs when α-PdD forms on those surfaces where the $2D^+ + 2e^- = D_2$ reaction occurs. In addition to this loss mechanism, some absorbed deuterium atoms recombine on the surface and leave directly as gas. Various surface poisons, such as thiourea,[72] affect this process—but not the loss through cracks.

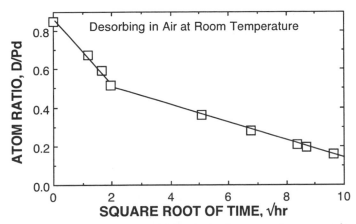

Figure 91. Composition of a typical PdD sample shown as a function of √time while deloading in air. (Storms)

As electrolysis continues, lithium dissolves in the β-PdD and forms various Pd-Li-D compounds of increasing Li content. A phase diagram for the Pd-Li system is shown in Figure 92. Compounds in the Li rich region are soluble in aqueous solutions and those in the palladium-rich region are insoluble. When the Li content reaches a critical value, a little of the alloy dissolves in the electrolyte, thereby adding Pd^{++} to the solution. In addition, Pt^{++} is slowly added to the solution as the black colored platinum oxide slowly dissolves from the platinum anode. These Pd^{++} and Pt^{++} ions slowly redeposit on the cathode surface to form a layer having a new structure and composition. As a result, the surface and the NAE, if present, are no longer pure palladium.

If the anode is gold, this element will rapidly transfer to the cathode[73]. Also, copper, nickel or stainless steel connecting wires can contaminate the surface with these elements no matter how well protected the wires might be. Impurities in commercially supplied heavy-water will also deposit on the surface unless they are removed by distillation or by pre-electrolysis using an expendable cathode. Some silicon and boron will dissolve from Pyrex glass and be deposited. Other elements, many of which have been observed to deposit,[74] might be required for the NAE to form or may even prevent its formation. To make the problem more complicated, these elements do not always deposit in a uniform layer, but instead form isolated islands having different structures and combinations of elements. As a result, general statements cannot be made about the surface composition or structure resulting from electrolysis.

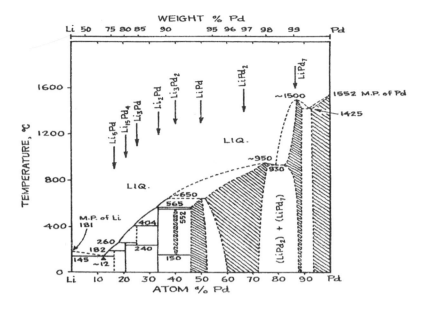

Figure 92. Phase diagram of the Li-Pd system. (From the Handbook of Binary Phase Diagrams, Volume III, W. G. Moffatt, General Electric Co., 1974.)

When electrodeposition is done on purpose within an operating cell, it is called co-deposition by Szpak. This technique has been studied by several workers[75-79] and has frequently given positive results. The deposited structure will be influenced by the structure of and impurities on the base material, which act as nucleation sites. Consequently, prior treatment of a surface has an important effect on where a deposit forms and its structure. For example, Rolison and Trzaskoma[80] examined the surface of palladium using a SEM after etching the surface and loading the sample with deuterium or hydrogen. Loading with H had a different effect on the morphology than did D. However, both loadings caused significant changes that depend on the nature of the local region. The surface is neither ideal nor uniform and the characteristics change with time. Success depends on the source and treatment of palladium when it is used as the substrate. For a cathode to become nuclear-active, it must have the proper surface properties, the proper ions must be present in the solution, and sufficient time must be taken to deposit these ions on the surface. In addition, the surface must be able to achieve a high deuterium content.

As the temperature of an electrolyzed cathode is increased, the average composition will decrease because all of the mechanisms leading to loss of hydrogen will increase, including diffusion rate and pressure within surface penetrating cracks. Consequently, the diffusion flux will increase resulting in a steeper concentration gradient between the surface and the interior. As long as sufficient current is applied, the active surface can retain a relatively constant composition. However, the magnitude of this composition is influenced by how rapidly hydrogen ions can enter the surface instead of reacting on the surface to form H_2 (D_2). This rate can be modified by application of surface impurities, including the unintended plating of lithium and platinum during extended electrolysis. As a result, the actual composition of the surface, hence the composition of the NAE, cannot be predicted or easily controlled when electrolysis is used.

C.6 Measurement of Surface Activity

Changes in surface composition and structure can be observed using the open-circuit voltage (OCV), which is the voltage measured between the cathode and a stable reference electrode when current applied to the cell is interrupted for a brief time, usually for a few milliseconds. The OCV is equal to the sum of the half-reaction voltage generated by the cathode combined with that generated by the reference electrode. If the reference electrode is one for which a half-reaction voltage is known, the absolute voltage of the cathode can be determined. Various reference electrodes have been used, including (Ag,AgCl, saturated KCl, E = 0.197 V), (Hg, saturated Hg_2Cl_2, E = 0.241 V), and (Pt, 1 atm D_2, E = 0.000 V). Each of these reference electrodes has a disadvantage. In the first two cases, silver or mercury can enter the cell and deposit on the cathode. The Pt + D_2 (SHE) electrode involves the nuisance of bubbling D_2 over the activated Pt reference electrode.

To avoid these problems, Fleischmann and Pons used a palladium reference electrode containing alpha+beta-PdD. This electrode will remain stable at a known half-voltage long enough for measurements to be made. Storms[81] simplified the design still further by using a piece of platinum having a high surface area as the reference electrode. This electrode has a half-reaction voltage determined by the effective activity of D_2 gas dissolved in the electrolyte. Although the pressure may be unknown and much less than one atmosphere, it will be relatively stable. The magnitude of this voltage can be determined by measuring the OCV when the cathode has acquired the two-phase mixture. This condition will produce an arrest in the OCV or just a break in slope, as shown in Figure 16. The activity of deuterium in this two-phase region can be calculated using Equation 1 (Section C.4). This value can then be inserted in Equation (3),

$$\Delta E = [RT/F] \ \ln a/a_r \qquad\qquad (3)$$

where ΔE is the voltage difference between the two electrodes, *i.e.* the OCV, "a" is the deuterium activity at the cathode surface and "a_r" is the deuterium activity of the reference electrode, T is temperature in K, R is the gas constant [8.314510 J/mol*K], and F [96485.309 C/mol] is the Faraday constant. Once a_r is known, any activity can be calculated from the measured OCV. By using this activity, the

D/Pd ratio at the surface can be calculated from Equation 2 (Section C.4) as long as the surface is pure β-PdD and ideal conditions exist. An example of how this equation can be used is shown in Figure 93 where the OCV is compared to the D_2 pressure and D/Pd ratio on the surface. Composition gradients exist between the surface and the bulk material causing the average surface composition to be significantly greater than the measured average D/Pd ratio of bulk material.

The approach can be applied, as shown in Figure 94, where the changing D/Pd ratio and OCV were measured over a period of time while palladium was electrolyzed in D_2O + LiOD. Beta phase starts forming at OCV = 0.49 V in this cell. By combining Equations [1], [2] and [3], the maximum measured OCV of 0.865 V gives an activity of 87500 atm, which would be produced by pure β-PdD having an average D/Pd = 0.90. The difference between the measured D/Pd of 0.78 and that calculated gives the magnitude of the average difference between the surface and bulk material. By measuring the OCV as a function of time, as shown in Figure 95, changes in the properties of the surface can be revealed. Absence of an arrest within the α–β region and additional breaks in slope at higher OCV are thought to be caused by changes in surface purity as lithium is absorbed. Consequently, when the surface is no longer pure PdD, the D/Pd ratio can not be calculated using this method.

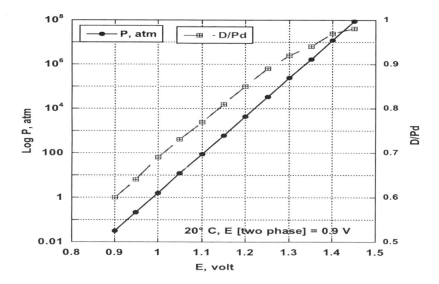

Figure 93. Relationship between D_2 pressure, D/Pd ratio, and OCV.

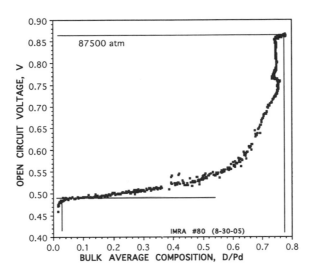

Figure 94. Open circuit voltage produced by the surface composition as a function of average bulk composition. The calculated final D_2 pressure at the surface is shown.

C.7 Proposed Complete Phase Diagram of the Pd-D System

The relationship between OCV and NAE is complex. Formation of another phase, such as PdD_2, might be required to create the NAE under electrolytic conditions. A proposed phase relationship for the pure Pd-D_2 system and a proposed composition for the NAE can be seen in Figure 96. On the other hand, the NAE might actually form at a composition much lower than the proposed composition, but the deuterium concentration in that region of the sample might be too small to produce detectable heat or nuclear products, thereby making the existence of the NAE invisible. This concept needs to be emphasized; unless the deuterium or hydrogen concentration is above a critical value, nuclear reactions will not occur even if the NAE is present. Because the surface has the highest composition in an electrolytic cell or during gas discharge, nuclear-activity will first occur at the surface and extend further into the sample as the critical composition moves deeper, provided the NAE extends below the surface. Until the actual conditions present in the surface region are understood and controlled, the effect cannot be expected to be reproducible. On the other hand, if the surface has the structure and composition needed to form a large amount of NAE, a high deuterium activity (pressure) may not be required to produce detectable nuclear products. These considerations make a study of surface deposits, their composition and structure, very important.

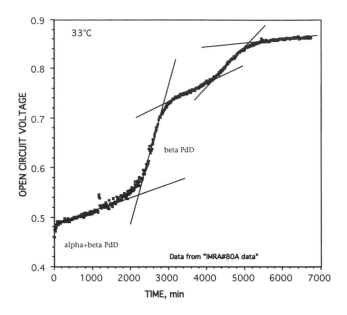

Figure 95. OCV change during loading. Sample had been electrolyzed for a sufficient time to cause significant lithium to dissolve in the surface.

C.8 Surface Characteristics and Deposits

Like palladium, the surface of platinum also has a very nonuniform character. Such a complex surface is shown in the SEM picture (Figure 97) of a platinum surface after it had been polished and then etched for 20 hr using Aqua Regia (HCl + HNO$_3$). Notice that some regions are smooth while other regions are rough to varying degrees. This irregularity occurs because different regions have a different chemical susceptibility to attack by Aqua Regia, which reveals the presence of a highly variable chemical susceptibility. This causes each region to accept a deposit in a different way. A thin deposit of palladium deposited on platinum from a very dilute solution of PdCl$_2$ forms islands (Figure 98) which give the surface a black appearance when viewed by eye. This type of deposit forms slowly on the cathode during extended electrolysis. A more concentrated solution produces a fern-like deposit having a very non-uniform structure (Figure 99). Many other morphologies can be formed depending on temperature, applied current density, and the chemical form and concentration of palladium in the plating solution, as described in detail by Julin and Bursill[83] and Bockris and Reddy.[84] Many other papers explore this behavior as part of the literature describing the electroplating of palladium as a commercial process. Another example of how variable the deposition process can be is provided by Lee et al.,[85] who found that platinum electrodeposited on graphite formed isolated

10 nm islands when 0.013 M H_2PtCl_6 and 0.104 M $HClO_4$ were used in the electrolyte and pulsed current of 1 μA was applied for 3 sec. The nature of the deposit was sensitive to concentration and applied current. This same behavior is expected during palladium deposition.

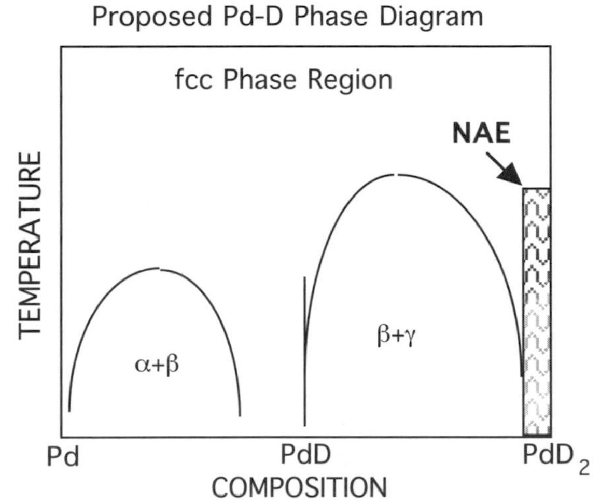

Proposed Pd-D Phase Diagram

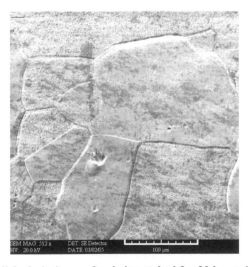

Figure 96. Proposed diagram of the Pd-PdD$_2$ phase region under equilibrium conditions.[82]

Figure 97. Polished platinum after being etched for 20 hours in Aqua Regia.

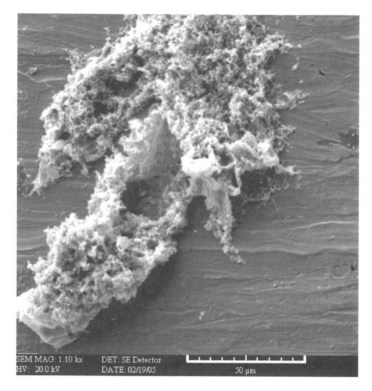

Figure 98. A typical island deposit of Pd formed on Pt from a very dilute solution of PdCl$_2$.

Besides these recognizable structures, a thin layer of palladium can form with sufficient thickness to hide the platinum substrate from EDX using 20 keV electrons, but this layer is frequently too thin to be noticed because it hide surface detail. Consequently, a thin deposit of palladium can be easily missed when it forms on palladium. Furthermore, active surfaces and those that are inert have a similar overall appearance, suggesting the NAE can not be located based on obvious surface structures.

Several studies have found various possible transmutation products only in small, isolated regions,[86-90] but the size of the active region is difficult to estimate. Iwamura *et al.*[91] examined samples of palladium in which they had converted Cs to Pr, as described previously, using X-ray fluorescence spectrometry. The Cs + 4D = Pr reaction was found to occur in isolated regions smaller than 100 μm. Arata and Zhang achieved good success by exposing very small particles (~5 nm) of palladium to D$_2$ gas, as described previously. Is success possible even when applied pressure (activity) is low because a large amount of NAE is present in the samples? If so, which of the millions of particles present in their sample are active and why? How can the active material be applied to a cathode where it can be exposed to a much higher concentration of deuterium, thereby increasing the nuclear reaction rate? These are questions worth exploring in future studies.

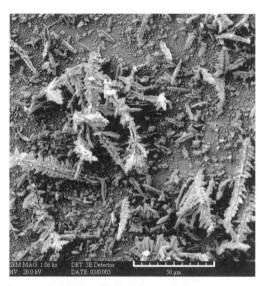

Figure 99. Palladium deposited on platinum from a concentrated solution of PdCl$_2$.

C.9 Crack Formation During Loading

The extra volume generated by crack formation can be determined by comparing the physical volume of the metal to that calculated from the lattice dimension based on X-ray diffraction.[92] At compositions up to the lower boundary of the beta phase (*i.e.* while two phases are present), the physical dimension agrees with that calculated using the lattice parameter—meaning no cracks have formed. As deuterium is added to the beta phase, extra volume grows. This extra volume is retained each time the sample is deloaded, as shown in the Figure 100. In other words, cracks and dislocations do not heal, but accumulate during each loading-deloading cycle of the β-PdD phase.

The amount of this extra volume is very small for some samples and large for others. Anomalous energy resulting from electrolysis is more likely to be produced by samples having only a small amount of extra volume (see Table 1), hence a small number of cracks, with an especially negative impact caused by those penetrating the surface.[81] Surface cracks allow deuterium gas to leave the sample without being influenced by the electrolytic process or by surface poisons, thereby limiting the maximum deuterium content (see Figure 12).

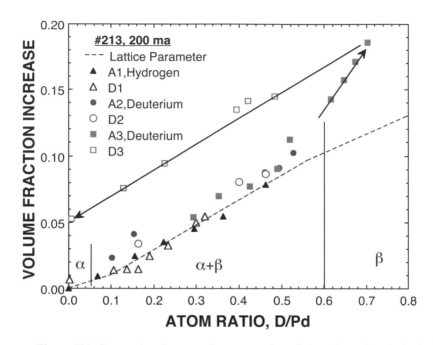

Figure 100. Comparison between the atom ratio and the volume based physical measurement and that calculated from the lattice parameter. "A" represents addition of hydrogen isotopes and "D" designates deloading.[93]

The concentration of surface penetrating cracks can be explored by measuring the deloading rate once current is stopped (see Figure 15). Samples able to reach a high D/Pd ratio generally have a smaller deloading rate than less well-loaded samples because they have fewer surface penetrating cracks.[94] This loss from cracks can be seen as bubbles, shown in Figure 64, that rise from isolated regions during deloading of palladium. The same behavior is seen during electrolysis, with bubbles being generated only from certain regions, usually the same regions from which gas is lost during deloading.

Conference Proceedings

Location, date, and source from which the proceedings might be ordered are shown.

ICCF CONFERENCE

ICCF-1, University Park Hotel, Salt Lake City, Utah, March 28-31, 1990, unavailable. (39 papers)

ICCF-2, Villa Olmo, Como, Italy, June 29-July 4, 1991, Italian Physical Soc. (Volume 33), Bologna, Italy. (56 papers)

ICCF-3, Nagoya, Japan, Oct. 21-25, 1992, Universal Academy Press, Inc., Tokyo, Japan. (102 papers)

ICCF-4, Lahaina, Maui, Hawaii, Dec. 6-9, 1994, Electric Power Research Institute, Palo Alto, CA (EPRI TR-104188, 4 volumes). (125 papers)

ICCF-5, Hotel Loews, Monte-Carlo, Monaco, April 9-13, 1995, unavailable. (76 papers)

ICCF-6, Lake Toya, Hokkaido, Japan, Oct. 13-18, 1996, unavailable. (110 papers)

ICCF-7, Vancouver Trade and Convention Centre, Vancouver, Canada, April 19-24, 1998, ENECO, Inc., Salt Lake City, Utah. (76 papers)

ICCF-8, Villa Marigola, Lerici (La Spezia), Italy, May 21-26, 2000, Italian Physical Soc. (Volume 70), Bologna, Italy. (68 papers)

ICCF-9, International. Convention Center, Tsinghua Univ., Beijing, China, May 19-24, 2002, Prof. X. Z. Li, Tsinghua Univ. (87 papers)

ICCF-10, Royal Sonesta Hotel, Cambridge, MA, Aug. 24-29, 2003, World Scientific Publishing Co. (93 papers)

ICCF-11, Hotel Mercure, Marseilles, France, Oct. 31-Nov. 5, 2004, World Scientific Publishing Co. (72 papers)

ICCF-12, Shin Yokoama Prince Hotel, Yokohama, Japan, Nov. 27-Dec. 2, 2005, World Scientific Publishing Co. (63 papers)

ICCF-13, Dagomys, Sochi, Russia, June 25-July 1, 2007

Many papers from these conferences are available at www.LENR-CANR.org in full text.

OTHER CONFERENCES

Workshop on Cold Fusion Phenomena, Santa Fe, NM, May 23-25, 1989, LANL, Los Alamos, NM (Report # LA-11686-C).

World Hydrogen Conference #8, Hawaii, USA, July 22-27, 1990, Univ. Of Hawaii, Honolulu, Hawaii.

Anomalous Nuclear Effects in Deuterium/Solid Systems, Brigham Young Univ., Provo, UT, Oct. 22-23, 1990, Prof. Steven Jones, BYU

Proceedings of the Japanese Cold Fusion Society:
http://wwwcf.elc.iwate-u.ac.jp/jcf/PAPER.HTML

International Symposium on Cold Fusion and Advanced Energy Sources, Belarusian State Univ., Minsk, Belarus, May 24-26, 1994, Hal Fox, Fusion Information Center, Salt Lake City, UT. (www.padrak.com/ine/products.html)

Asti Workshop on Anomalies in Hydrogen/Deuterium Loaded Metals, Villa Riccardi, Rocca d'Arazzo, Italy, Nov. 27-30, 1994, Italian Physical Soc. (Vol. 64), Bologna, Italy.

8th International Workshop on Anomalies in Hydrogen/Deuterium Loaded Metals, Sicily, 13-18 October 2007, Fulvio Frisone Foundation.

Appendix E

Enrichment of Tritium During Electrolysis

A chemical reaction involving hydrogen will change the isotopic ratio between the three isotopes, designated here as H (protium), D (deuterium) and T (tritium) in the products of that reaction. This allows H and D to be almost completely separated in the commercial production of heavy-water (D_2O). During electrolysis, H goes into the gas phase more rapidly than does D, which leaves faster than tritium (T). Likewise, hydrogen dissolves in palladium more easily than does deuterium, which dissolves more easily than tritium. In other words, tritium is the least reactive of the three isotopes.

All heavy-water contains some tritium. As water is decomposed during electrolysis, the amount of tritium remaining in the cell will gradually increase.[1,2,3-11] Various efforts have been made to derive an equation for calculating the tritium increase caused by this process. Anomalous tritium is then assumed to be the difference between the measured amount and that calculated using the derived equation. To make the situation confusing, several different equations have been published and used to make this correction. Publications seldom provide enough information to test the accuracy of the employed equation or to apply different corrections. Several examples of equations used are provided below. Each assumes the volume (mass) of the electrolyte is constant, with any loss replaced by D_2O before a sample is taken. The published equations have been modified to use the same designations for the variables. These designations are:

> Separation factor = S = (T/D)gas/(T/D)liquid
> Faraday constant = 96489 [coulomb/electron mol]
> Rate of mass change = R = A*20/2*F [gm/sec]
> Mass of electrolyte = M [gram]
> Volume of electrolyte = V [ml]
> Number of moles D_2O in electrolyte = N [mole]
> Time since start of electrolysis = t [sec]
> Applied current = A [amp]
> Initial tritium concentration in electrolyte = Ti = decay rate/ml
> Tritium concentration in electrolyte at time (t) = Tt = decay rate/ml
> Rate of anomalous tritium production = q
> Ratio of sampling rate to electrolysis rate = a
> Ratio of evaporation rate to electrolysis rate = g

$$Tt/Ti=[(M-R*t)/M]^{S-1} + q/(Ti*(S-1)*R)*[1-[(Ti-R*t)/Ti]^{S-1}]$$
Szpak and Mosier-Boss[11]

$$Tt/Ti=[1/(S+a+g)]*[1+a+g-(1-S)*exp[-(S+a+g)*R*t)/N]]$$
Williams et al.[1]

$$Tt/Ti=S-(S-1)*exp[-A*t/(S*M*2*F)]$$
Hodko and Bockris[8]

$$Tt/Ti=S*(1-exp[-R/(1.1*V*S)]) + exp[-R/(1.1*V*S)]$$
Sona et al.[2]

$$Tt/Ti=(V(initial)/V(final))^{(1-1/S)}$$
No addition of makeup D_2O or withdrawal for sampling.
Corrigan and Schneider[10]

Rather than using an analytical representation, a series of corrections can be calculated over short intervals of time, which are summed to give a value for the expected tritium content as a function of time. For this method, Szpak et al.[6] used individual equations to account for tritium lost to the gas, removed by sampling, and added in the makeup D_2O. The following equation describes the method.

$$Tt /Ti = \sum(V*Ti – G + L – S) = dpm(at\ total\ time)/dpm(initial)$$

G = moles T lost to gas in time Δt = $S*\Delta t*A/F$
L = moles T gain by makeup D_2O = $Ti*\Delta t*A*20/(1.1*2*F)$
S = moles T removed in each sample after the D_2O is replaced = $Vs*(Tt-Ti)$

where Ti is the initial tritium concentration in D_2O [mole/ml], Tt is tritium concentration after each summation is applied for the time interval Δt, V is total volume of electrolyte, Va is the volume of makeup D_2O added, Vs is the volume of each sample removed. Liquid lost with the gas as vapor is not included. Makeup liquid is assumed to be added just as fast as D_2O is lost, not as a periodic batch addition. If batch addition is used, the amount added each time needs to be used to calculate the amount of T, rather than using applied current to calculate the amount of D_2O. No pickup of H_2O from the atmosphere is assumed. This method avoids the complex and perhaps dubious published equations.

As Williams et al.[1] and others point out[7], the separation factor is sensitive to many conditions within the cell. Nevertheless, Roy[12] measured the separation factors of all species as a function of temperature, from which the values in Figure 101 were obtained. In the figure, $\alpha=(H/D)_{gas}/(H/D)_{liquid}$, $\beta=(H/T)_{gas}/(H/T)_{liquid}$, and $\gamma=(D/T)_{gas}/(D/T)_{liquid}$. These values for D/T agree with other measurements as shown above.

Calculations based on reported values of S can only be used to obtain an approximate correction. Only by preventing gas loss or by keeping a complete inventory of tritium addition and loss can reliable information be obtained.

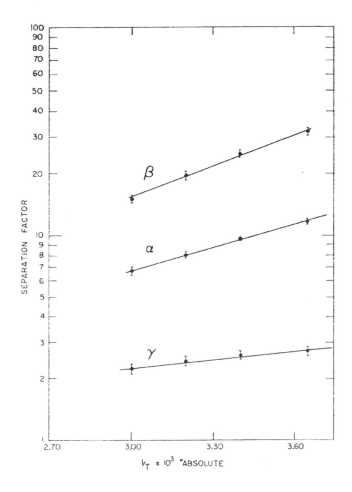

Figure 101. Separation factors for Pd+H,D,T.

Appendix F

Methods for Measuring the Amount of Hydrogen (Deuterium) in Palladium

F.1 Introduction

The average D/Pd ratio has been measured several different ways. Each of these methods measures the average deuterium content within the entire sample, not the deuterium content of the NAE. Because the deuterium content is very nonuniform,[1] the average value will depend strongly on the size and shape of the sample. In general, the smaller the sample, the larger the measured D/Pd ratio for the same conditions. This effect must be taken into account when a comparison is made between studies using samples of different shape and size.

F.2 Weight-Gain Method

The amount of deuterium contain within palladium can be measured by weighing the sample before and after loading. A sample of palladium weighing 1 gm with a D/Pd ratio of 1.0 contains 0.01879g of deuterium. Consequently, a four-place balance is required. Because the contained deuterium is rapidly lost from the sample at room temperature, the composition existing during an experiment must be determined by extrapolating the weight back to the time when loading stopped. As plotted in Figure 102, this extrapolation is based on the square root of time. The more quickly the sample can be weighed after loading is stopped, the more accurately the composition can be determined. This method can be made accurate to ±0.01 in the atom ratio and applied to any method used to initiate the cold fusion effect, provided weight can be determine quickly. However, an unknown amount of protium in the sample can make the value uncertain.

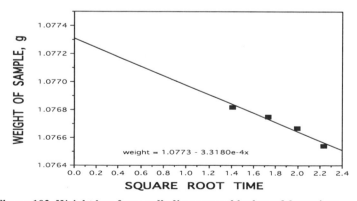

Figure 102. Weight loss from palladium caused by loss of deuterium.

F.3 Orphan Oxygen Method

When D_2O is split into D_2 and O_2 by electrolysis and the D_2 reacts with palladium, O_2 remains behind to accumulate in the cell. The amount of D_2 reacted with Pd can be determined by measuring the amount of this oxygen. Of course, the cell must be sealed and contain a recombining catalyst to eliminate unreacted D_2 and O_2. The amount of orphaned O_2 can be determined several different ways. A volume measurement can be made by bubbling gas into a gas burette containing oil. The level of the oil can be made equal to the level of the electrolyte in the cell to eliminate the effect of pressure. Another method uses an oil reservoir from which mineral oil is transferred to a balance as oxygen accumulates. A drawing in Figure 103 shows the method to be simple and the weight can be continuously recorded to provide a continuous value for the composition. A sensitivity of about 7600 g oil/mol D in Pd is typical. This number can be determined after each experiment by measuring how much deuterium is in the palladium using the Weight Method. The method can be made accurate to ±0.001 in atom ratio, but it must be calibrated regularly because it is affected by changes in atmospheric pressure and room temperature.

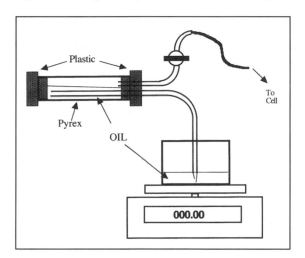

Figure 103. Oil reservoir used for measuring D/Pd ratio.

F.4 Pressure Change Method

The cell can be flushed and filled with deuterium gas before the experiment begins. As O_2 accumulates during electrolysis, it reacts with the extra D_2 at a recombining catalyst, causing a reduction in pressure. This method is relatively insensitive to changes in atmospheric pressure and does not need to be calibrated if the volume is known. The moles of D absorbed by the palladium can be calculated by the formula

$$N = 2*\Delta P*V/R*T,$$

where ΔP (atm) is the pressure change, V (liter) is the volume of gas, T (K) is temperature, and R = 0.082054 liter-atm/deg-mole.

A variation of this method for use in the absence of a recombiner has been described by Algueró et al.[2] When pressure within the sealed cell reaches a preset value, gas is vented. By knowing the volume of the system, the amount of $D_2 + O_2$ released each time can be determined. The difference between this volume and the volume expected to be produced by applied current gives the amount of deuterium that has reacted with the palladium. Complete absence of recombination must be assumed.

F.5 Resistance Change Method

The resistance of palladium changes as it takes up deuterium, as shown in Figure 104. Notice that the resistance ratio, R(PdD)/R(Pd), goes through a maximum near the lower phase boundary of the beta phase and decreases as the D/Pd ratio increases within the single-phase region. This relationship will change as a result of repeated loading and deloading.[4]

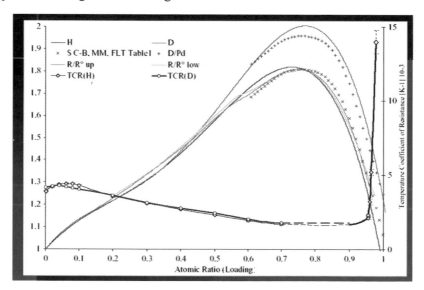

Figure 104. Resistance ratio and temperature coefficient for Pd-H$_2$ and Pd-D$_2$ systems. (McKubre et al.[3])

The temperature coefficient, TCR, shows an abrupt increase near D/Pd=0.96. This change suggests, among other things, formation of another phase mixed with β-PdD above this composition has occurred.

This measurement is made by attaching four wires to the sample, with a small AC current passing between two of the wires and a voltage drop measured between the other two. Composition gradients and unknown amounts of protium and lithium will cause an unknown error. In addition, use of a single function to describe behavior from Pd to $PdD_{1.0}$ is not correct because between approximately $PdD_{0.05}$ and $PdD_{0.68}$ the material contains two phases. Consequently, the resistance will be a linear average between the resistance of α-$PdD_{0.05}$ and β-$PdD_{0.68}$, with minor variations caused by non-uniform mixing of grains having the two different compositions—a condition difficult to control. According to Fazlekibria et al.,[5] the resistances of these two phases are almost equal. When this is true, the resistance ratio in the two-phase region should be constant under equilibrium conditions. However, this detail is not important if the values are used to calculate composition only in the β-phase region as is normally the case.

Errors are introduced by changes in lithium content, the presence of parallel resistance caused by the electrolyte when long thin wires are used, uncorrected resistance change caused by unknown and non-uniform temperature, non-uniform composition, and changes in the size of the sample caused by repeated loading-deloading cycles. These problems have been addressed by Zhang et al.[4]

F.6 Electrolytic Deloading Method

The PdD can be deloaded in an electrolytic cell by applying a voltage less than 1.52 V. Current will flow as long as D^+ is leaving the PdD structure. The total number of Coulombs (integrated current * time) measured during this process is proportional to the number of moles of deuterium atoms removed. However, removal becomes very slow as the last deuterium tries to leave, making this method impractical for most studies.

F.7 X-ray and Neutron Diffraction Method

X-rays and neutrons have been used on a number of occasions to study PdD and PdH.[6-13] According to Schirber and Morosin,[14] the lattice parameter of the beta phase is linear over its entire range. Felici et al.[9] used the following relationship to determine the near-surface composition of β-PdD:

$$D/Pd = -23.502 + 59.904 *a,$$

where a = lattice parameter in nanometers. Addition of lithium during electrolysis using lithium salts in the electrolyte will cause the lattice parameter of the surface to decrease.

BIBLIOGRAPHY

Preface

[1] Holley, C. E. J., Rand, M. H., and Storms, E., *The chemical thermodynamics of actinide elements and compounds* International Atomic Energy Agency, Vienna, 1984.

[2] Storms, E., *The refractory carbides* Academic Press, New York, 1967.

Chapter 1

[1] Normile, D., Waiting for ITER, fusion jocks look east, *Science* 312, 992, 2006.

[2] Nagel, D. J., Program strategy for low-energy nuclear reactions, *Infinite Energy* 12 (69), 13, 2006.

[3] Chery, D., ITER's $12 billion gamble, *Science* 214 (5797), 238, 2006.

[4] Peat, F. D., *Cold fusion: The making of a scientific controversy.* Contempory Books, Chicago, 1989.

[5] Mallove, E., *Fire From Ice* John Wiley, NY, 1991.

[6] Fox, H., *Cold fusion impact in the enhanced energy age*, 2 ed. Fusion Information Center, Inc, Salt Lake City, 1992.

[7] Close, F., *Too hot to handle. The race for cold fusion*, second ed. Penguin, paperback, New York, 1992.

[8] Huizenga, J. R., *Cold fusion: The scientific fiasco of the century*, first ed. University of Rochester Press, Rochester, NY, 1992.

[9] Taubes, G., *Bad science. The short life and weird times of cold fusion* Random House, NY, 1993.

[10] Hoffman, N., *A dialogue on chemically induced nuclear effects. A guide for the perplexed about cold fusion* American Nuclear Society, La Grange Park, Ill, 1995.

[11] Beaudette, C. G., *Excess heat. Why cold fusion research prevailed* Oak Grove Press (Infinite Energy, Distributor), Concord, NH, 2000.

[12] Simon, B., *Undead science: Science studies and the afterlife of cold fusion* Rutgers University Press, New Brunswick, NJ, 2002.

[13] Krivit, S. B. and Winocur, N., *The rebirth of cold fusion; Real science, real hope, real energy* Pacific Oaks Press, Los Angeles, CA, 2004.

[14] Germano, R., *Fusione fredda. Moderna storia d'inquisizione e d'alchimia* Bibliopolis, Napoli, Italy, 2003.

[15] Stolper, T. E., *Genius inventor: The controversy about the work of Randell Mills, America's Newton, in historical and contemporary context* www.amazon.com, 2006.

[16] Kozima, H., *The science of the cold fusion phenomenon* Elsevier Science, 2006.

[17] Kozima, H., *Discovery of the cold fusion phenomenon* Ohotake Shuppan, Inc, Tokyo, Japan, 1998.

[18] Mizuno, T., *Nuclear transmutation: The reality of cold fusion* Infinite Energy Press, Concord, NH, 1998.

[19] Champion, J., *Producing precious metals at home* Discovery Publishing, Westboro, WI, 1994.

[20] Rothwell, J., *Cold Fusion and the Future* www.LENR-CANR.org, 2005.

[21] Lewenstein, B., From fax to facts: Communication in the cold fusion saga, *Social Studies of Science* 25 (3), 403, 1995.

[22] Lewenstein, B., La saga de la fusion froide" (The cold fusion saga), *Recherche* 25, 636 (in French), 1994.

[23] Lewenstein, B. V. and Baur, W., A cold fusion chronology, *J. Radioanal. Nucl. Chem.* 152, 273, 1991.

[24] Lewenstein, B., Energy in a Jar, *The Sciences* July/August, 44, 1991.

[25] Lewenstein, B., Cold fusion saga: Lesson in science, *Forum Appl. Res. Public Policy* 7 (4), 67, 1992.

[26] Lewenstein, B. V., Cold fusion and hot history, *Osiris* 7, 135, 1992.
[27] Lewenstein, B. V., Do public electronic bulletin boards help create scientific knowledge? The cold fusion case, *Sci., Technol. Human Values* 20, 123, 1995.

Chapter 2

[1] Gottesfeld, S., Anderson, R. E., Baker, D. A., Bolton, R. D., Butterfield, K. B., Garzon, F. H., Goulding, C. A., Johnson, M. W., Leonard, E. M., Springer, T. E., and Zawodzinski, T., Experiments and nuclear measurements in search of cold fusion processes, *J. Fusion Energy* 9 (3), 287, 1990.

[2] Fleischmann, M., Pons, S., and Hawkins, M., Electrochemically Induced Nuclear Fusion of Deuterium, *J. Electroanal. Chem.* 261, 301 and errata in Vol. 263, 1989.

[3] Deryagin, B. V., Klyuev, V. A., Lipson, A. G., and Toporov, Y. P., Possibility of nuclear reactions during the fracture of solids, *Colloid J. USSR* 48, 8, 1986.

[4] Preparata, G., Fractofusion revisted, in *Anomalous Nuclear Effects in Deuterium/Solid Systems, "AIP Conference Proceedings 228"*, Jones, S., Scaramuzzi, F., andWorledge, D. American Institute of Physics, New York, Brigham Young Univ., Provo, UT, 1990, pp. 840.

[5] Menlove, H. O., Fowler, M. M., Garcia, E., Miller, M. C., Paciotti, M. A., Ryan, R. R., and Jones, S. E., Measurement of neutron emission from Ti and Pd in pressurized D_2 gas and D_2O electrolysis cells, *J. Fusion Energy* 9 (4), 495, 1990.

[6] De Ninno, A., Frattolillo, A., Lollobattista, G., Martinis, L., Martone, M., Mori, L., Podda, S., and Scaramuzzi, F., Emission of neutrons as a consequence of titanium-deuterium interaction, *Nuovo Cimento Soc. Ital. Fis. A* 101, 841, 1989.

[7] Preparata, G., A New Look at Solid-State Fractures, Particle Emission and 'Cold' Nuclear Fusion, *Nuovo Cimento A* 104, 1259, 1991.

[8] Jones, S. E., Palmer, E. P., Czirr, J. B., Decker, D. L., Jensen, G. L., Thorne, J. M., Taylor, S. F., and Rafelski, J., Observation of cold nuclear fusion in condensed matter, *Nature (London)* 338, 737, 1989.

[9] Menlove, H. O., Paciotti, M. A., Claytor, T. N., Maltrud, H. R., Rivera, O. M., Tuggle, D. G., and Jones, S. E., Reproducible neutron emission measurements from Ti metal in pressurized D_2 gas, in *Anomalous Nuclear Effects in Deuterium/Solid Systems, "AIP Conference Proceedings 228"*, Jones, S., Scaramuzzi, F., andWorledge, D. American Institute of Physics, New York, Brigham Young Univ., Provo, UT, 1990, pp. 287.

[10] De Ninno, A., Frattolillo, A., Lollobattista, G., Martinis, L., Martone, M., Mori, L., Podda, S., and Scaramuzzi, F., Evidence of emission of neutrons from a titanium-deuterium system, *Europhys. Lett.* 9, 221, 1989.

[11] Lewis, N. S., Barnes, C. A., Heben, M. J., Kumar, A., Lunt, S. R., McManis, G. E., Miskelly, G. M., Penner, R. M., Sailor, M. J., Santangelo, P. G., Shreve, G. A., Tufts, B. J., Youngquist, M. G., Kavanagh, R. W., Kellogg, S. E., Vogelaar, R. B., Wang, T. R., Kondrat, R., and New, R., Searches for low-temperature nuclear fusion of deuterium in palladium, *Nature (London)* 340 (6234), 525, 1989.

[12] Miskelly, G. M., Heben, M. J., Kumar, A., Penner, R. M., Sailor, M. J., and Lewis, N. S., Analysis of the published calorimetric evidence for electrochemical fusion of deuterium in palladium, *Science* 246, 793, 1989.

[13] Koonin, S. E. and Nauenberg, M., Calculated fusion rates in isotopic hydrogen molecules, *Nature (London)* 339, 690, 1989.

[14] Noninski, V. C. and Noninski, C. I., Notes on two papers claiming no evidence for the existence of excess energy during the electrolysis of 0.1M LiOD/D_2O with palladium cathodes, *Fusion Technol.* 23, 474, 1993.

[15] Miles, M. H., Bush, B. F., and Stilwell, D. E., Calorimetric principles and problems in measurements of excess power during Pd-D_2O electrolysis, *J. Phys. Chem.* 98, 1948, 1994.

16. Goodstein, D., Pariah science. Whatever happened to cold fusion?, *The American Scholar* 63 (4), 527, 1994.

17. Goodstein, D., Whatever happened to cold fusion?, *Accountability Res.* 8, 2000.

18. Park, R., *Voodoo science* Oxford University Press, New York, NY, 2000.

19. Storms, E. and Talcott, C. L., Electrolytic tritium production, *Fusion Technol.* 17, 680, 1990.

20. Claytor, T. N., Seeger, P., Rohwer, R. K., Tuggle, D. G., and Doty, W. R., Tritium and neutron measurements of a solid state cell, in *NSF/EPRI Workshop on Anomalous Effects in Deuterated Materials* LA-UR-89-39-46, Washington, DC, 1989.

21. Claytor, T. N., Jackson, D. D., and Tuggle, D. G., Tritium production from low voltage deuterium discharge on palladium and other metals, *Infinite Energy* 2 (7), 39, 1996.

22. Huizenga, J. R., *Cold fusion: The scientific fiasco of the century*, first ed. University of Rochester Press, Rochester, NY, 1992.

23. Holst-Hansen, P. and Britz, D., Can current fluctuations account for the excess heat claims of Fleischmann and Pons?, *J. Electroanal. Chem.* 388, 11, 1995.

24. Schwinger, J., Nuclear energy in an atomic lattice, *Prog. Theor. Phys.* 85, 711, 1991.

25. Schwinger, J., Cold fusion, A brief history of mine, *Trans. Fusion Technol.* 26 (4T), xiii, 1994.

26. Bockris, J. O. M., Accountability and academic freedom: The battle concerning research on cold fusion at Texas A&M University, *Accountability in Research* 8, 103, 2000.

27. Bockris, J., History of the discovery of transmutation at Texas A&M University, in *11th International Conference on Cold Fusion*, Biberian, J.-P. World Scientific Co., Marseilles, France, 2004, pp. 562.

28. Cedzynska, K., Barrowes, S. C., Bergeson, H. E., Knight, L. C., and Will, F. G., Tritium analysis in palladium with an open system analytical procedure, in *Anomalous Nuclear Effects in Deuterium/Solid Systems, "AIP Conference Proceedings 228"*, Jones, S., Scaramuzzi, F., andWorledge, D. American Institute of Physics, New York, Brigham Young Univ., Provo, UT, 1990, pp. 463.

29. Will, F. G., Cedzynska, K., Yang, M. C., Peterson, J. R., Bergeson, H. E., Barrowes, S. C., West, W. J., and Linton, D. C., Studies of electrolytic and gas phase loading of palladium with deuterium, in *Second Annual Conference on Cold Fusion, "The Science of Cold Fusion"*, Bressani, T., Giudice, E. D., andPreparata, G. Societa Italiana di Fisica, Bologna, Italy, Como, Italy, 1991, pp. 373.

30. Will, F. G., Cedzynska, K., and Linton, D. C., Tritium generation in palladium cathodes with high deuterium loading, in *Fourth International Conference on Cold Fusion* Electric Power Research Institute 3412 Hillview Ave., Palo Alto, CA 94304, Lahaina, Maui, 1993, pp. 8.

31. Will, F., Hydrogen + oxygen recombination and related heat generation in undivided electrolysis cells, *J. Electroanal. Chem.* 426, 177, 1997.

32. Langmuir, I., Pathological science, *Physics Today* October, 36, 1989.

33. Storms, E., Review of experimental observations about the cold fusion effect, *Fusion Technol.* 20, 433, 1991.

34. Tsarev, V. A. and Worledge, D. H., New results on cold nuclear fusion: a review of the conference on anomalous nuclear effects in deuterium/solid systems, Provo, Utah, October 22-24, 1990, *Fusion Technol.* 20, 484, 1991.

35. Tsarev, V. A., Cold fusion studies in the USSR, in *Second Annual Conference on Cold Fusion, "The Science of Cold Fusion"*, Bressani, T., Del Giudice, E., andPreparata, G. Societa Italiana di Fisica, Bologna, Italy, Como, Italy, 1991, pp. 319.

36. Srinivasan, M., Nuclear fusion in an atomic lattice: An update on the international status of cold fusion research, *Curr. Sci.* 60, 417, 1991.

37. Scaramuzzi, F., Survey of gas loading experiments, in *Second Annual Conference on Cold Fusion, "The Science of Cold Fusion"*, Bressani, T., Del Giudice, E., andPreparata, G. Societa Italiana di Fisica, Bologna, Italy, Como, Italy, 1991, pp. 445.

38. Fleischmann, M., The present status of research in cold fusion, in *Second Annual Conference on Cold Fusion, "The Science of Cold Fusion"*, Bressani, T., Del Giudice, E., andPreparata, G. Societa Italiana di Fisica, Bologna, Italy, Como, Italy, 1991, pp. 475.

[39.] Takahashi, A., Mega, A., Takeuchi, T., Miyamaru, H., and Iida, T., Anomalous excess heat by D_2O/Pd cell under L-H mode electrolysis, in *Third International Conference on Cold Fusion, "Frontiers of Cold Fusion"*, Ikegami, H. Universal Academy Press, Inc., Tokyo, Japan, Nagoya Japan, 1992, pp. 79.

[40.] Takahashi, A., Iida, T., Takeuchi, T., and Mega, A., Excess heat and nuclear products by D_2O/Pd electrolysis and multibody fusion, *Int. J. Appl. Electromagn. Mater.* 3, 221, 1992.

[41.] Miyamaru, H. and Takahashi, A., Periodically current-controlled electrolysis of D_2O/Pd system for excess heat production, in *Third International Conference on Cold Fusion, "Frontiers of Cold Fusion"*, Ikegami, H. Universal Academy Press, Inc., Tokyo, Japan, Nagoya Japan, 1992, pp. 393.

[42.] Kobayashi, M., Imai, N., Hasegawa, N., Kubota, A., and Kunimatsu, K., Measurements of D/Pd and excess heat during electrolysis of LiOD in a fuel-cell type closed cell using a palladium sheet cathode, in *Third International Conference on Cold Fusion, "Frontiers of Cold Fusion"*, Ikegami, H. Universal Academy Press, Inc., Tokyo, Japan, Nagoya Japan, 1992, pp. 385.

[43.] Celani, F., Spallone, A., Tripodi, P., and Nuvoli, A., Measurement of excess heat and tritium during self-biased pulsed electrolysis of Pd-D_2O, in *Third International Conference on Cold Fusion, "Frontiers of Cold Fusion"*, Ikegami, H. Universal Academy Press, Inc., Tokyo, Japan, Nagoya Japan, 1992, pp. 93.

[44.] Bockris, J. O. M., Sundaresan, R., Minevski, Z., and Letts, D., Triggering of heat and sub-surface changes in Pd-D systems, *Trans. Fusion Technol.* 26 (#4T), 267, 1994.

[45.] Storms, E., Measurements of excess heat from a Pons-Fleischmann-type electrolytic cell using palladium sheet, *Fusion Technol.* 23, 230, 1993.

[46.] Storms, E., Statement of Dr. Edmund Storms, in *Hearing Before the Subcommittee on Energy of the Committee on Science, Space, and Technology, U. S. House of Representatives, One Hundred Third Congress, First Session* U.S. Government Printing Office, Washington, C.D., 1993, pp. 114.

[47.] Taubes, G., *Bad science. The short life and weird times of cold fusion* Random House, NY, 1993.

[48.] Huizenga, J. R., *Cold fusion: The scientific fiasco of the century*, second ed. Oxford University Press, New York, 1993.

[49.] Milton, R., *Forbidden science. Suppressed research that could change our lives* Fourth Estate, London, 1994.

[50.] Cohen, I. B., *Revolution in science* The Belknap Press, Harvard Univ. Press, Cambridge, Mass., 1985.

[51.] Platt, C., What if cold fusion is real?, in *Wired* 1998, pp. 170.

[52.] Krivit, S. B. and Winocur, N., *The rebirth of cold fusion; Real science, real hope, real energy* Pacific Oaks Press, Los Angeles, CA, 2004.

[53.] Beaudette, C. G., *Excess heat. Why cold fusion research prevailed* Oak Grove Press (Infinite Energy, Distributor), Concord, NH, 2000.

[54.] DoE, Report of the review of low energy nuclear reactions, in *Review of Low Energy Nuclear Reactions* Department of Energy, Office of Science, Washington, DC, 2004.

[55.] DoE, Report No. www.LENR-CANR.org, 2004.

[56.] Storms, E., Report No. http://lenr-canr.org/acrobat/StormsEaresponset.pdf, 2005.

Chapter 3

[1.] Storms, E. and Talcott, C. L., Electrolytic tritium production, *Fusion Technol.* 17, 680, 1990.

[2.] Storms, E. and Talcott-Storms, C., The effect of hydriding on the physical structure of palladium and on the release of contained tritium, *Fusion Technol.* 20, 246, 1991.

[3.] Storms, E., Measurements of excess heat from a Pons-Fleischmann-type electrolytic cell using palladium sheet, *Fusion Technol.* 23, 230, 1993.

[4.] Champion, J., *Producing precious metals at home* Discovery Publishing, Westboro, WI, 1994.

[5.] Bockris, J., History of the discovery of transmutation at Texas A&M University, in *11th International Conference on Cold Fusion*, Biberian, J.-P. World Scientific Co., Marseilles, France, 2004, pp. 562.

[6.] Bockris, J. O. M., Accountability and academic freedom: The battle concerning research on cold fusion at Texas A&M University, *Accountability in Research* 8, 103, 2000.

[7.] Storms, E., A study of those properties of palladium that influence excess energy production by the Pons-Fleischmann effect, *Infinite Energy* 2 (8), 50, 1996.

[8.] Storms, E., Formation of b-PdD containing high deuterium concentration using electrolysis of heavy-water, *J. Alloys Comp.* 268, 89, 1998.

[9.] Storms, E., Relationship between open-circuit-voltage and heat production in a Pons-Fleischmann cell, in *The Seventh International Conference on Cold Fusion*, Jaeger, F. ENECO, Inc., Salt Lake City, UT, Vancouver, Canada, 1998, pp. 356.

[10.] Pons, S. and Fleischmann, M., Heat after death, *Trans. Fusion Technol.* 26 (4T), 97, 1994.

[11.] Storms, E., The nature of the energy-active state in Pd-D, in *II Workshop on the Loading of Hydrogen/Deuterium in Metals, Characterization of Materials and Related Phenomena*, Asti, Italy, 1995.

[12.] Storms, E., The nature of the energy-active state in Pd-D, *Infinite Energy* (#5 and #6), 77, 1995.

[13.] Storms, E., A review of the cold fusion effect, *J. Sci. Exploration* 10 (2), 185, 1996.

[14.] Storms, E., Cold fusion, a challenge to modern science, *J. Sci. Expl.* 9, 585, 1995.

[15.] Storms, E., Cold fusion revisited, *Infinite Energy* 4 (21), 16, 1998.

[16.] Lewis, N. S., Barnes, C. A., Heben, M. J., Kumar, A., Lunt, S. R., McManis, G. E., Miskelly, G. M., Penner, R. M., Sailor, M. J., Santangelo, P. G., Shreve, G. A., Tufts, B. J., Youngquist, M. G., Kavanagh, R. W., Kellogg, S. E., Vogelaar, R. B., Wang, T. R., Kondrat, R., and New, R., Searches for low-temperature nuclear fusion of deuterium in palladium, *Nature (London)* 340 (6234), 525, 1989.

[17.] Storms, E., Description of a dual calorimeter, *Infinite Energy* 6 (34), 22, 2000.

[18.] Shanahan, K., A systematic error in mass flow calorimetery demonstrated, *Thermochim. Acta* 382 (2), 95-101, 2002.

[19.] Szpak, S., Mosier-Boss, P. A., and Smith, J. J., On the behavior of Pd deposited in the presence of evolving deuterium, *J. Electroanal. Chem.* 302, 255, 1991.

[20.] Storms, E., Use of a very sensitive Seebeck calorimeter to study the Pons-Fleischmann and Letts effects, in *Tenth International Conference on Cold Fusion*, Hagelstein, P. L. andChubb, S. R. World Scientific Publishing Co., Cambridge, MA, 2003, pp. 183.

[21.] Case, L. C., Catalytic fusion of deuterium into helium-4, in *The Seventh International Conference on Cold Fusion*, Jaeger, F. ENECO, Inc., Salt Lake City, UT, Vancouver, Canada, 1998, pp. 48.

[22.] Letts, D. and Cravens, D., Laser stimulation of deuterated palladium, *Infinite Energy* 9 (50), 10, 2003.

[23.] Letts, D. and Cravens, D., Laser stimulation of deuterated palladium: past and present, in *Tenth International Conference on Cold Fusion*, Hagelstein, P. L. andChubb, S. R. World Scientific Publishing Co., Cambridge, MA, 2003, pp. 159.

[24.] Letts, D. and Cravens, D., Cathode frabrication methods to reproduce the Letts-Cravens effect, in *ASTI-5* www.iscmns.org/, Asti, Italy, 2004.

[25.] Claytor, T. N., 2003.

Chapter 4

[1.] Langmuir, I., Pathological science, *Physics Today* October, 36, 1989.

[2.] Petrasso, R. D., Chen, X., Wenzel, K. W., Parker, R. R., Li, C. K., and Fiore, C., Problems with the gamma-ray spectrum in the Fleischmann *et al.* experiments, *Nature* 339 (6221), 667, 1989.

[3] Fleischmann, M., Pons, S., Hawkins, M., and Hoffman, R. J., Measurements of gamma-rays from cold fusion., *Nature (London)* 339 (622), 667, 1989.

[4] Wilson, R. H., Bray, J. W., Kosky, P. G., Vakil, H. B., and Will, F. G., Analysis of experiments on the calorimetry of LiOD-D2O electrochemical cells, *J. Electroanal. Chem.* 332, 1, 1992.

[5] Morrison, D. R. O., Comments on claims of excess enthalpy by Fleischmann and Pons using simple cells made to boil, *Phys. Lett. A* 185, 498, 1994.

[6] Miskelly, G. M., Heben, M. J., Kumar, A., Penner, R. M., Sailor, M. J., and Lewis, N. S., Analysis of the published calorimetric evidence for electrochemical fusion of deuterium in palladium, *Science* 246, 793, 1989.

[7] Saito, T., Sumi, M., Asami, N., and Ikegami, H., Studies on Fleishmann-Pons calorimetry with ICARUS 1., in *5th International Conference on Cold Fusion*, Pons, S. IMRA Europe, Sophia Antipolis Cedex, France, Monte-Carlo, Monaco, 1995, pp. 105.

[8] Fleischmann, M. and Pons, S., Reply to the critique by Morrison entitled 'Comments on claims of excess enthalpy by FLeischmann and Pons using simple cells made to boil, *Phys. Lett. A* 187, 276, 1994.

[9] Fleischmann, M. and Pons, S., Some comments on the paper "Analysis of experiments on the calorimetry of LiOD-D_2O electrochemical cells", R.H. Wilson *et al.*, J. Electroanal. Chem. 332 [1992] 1, *J. Electroanal. Chem.* 332, 33, 1992.

[10] Pons, S. and Fleischmann, M., Calorimetric measurements of the palladium/deuterium system: fact and fiction, *Fusion Technol.* 17, 669, 1990.

[11] Hansen, W. N. and Melich, M. E., Pd/D calorimetry- The key to the F/P effect and a challenge to science, in *Fourth International Conference on Cold Fusion* Electric Power Research Institute 3412 Hillview Ave., Palo Alto, CA 94304, Lahaina, Maui, 1993, pp. 11.

[12] Hansen, W. N. and Melich, M. E., Pd/D calorimetry- The key to the F/P effect and a challenge to science, *Trans. Fusion Technol.* 26 (4T), 355, 1994.

[13] Hansen, W. N., Report to the Utah State Fusion/Energy Council on the analysis of selected Pons Fleischmann calorimetric data, in *Second Annual Conference on Cold Fusion, "The Science of Cold Fusion"*, Bressani, T., Del Giudice, E., andPreparata, G. Societa Italiana di Fisica, Bologna, Italy, Como, Italy, 1991, pp. 491.

[14] Nissant, M., The plight of the obscure innovator in science, *Soc. Studies Sci.* 25, 165, 1995.

[15] Jones, S. E., Palmer, E. P., Czirr, J. B., Decker, D. L., Jensen, G. L., Thorne, J. M., Taylor, S. F., and Rafelski, J., Observation of cold nuclear fusion in condensed matter, *Nature (London)* 338, 737, 1989.

[16] Williams, D. E. G., Findlay, D. J. S., Craston, D. H., Sene, M. R., Bailey, M., Croft, S., Hooton, B. W., Jones, C. P., Kucernak, A. R. J., Mason, J. A., and Taylor, R. I., Upper bounds on 'cold fusion' in electrolytic cells, *Nature (London)* 342, 375, 1989.

[17] Lewis, N. S., Barnes, C. A., Heben, M. J., Kumar, A., Lunt, S. R., McManis, G. E., Miskelly, G. M., Penner, R. M., Sailor, M. J., Santangelo, P. G., Shreve, G. A., Tufts, B. J., Youngquist, M. G., Kavanagh, R. W., Kellogg, S. E., Vogelaar, R. B., Wang, T. R., Kondrat, R., and New, R., Searches for low-temperature nuclear fusion of deuterium in palladium, *Nature (London)* 340 (6234), 525, 1989.

[18] Albagli, D., Ballinger, R., Cammarata, V., Chen, X., Crooks, R. M., Fiore, C., Gaudreau, M. P. J., Hwang, I., Li, C. K., Linsay, P., Luckhardt, S. C., Parker, R. R., Petrasso, R. D., Schloh, M. O., Wenzel, K. W., and Wrighton, M. S., Measurement and analysis of neutron and gamma-ray emission rates, other fusion products, and power in electrochemical cells having Pd cathodes, *J. Fusion Energy* 9, 133, 1990.

[19] Dardik, I., Zilov, T., Branover, H., El-Boher, A., Greenspan, E., Khachaturov, B., Krakov, V., S., L., and Tsirlin, M., Progress in electrolysis experiments at Energetics Technologies, in *12th International Conference on Condensed Matter Nuclear Science*, Yokohama, Japan, 2005.

[20] Stringham, R., Low mass 1.6 MHz sonofusion reactor, in *11th International Conference on Cold Fusion*, Biberian, J.-P. World Scientific Co., Marseilles, France, 2004, pp. 238.

[21.] Stringham, R., Cavitation and fusion, in *Tenth International Conference on Cold Fusion*, Hagelstein, P. L. andChubb, S. R. World Scientific Publishing Co., Cambridge, MA, 2003, pp. 233.

[22.] Savvatimova, I. and Gavritenkov, D. V., Results of analysis of Ti foil after glow discharge with deuterium, in *11th International Conference on Cold Fusion*, Biberian, J.-P. World Scientific Co., Marseilles, France, 2004, pp. 438.

[23.] Mizuno, T., Aoki, Y., Chung, D. Y., and Sesftel, F., Generation of heat and products during plasma electrolysis, in *11th International Conference on Cold Fusion*, Biberian, J.-P. World Scientific Co., Marseilles, France, 2004, pp. 161.

[24.] Mizuno, T., Ohmori, T., and Akimoto, T., Generation of heat and products during plasma electrolysis, in *Tenth International Conference on Cold Fusion*, Chubb, P. I. H. a. S. R. World Scientific Publishing Co., Cambridge, MA, 2003, pp. 73.

[25.] Tian, J., Jin, L. H., Weng, Z. K., Song, B., Zhao, X. L., Xiao , Z. J., Chen, G., and Du, B. Q., "Excess heat" during electrolysis in platinum/K_2CO_3/nickel light water system, in *11th International Conference on Cold Fusion*, Biberian, J.-P. World Scientific Co., Marseilles, France, 2004, pp. 102.

[26.] Szpak, S., Mosier-Boss, P. A., Miles, M., and Fleischmann, M., Thermal behavior of polarized Pd/D electrodes prepared by co-deposition, *Thermochim. Acta* 410, 101, 2004.

[27.] Campari, E. G., Fasano, G., Focardi, S., Lorusso, G., Gabbani, V., Montalbano, V., Piantelli, F., Stanghini, C., and Veronesi, S., Photon and particle emission, heat production and surface transformation in Ni-H system, in *11th International Conference on Cold Fusion*, Biberian, J.-P. World Scientific Co., Marseilles, France, 2004, pp. 405.

[28.] Dash, J. and Ambadkar, A., Co-deposition of palladium with hydrogen isotopes, in *11th International Conference on Cold Fusion*, Biberian, J.-P. World Scientific Co., Marseilles, France, 2004, pp. 477.

[29.] Dardik, I., Zilov, T., Branover, H., El-Boher, A., Greenspan, E., Khachatorov, B., Krakov, V., Lesin, S., and Tsirlin, M., Excess heat in electrolysis experiments at Energetics Technologies, in *11th International Conference on Cold Fusion*, Biberian, J.-P. World Scientific Co., Marseilles, France, 2004, pp. 84.

[30.] Wei, Q., Li, X. Z., and Cui, Y. O., Excess heat in heavy water-Pd/C catalyst cathode (Case-Type) electrolysis at the temperature near the boiling point, in *Tenth International Conference on Cold Fusion*, Chubb, P. I. H. a. S. R. World Scientific Publishing Co., Cambridge, MA, 2003, pp. 55.

[31.] Wei, Q. M., Cui, Y. O., Pan, G. H., Q., D. X., and Li, X. Z., Excess heat in Pd/C catalyst electrolysis experiment (Case-type cathode), in *The 9th International Conference on Cold Fusion, Condensed Matter Nuclear Science*, Li, X. Z. Tsinghua Univ. Press, Tsinghua Univ., Beijing, China, 2002, pp. 408.

[32.] Tsvetkov, S. A., Filatov, E. S., and Khokhlov, V. A., Excess heat in molten salts of (LiCl-KCl)+(LiD+LiF) at the titanium anode during electrolysis, in *Tenth International Conference on Cold Fusion*, Hagelstein, P. L. andChubb, S. R. World Scientific Publishing Co., Cambridge, MA, 2003, pp. 89.

[33.] Swartz, M. R. and Verner, G., Excess heat from low-electrical conducting heavy water spiral-wound Pd/D_2O/Pt and Pd/D_2O-PdCl$_2$/Pt devices, in *Tenth International Conference on Cold Fusion*, Chubb, P. I. H. a. S. R. World Scientific Publishing Co., Cambridge, MA, 2003, pp. 29.

[34.] Storms, E., Use of a very sensitive Seebeck calorimeter to study the Pons-Fleischmann and Letts effects, in *Tenth International Conference on Cold Fusion*, Hagelstein, P. L. andChubb, S. R. World Scientific Publishing Co., Cambridge, MA, 2003, pp. 183.

[35.] Miles, M., Fluidized bed experiments using platinum and palladium particles In heavy water, in *Tenth International Conference on Cold Fusion*, Hagelstein, P. L. andChubb, S. R. World Scientific Publishing Co., Cambridge, MA, 2003, pp. 23.

[36.] Li, X. Z., Liu, B., Cai, N., Wei, Q., Tian, J., and Cao, D. X., Progress in gas-loading D/Pd system: The feasibility of a self-sustaining heat generator, in *Tenth International Conference on*

Cold Fusion, Hagelstein, P. L. andChubb, S. R. World Scientific Publishing Co., Cambridge, MA, 2003, pp. 113.

[37.] Letts, D. and Cravens, D., Laser stimulation of deuterated palladium: past and present, in *Tenth International Conference on Cold Fusion*, Hagelstein, P. L. andChubb, S. R. World Scientific Publishing Co., Cambridge, MA, 2003, pp. 159.

[38.] Letts, D. and Cravens, D., Laser stimulation of deuterated palladium, *Infinite Energy* 9 (50), 10, 2003.

[39.] Karabut, A. B., Production of excess heat, impurity elements and unnatural isotopic ratios in high-current glow discharge experiments, in *Tenth International Conference on Cold Fusion*, Hagelstein, P. L. andChubb, S. R. World Scientific Publishing Co., Cambridge, MA, 2003, pp. 99.

[40.] Karabut, A., Excess heat production in Pd/D during periodic pulse discharge current in various conditions, in *11th International Conference on Cold Fusion*, Biberian, J.-P. World Scientific Co., Marseilles, France, 2004, pp. 178.

[41.] De Ninno, A., Frattolillo, A., Rizzo, A., and Del Gindice, E., ^{4}He detection In a cold fusion experiment, in *Tenth International Conference on Cold Fusion*, Hagelstein, P. L. andChubb, S. R. World Scientific Publishing Co., Cambridge, MA, 2003, pp. 133.

[42.] Dardik, I., Branover, H., El-Boher, A., Gazit, D., Golbreich, E., Greenspan, E., Kapusta, A., Khachatorov, B., Krakov, V., Lesin, S., Michailovitch, B., Shani, G., and Zilov, T., Intensification of low energy nuclear reactions using superwave excitation, in *Tenth International Conference on Cold Fusion*, Chubb, P. I. H. a. S. R. World Scientific Publishing Co., Cambridge, MA, 2003, pp. 61.

[43.] Celani, F., Spallone, A., Righi, E., Trenta, G., Catena, C., D'Agostaro, G., Quercia, P., Andreassi, V., Marini, P., Di Stefano, V., Nakamura, M., Mancini, A., Sona, P. G., Fontana, F., Gamberale, L., Garbelli, D., Falcioni, F., Marchesini, M., Novaro, E., and Mastromatteo, U., Thermal and isotopic anomalies when Pd cathodes are electrolyzed in electrolytes containing Th-Hg salts dissolved at micromolar concentration in C_2H_5OD/D_2O mixtures, in *Tenth International Conference on Cold Fusion*, Hagelstein, P. L. andChubb, S. R. World Scientific Publishing Co., Cambridge, MA, 2003, pp. 379.

[44.] Warner, J., Dash, J., and Frantz, S., Electrolysis of D_2O with titanium cathodes: enhancement of excess heat and further evidence of possible transmutation, in *The Ninth International Conference on Cold Fusion*, Li, X. Z. Tsinghua University, Beijing, China: Tsinghua University, 2002, pp. 404.

[45.] Tian, J., Li, X. Z., Yu, W. Z., Mei, M. Y., Cao, D. X., Li, A. L., Li, J., Zhao, Y. G., and Zhang, C., "Excess heat" and "heat after death" in a gas loading hydrogen/palladium system, in *The 9th International Conference on Cold Fusion, Condensed Matter Nuclear Science*, Li, X. Z. Tsinghua Univ. Press, Tsinghua Univ., Beijing, China, 2002, pp. 360.

[46.] Tian, J., Li, X. Z., Yu, W. Z., Cao, D. X., Zhou, R., Yu, Z. W., Ji, Z. F., Liu, Y., T., H. J., and Zhou, R. X., Anomanous heat flow and its correlation with deuterium flux in a gas-loading deuterium-palladium system, in *The 9th International Conference on Cold Fusion, Condensed Matter Nuclear Science*, Li, X. Z. Tsinghua Univ. Press, Tsinghua Univ., Beijing, China, 2002, pp. 353.

[47.] Li, X. Z., Liu, B., Ren, X. Z., Tian, J., Cao, D. X., Chen, S., Pan, G. H., Ho, D. l., and Deng, Y., "Super-absorption" - correlation between deuterium flux and excess heat, in *The 9th International Conference on Cold Fusion, Condensed Matter Nuclear Science*, Li, X. Z. Tsinghua Univ. Press, Tsinghua Univ., Beijing, China, 2002, pp. 202.

[48.] Swartz, M. R., Verner, G. M., and Frank, A. H., The impact of heavy water (D_2O) on nickel-light water cold fusion systems, in *The 9th International Conference on Cold Fusion, Condensed Matter Nuclear Science*, Li, X. Z. Tsinghua Univ. Press, Tsinghua Univ., Beijing, China, 2002, pp. 335.

[49.] Sun, Y., Zhang, Q., and Gou, Q., The crystal change and "excess heat" production by long time electrolysis of heavy water with titanium cathode due to deuterium atom entering the lattice of titanium, in *The 9th International Conference on Cold Fusion, Condensed Matter Nuclear Science*, Li, X. Z. Tsinghua Univ. Press, Tsinghua Univ., Beijing, China, 2002, pp. 329.

[50.] Storms, E., Ways to initiate a nuclear reaction in solid environments, *Infinite Energy* 8 (45), 45, 2002.

[51.] Miles, M. H., Szpak, S., Mosier-Boss, P. A., and Fleischmann, M., Thermal behavior of polarized Pd/D electrodes prepared by co-deposition, in *The Ninth International Conference on Cold Fusion*, Li, X.-Z. Tsinghua University Press, Tsinghua University, Beijing, China, 2002, pp. 250.

[52.] Li, X. Z., Liu, B., Ren, X. Z., Tian, J., Yu, W. Z., Cao, D. X., Chen, S., Pan, G. H., and Zheng, S. X., "Pumping effect", reproducible excess heat in a gas-loading D/Pd system, in *The 9th International Conference on Cold Fusion, Condensed Matter Nuclear Science*, Li, X. Z. Tsinghua Univ. Press, Tsinghua Univ., Beijing, China, 2002, pp. 197.

[53.] Kirkinskii, V. A., Drebushchak, V. A., and Khmelnikov, A. I., Experimental evidence of excess heat output during deuterium sorption-desorption in palladium deuteride, in *The 9th International Conference on Cold Fusion, Condensed Matter Nuclear Science*, Li, X. Z. Tsinghua Univ. Press, Tsinghua Univ., Beijing, China, 2002, pp. 170.

[54.] Kirkinskii, V. A., Drebushchak, V. A., and Khmelnikov, A. I., Excess heat release during deuterium sorption-desorption by finely powdered palladium deuteride, *Europhys. Lett.* 58 (3), 462, 2002.

[55.] Karabut, A. B., Excess heat power, nuclear products and X-ray emission in relation to the high current glow discharge experimental parameters, in *The 9th International Conference on Cold Fusion, Condensed Matter Nuclear Science*, Li, X. Z. Tsinghua Univ. Press, Tsinghua Univ., Beijing, China, 2002, pp. 151.

[56.] Fujii, M., Mitsushima, S., Kamiya, N., and Ota, K., Heat measurement during light water electrolysis using Pd/Ni rod cathodes, in *The 9th International Conference on Cold Fusion, Condensed Matter Nuclear Science*, Li, X. Z. Tsinghua Univ. Press, Tsinghua Univ., Beijing, China, 2002, pp. 96.

[57.] Inoue, O. H., Fujii, M., Mitsushima, S., Kamiya, N., and Ota, K., Heat measurement during light water electrolysis, in *The 4th Meeting of Japan CF-Research Soc.* http:/wwwcf.elc.iwate-u.ac.jp/jcf/papers.html, Iwate Univ, Japan, 2002, pp. JCF4-2.

[58.] Del Giudice, E., De Ninno, A., Frattolillo, A., Porcu, M., and Rizzo, A., Production of excess enthalpy in the electrolysis of D_2O on Pd cathodes, in *The 9th International Conference on Cold Fusion, Condensed Matter Nuclear Science*, Li, X. Z. Tsinghua Univ. Press, Tsinghua Univ., Beijing, China, 2002, pp. 82.

[59.] Del Giudice, E., De Ninno, A., Franttolillo, A., Preparata, G., Scaramuzzi, F., Bulfone, A., Cola, M., and Giannetti, C., The Fleischmann-Pons effect in a novel electrolytic configuration, in *8th International Conference on Cold Fusion*, Scaramuzzi, F. Italian Physical Society, Bologna, Italy, Lerici (La Spezia), Italy, 2000, pp. 47.

[60.] Chicea, D., On current density and excess power density in electrolysis experiments, in *The 9th International Conference on Cold Fusion, Condensed Matter Nuclear Science*, Li, X. Z. Tsinghua Univ. Press, Tsinghua Univ., Beijing, China, 2002, pp. 49.

[61.] Castano, C. H., Lipson, A. G., Kim, S. O., and Miley, G. H., Calorimetric measurements during Pd-Ni thin film-cathodes electrolysis in Li_2SO_4/H_2O solution, in *The 9th International Conference on Cold Fusion, Condensed Matter Nuclear Science*, Li, X. Z. Tsinghua Univ. Press, Tsinghua Univ., Beijing, China, 2002, pp. 24.

[62.] Isobe, Y., Uneme, S., Yabuta, K., Katayama, Y., Mori, H., Omote, T., Ueda, S., Ochiai, K., Miyamaru, H., and Takahashi, A., Search for multibody nuclear reactions in metal deuteride induced with ion beam and electrolysis methods, *Jpn. J. Appl. Phys.* 41 (3), 1546, 2002.

[63.] Arata, Y. and Zhang, Y. C., Picnonuclear fusion generated in "lattice-reactor" of metallic deuterium lattice within metal atom-clusters. II Nuclear fusion reacted inside a metal by intense sonoimplantion effect, in *The 9th International Conference on Cold Fusion, Condensed Matter Nuclear Science*, Li, X. Z. Tsinghua Univ. Press, Tsinghua Univ., Beijing, China, 2002, pp. 5.

[64.] Dufour, J., Murat, D., Dufour, X., and Foos, J., Experimental observation of nuclear reactions in palladium and uranium - possible explanation by Hydrex mode, *Fusion Sci. & Technol.* 40, 91, 2001.

[65.] Zhang, Z. L., Zhang, W. S., Zhong, M. H., and Tan, F., Measurements of excess heat in the open Pd/D_2O electrolytic system by the Calvet calorimetry, in *8th International Conference on Cold Fusion*, Scaramuzzi, F. Italian Physical Society, Bologna, Italy, Lerici (La Spezia), Italy, 2000, pp. 91.

[66.] Warner, J. and Dash, J., Heat produced during the electrolysis of D_2O with titanium cathodes, in *8th International Conference on Cold Fusion*, Scaramuzzi, F. Italian Physical Society, Bologna, Italy, Lerici (La Spezia), Italy, 2000, pp. 161.

[67.] Storms, E., Excess power production from platinum cathodes using the Pons-Fleischmann effect, in *8th International Conference on Cold Fusion*, Scaramuzzi, F. Italian Physical Society, Bologna, Italy, Lerici (La Spezia), Italy, 2000, pp. 55.

[68.] Mizuno, T., Ohmori, T., Akimoto, T., and Takahashi, A., Production of heat during plasma electrolysis, *Jpn. J. Appl. Phys. A* 39, 6055, 2000.

[69.] Mizuno, T., Ohmori, T., Azumi, K., Akimoto, T., and Takahashi, A., Confirmation of heat generation and anomalous element caused by plasma electrolysis in the liquid, in *8th International Conference on Cold Fusion*, Scaramuzzi, F. Italian Physical Society, Bologna, Italy, Lerici (La Spezia), Italy, 2000, pp. 75.

[70.] Miles, M. H., Calorimetric studies of palladium alloy cathodes using Fleischmann-Pons Dewar type cells, in *8th International Conference on Cold Fusion*, Scaramuzzi, F. Italian Physical Society, Bologna, Italy, Lerici (La Spezia), Italy, 2000, pp. 97.

[71.] Miles, M., Imam, M. A., and Fleischmann, M., Excess heat and helium production in the palladium-boron system, *Trans. Am. Nucl. Soc.* 83, 371, 2000.

[72.] Miles, M. H., Imam, M. A., and Fleischmann, M., "Case studies" of two experiments carried out with the ICARUS systems, in *8th International Conference on Cold Fusion*, Scaramuzzi, F. Italian Physical Society, Bologna, Italy, Lerici (La Spezia), Italy, 2000, pp. 105.

[73.] Miles, M. H., Calorimetric studies of Pd/D_2O+LiOD electrolysis cells, *J. Electroanal. Chem.* 482, 56, 2000.

[74.] McKubre, M. C. H., Tanzella, F. L., Tripodi, P., and Hagelstein, P. L., The emergence of a coherent explanation for anomalies observed in D/Pd and H/Pd system: evidence for ^{4}He and ^{3}He production, in *8th International Conference on Cold Fusion*, Scaramuzzi, F. Italian Physical Society, Bologna, Italy, Lerici (La Spezia), Italy, 2000, pp. 3.

[75.] Dufour, J., Murat, D., Dufour, X., and Foos, J., Hydrex catallyzed transmutation of uranium and palladium: experimental part, in *8th International Conference on Cold Fusion*, Scaramuzzi, F. Italian Physical Society, Bologna, Italy, Lerici (La Spezia), Italy, 2000, pp. 153.

[76.] Campari, E. G., Focardi, S., Gabbani, V., Montalbano, V., Piantelli, F., Porcu, E., Tosti, E., and Veronesi, S., Ni-H systems, in *8th International Conference on Cold Fusion*, Scaramuzzi, F. Italian Physical Society, Bologna, Italy, Lerici (La Spezia), Italy, 2000, pp. 69.

[77.] Isobe, Y., Uneme, S., Yabuta, K., Mori, H., Omote, T., Ucda, S., Ochiai, K., Miyadera, H., and Takahashi, A., Search for coherent deuteron fusion by beam and electrolysis experiments, in *8th International Conference on Cold Fusion*, Scaramuzzi, F. Italian Physical Society, Bologna, Italy, Lerici (La Spezia), Italy, 2000, pp. 17.

[78.] Bernardini, M., Manduchi, C., Mengoli, G., and Zannoni, G., Anomalous effects induced by D_2O electrolysis at titanium, in *8th International Conference on Cold Fusion*, Scaramuzzi, F. Italian Physical Society, Bologna, Italy, Lerici (La Spezia), Italy, 2000, pp. 39.

[79.] Arata, Y. and Zhang, Y. C., Definite difference amoung [DS-D_2O], [DS-H_2O] and [Bulk-D_2O] cells in the deuterization and deuterium-reaction, in *8th International Conference on Cold Fusion*, Scaramuzzi, F. Italian Physical Society, Bologna, Italy, Lerici (La Spezia), Italy, 2000, pp. 11.

[80.] Arata, Y. and Zhang, Y. C., Observation of anomalous heat release and helium-4 production from highly deuterated fine particles, *Jpn. J. Appl. Phys. Part 2* 38 (7A), L774, 1999.

[81.] Arata, Y. and Zhang, Y. C., Definitive difference between [DS-D$_2$O] and [Bulk-D$_2$O] cells in 'deuterium-reaction', *Proc. Jpn. Acad., Ser. B* 75 Ser. B, 71, 1999.

[82.] Arata, Y. and Zhang, Y. C., Critical condition to induce 'excess energy' within [DS-H2O] cell, *Proc. Jpn. Acad., Ser. B* 75 Ser. B, 76, 1999.

[83.] Szpak, S., Mosier-Boss, P. A., and Miles, M. H., Calorimetry of the Pd+D co-deposition, *Fusion Technol.* 36, 234, 1999.

[84.] Takahashi, A., Results of experimental studies of excess heat vs nuclear products correlation and conceivable reaction model, in *The Seventh International Conference on Cold Fusion*, Jaeger, F. ENECO, Inc., Salt Lake City, UT., Vancouver, Canada, 1998, pp. 378.

[85.] Takahashi, A., Fukuoka, H., Yasuda, K., and Taniguchi, M., Experimental study on correlation between excess heat and nuclear products by D2O/Pd electrolysis, *Int. J. Soc. Mat. Eng. Resources* 6 (1), 4, 1998.

[86.] Stringham, R., Chandler, J., George, R., Passell, T. O., and Raymond, R., Predictable and reproducible heat, in *The Seventh International Conference on Cold Fusion*, Jaeger, F. ENECO, Inc., Salt Lake City, UT, Vancouver, Canada, 1998, pp. 361.

[87.] Savvatimova, I. B. and Korolev, V. U., Comparative analysis of heat effect in various cathode materials exposed to glow Ddscharge, in *The Seventh International Conference on Cold Fusion*, Jaeger, F. ENECO, Inc., Salt Lake City, UT, Vancouver, Canada, 1998, pp. 335.

[88.] Oya, Y., Ogawa, H., Aida, M., Iinuma, K., and Okamoto, M., Material conditions to replicate the generation of excess energy and the emission of excess neutrons, in *The Seventh International Conference on Cold Fusion*, Jaeger, F. ENECO, Inc., Salt Lake City, UT, Vancouver, Canada, 1998, pp. 285.

[89.] Ota, K., Kobayashi, T., Motohira, N., and Kamiya, N., Heat measurement during the heavy water electrolysis using Pd cathode, in *The Seventh International Conference on Cold Fusion*, Jaeger, F. ENECO, Inc., Salt Lake City, UT, Vancouver, Canada, 1998, pp. 297.

[90.] Ota, K., Kobayashi, T., Motohira, N., and Kamiya, N., Effect of boron for the heat production during the heavy water electrolysis using palladium cathode, *Int. J. Soc. Mat. Eng. Resources* 6 (1), 26, 1998.

[91.] Ohmori, T. and Mizuno, T., Strong excess energy evolution, new element production, and electromagnetic wave and/or neutron emission in the light water electrolysis with a tungsten cathode, in *The Seventh International Conference on Cold Fusion*, Jaeger, F. ENECO, Inc., Salt Lake City, UT, Vancouver, Canada, 1998, pp. 279.

[92.] Ohmori, T. and Mizuno, T., Excess energy evolution and transmutation, *Infinite Energy* 4 (20), 14, 1998.

[93.] Mengoli, G., Bernardini, M., Manducchi, C., and Zannoni, G., Anomalous heat effects correlated with electrochemical hydriding of nickel, *Nuovo Cimento Soc. Ital. Fis. D* 20 (3), 331, 1998.

[94.] Mengoli, G., Bernardini, M., Comisso, N., Manduchi, C., and Zannoni, G., The nickel-K$_2$CO$_3$, H$_2$O system: an electrochemical and calorimetric investigation, in *Asti Workshop on Anomalies in Hydrogen/Deuterium Loaded Metals*, Collis, W. J. M. F. Italian Phys. Soc., Riccardi, Rocca d'Arazzo, Italy, 1997, pp. 71.

[95.] Mengoli, G., Bernardini, M., Manduchi, C., and Zannoni, G., Calorimetry close to the boiling temperature of the D$_2$O/Pd electrolytic system, *J. Electroanal. Chem.* 444, 155, 1998.

[96.] Lonchampt, G., Biberian, J.-P., Bonnetain, L., and Delepine, J., Excess heat measurement with Pons and Fleischmann Type cells, in *The Seventh International Conference on Cold Fusion*, Jaeger, F. ENECO, Inc., Salt Lake City, UT, Vancouver, Canada, 1998, pp. 202.

[97.] Lonchampt, G., Biberian, J.-P., Bonnetain, L., and Delepine, J., Excess heat measurement with Patterson Type cells, in *The Seventh International Conference on Cold Fusion*, Jaeger, F. ENECO, Inc., Salt Lake City, UT, Vancouver, Canada, 1998, pp. 206.

[98.] Li, X. Z., Zheng, S. X., Huang, H. F., Huang, G. S., and Yu, W. Z., New measurements of excess heat in a gas loaded D-Pd system, in *The Seventh International Conference on Cold Fusion*, Jaeger, F. ENECO, Inc., Salt Lake City, UT, Vancouver, Canada, 1998, pp. 197.

[99.] Iwamura, Y., Itoh, T., Gotoh, N., Sakano, M., Toyoda, I., and Sakata, H., Detection of anomalous elements, X-ray and excess heat induced by continous diffusion of deuterium through multi-layer cathode (Pd/CaO/Pd), in *The Seventh International Conference on Cold Fusion*, Jaeger, F. ENECO, Inc., Salt Lake City, UT, Vancouver, Canada, 1998, pp. 167.

[100.] Iwamura, Y., Itoh, H., Gotoh, N., Sakano, M., Toyoda, I., and Sakata, H., Detection of anomalous elements, X-ray and excess heat induced by continuous diffusion of deuterium through multi-layer cathode (Pd/CaO/Pd), *Infinite Energy* 4 (20), 56, 1998.

[101.] Iwamura, Y., Itoh, T., Gotoh, N., and Toyoda, I., Detection of anomalous elements, X-ray, and excess heat in a D_2-Pd system and its interpretation by the electron-induced nuclear reaction model, *Fusion Technol.* 33, 476, 1998.

[102.] Gozzi, D., Cellucci, F., Cignini, P. L., Gigli, G., Tomellini, M., Cisbani, E., Frullani, S., and Urciuoli, G. M., Erratum to "X-ray, heat excess and ^{4}He in the D/Pd system", *J. Electroanal. Chem.* 452, 251, 1998.

[103.] Focardi, S., Gabbani, V., Montalbano, V., Piantelli, F., and Veronesi, S., Large excess heat production in Ni-H systems, *Nuovo Cimento* 111A (11), 1233, 1998.

[104.] Cain, B. L., Cheney, A. B., Rigsbee, J. M., Cain, R. W., and McMillian, L. S., Thermal power produced using thin-film palladium cathodes in concentrated lithium salt electrolyte, in *The Seventh International Conference on Cold Fusion*, Jaeger, F. ENECO, Inc., Salt Lake City, UT., Vancouver, Canada, 1998, pp. 43.

[105.] Bush, B. F. and Lagowski, J. J., Methods of generating excess heat with the Pons and Fleischmann effect: rigorous and cost effective calorimetry, nuclear products analysis of the cathode and helium analysis, in *The Seventh International Conference on Cold Fusion*, Jaeger, F. ENECO, Inc., Salt Lake City, UT., Vancouver, Canada, 1998, pp. 38.

[106.] Biberian, J.-P., Lonchampt, G., Bonnetain, L., and Delepine, J., Electrolysis of LaAlO₃ single crystals and ceramics in a deuteriated atmosphere, in *The Seventh International Conference on Cold Fusion*, Jaeger, F. ENECO, Inc., Salt Lake City, UT, Vancouver, Canada, 1998, pp. 27.

[107.] Arata, Y. and Zhang, Y.-C., Anomalous 'deuterium-reaction energies' within solid, *Proc. Japan. Acad.* 74 B, 155, 1998.

[108.] Arata, Y. and Zhang, Y.-C., Anomalous difference between reaction energies generated within D_2O-cell and H_2O-cell, *Jpn. J. Appl. Phys. Pt.2* 37 (11A), L1274, 1998.

[109.] Arata, Y. and Zhang, Y.-C., Solid-state plasma fusion ('cold fusion'), *J. High Temp. Soc.* 23 (special volume), 1-56, 1997.

[110.] Swartz, M. R., Consistency of the biphasic nature of excess enthalpy in solid-state anomalous phenomena with the quasi-one-dimensional model of isotope loading into a material, *Fusion Technol.* 31, 63, 1997.

[111.] Ohmori, T., Enyo, M., Mizuno, T., Nodasaka, Y., and Minagawa, H., Transmutation in the electrolysis of light water - excess energy and iron production in a gold electrode, *Fusion Technol.* 31, 210, 1997.

[112.] Dufour, J., Foos, J., Millot, J. P., and Dufour, X., Interaction of palladium/hydrogen and palladium/deuterium to measure the excess energy per atom for each isotope, *Fusion Technol.* 31, 198, 1997.

[113.] Numata, H. and Fukuhara, M., Low-temperature elastic anomalies and heat generation of deuterated palladium, *Fusion Technol.* 31, 300, 1997.

[114.] Focardi, S., Gabbani, V., Montalbano, V., Piantelli, F., and Veronesi, S., On the Ni-H system, in *Asti Workshop on Anomalies in Hydrogen/Deuterium Loaded Metals*, Collins, W. J. M. F. Societa Italiana Di Fisica, Villa Riccardi, Italy, 1997, pp. 35.

[115.] Cammarota, G., Collis, W. J. M. F., Rizzo, A., and Stremmenos, C., A flow calorimeter study of the Ni/H system, in *Asti Workshop on Anomalies in Hydrogen/Deuterium Loaded Metals*, Collis, W. J. M. F. Italian Phys. Soc., Villa Riccardi, Rocca d'Arazzo, Italy, 1997, pp. 1.

[116.] Kopecek, R. and Dash, J., Excess heat and unexpected elements from electrolysis of heavy water with titanium cathodes, *J. New Energy* 1 (3), 46, 1996.

[117.] Li, X. Z., Yue, W. Z., Huang, G. S., Shi, H., Gao, L., Liu, M. L., and Bu, F. S., "Excess heat" measurement in gas-loading D/Pd system, *J. New Energy* 1 (4), 34, 1996.

[118.] Yasuda, K., Nitta, Y., and Takahashi, A., Study of excess heat and nuclear products with closed D_2O electrolysis systems, in *Sixth International Conference on Cold Fusion, Progress in New Hydrogen Energy*, Okamoto, M. New Energy and Industrial Technology Development Organization, Tokyo Institute of Technology, Tokyo, Japan, Lake Toya, Hokkaido, Japan, 1996, pp. 36.

[119.] Celani, F., Spallone, A., Tripodi, P., Di Gioacchino, D., Marini, P., Di Stefano, V., Mancini, A., and Pace, S., New kinds of electrolytic regimes and geometrical configurations to obtain anomalous results in Pd(M)-D systems, in *Sixth International Conference on Cold Fusion, Progress in New Hydrogen Energy*, Okamoto, M. New Energy and Industrial Technology Development Organization, Tokyo Institute of Technology, Tokyo, Japan, Lake Toya, Hokkaido, Japan, 1996, pp. 93.

[120.] Roulette, T., Roulette, J., and Pons, S., Results of ICARUS 9 experiments run at IMRA Europe, in *Sixth International Conference on Cold Fusion, Progress in New Hydrogen Energy*, Okamoto, M. New Energy and Industrial Technology Development Organization, Tokyo Institute of Technology, Tokyo, Japan, Lake Toya, Hokkaido, Japan, 1996, pp. 85.

[121.] Preparata, G., Scorletti, M., and Verpelli, M., Isoperibolic calorimetry on modified Fleischmann-Pons cells, *J. Electroanal. Chem.* 411, 9, 1996.

[122.] Oyama, N., Ozaki, M., Tsukiyama, S., Hatozaki, O., and Kunimatsu, K., In situ interferometric microscopy of Pd electrode surfaces and calorimetry during electrolysis of D_2O solution containing sulfur ion, in *Sixth International Conference on Cold Fusion, Progress in New Hydrogen Energy*, Okamoto, M. New Energy and Industrial Technology Development Organization, Tokyo Institute of Technology, Tokyo, Japan, Lake Toya, Hokkaido, Japan, 1996, pp. 234.

[123.] Oya, Y., Ogawa, H., Ono, T., Aida, M., and Okamoto, M., Hydrogen isotope effect induced by neutron irradiation in Pd-LiOD(H) electrolysis, in *Sixth International Conference on Cold Fusion, Progress in New Hydrogen Energy*, Okamoto, M. New Energy and Industrial Technology Development Organization, Tokyo Institute of Technology, Tokyo, Japan, Lake Toya, Hokkaido, Japan, 1996, pp. 370.

[124.] Oriani, R. A., An investigation of anomalous thermal power generation from a proton-conducting oxide, *Fusion Technol.* 30, 281, 1996.

[125.] Niedra, J. M. and Myers, I. T., Replication of the apparent excess heat effect in light water-potassium carbonate-nickel-electrolytic cell, *Infinite Energy* 2 (7), 62, 1996.

[126.] Mizuno, T., Akimoto, T., Azumi, K., Kitaichi, M., and Kurokawa, K., Anomalous heat evolution from a solid-state electrolyte under alternating current in high-temperature D_2 gas, *Fusion Technol.* 29, 385, 1996.

[127.] Miles, M. H., Johnson, K. B., and Imam, M. A., Electrochemical loading of hydrogen and deuterium into palladium and palladium-boron alloys, in *Sixth International Conference on Cold Fusion, Progress in New Hydrogen Energy*, Okamoto, M. New Energy and Industrial Technology Development Organization, Tokyo Institute of Technology, Tokyo, Japan, Lake Toya, Hokkaido, Japan, 1996, pp. 208.

[128.] Miles, M. H. and Johnson, K. B., Improved, open cell, heat conduction, isoperibolic calorimetry, in *Sixth International Conference on Cold Fusion, Progress in New Hydrogen Energy*, Okamoto, M. New Energy and Industrial Technology Development Organization, Tokyo Institute of Technology, Tokyo, Japan, Lake Toya, Hokkaido, Japan, 1996, pp. 496.

[129.] Lonchampt, G., Bonnetain, L., and Hieter, P., Reproduction of Fleischmann and Pons experiments, in *Sixth International Conference on Cold Fusion, Progress in New Hydrogen Energy*, Okamoto, M. New Energy and Industrial Technology Development Organization, Tokyo Institute of Technology, Tokyo, Japan, Lake Toya, Hokkaido, Japan, 1996, pp. 113.

[130.] Kamimura, H., Senjuh, T., Miyashita, S., and Asami, N., Excess heat in fuel cell type cells from pure Pd cathodes annealed at high temperatures, in *Sixth International Conference on Cold Fusion, Progress in New Hydrogen Energy*, Okamoto, M. New Energy and Industrial Technology

Development Organization, Tokyo Institute of Technology, Tokyo, Japan, Lake Toya, Hokkaido, Japan, 1996, pp. 45.

[131.] Iwamura, Y., Itoh, T., Gotoh, N., and Toyoda, I., Correlation between behavior of deuterium in palladium and occurance of nuclear reactions observed by simultaneous measurement of excess heat and nuclear products, in *Sixth International Conference on Cold Fusion,Progress in New Hydrogen Energy*, Okamoto, M. Lake Toya, Hokkaido, Japan, Lake Toya, Hokkaido, Japan, 1996, pp. 274.

[132.] Isagawa, S. and Kanda, Y., Mass spectroscopic search for helium in effluent gas and palladium cathodes of D_2O electrolysis cells involving excess power, in *Sixth International Conference on Cold Fusion, Progress in New Hydrogen Energy*, Okamoto, M. New Energy and Industrial Technology Development Organization, Tokyo Institute of Technology, Tokyo, Japan, Lake Toya, Hokkaido, Japan, 1996, pp. 12.

[133.] Dufour, J., Foos, J., Millot, J. P., and Dufour, X., From "cold fusion" to "Hydrex" and "Deutex" states of hydrogen, in *Sixth International Conference on Cold Fusion, Progress in New Hydrogen Energy*, Okamoto, M. New Energy and Industrial Technology Development Organization, Tokyo Institute of Technology, Tokyo, Japan, Lake Toya, Hokkaido, Japan, 1996, pp. 482.

[134.] De Marco, F., De Ninno, A., Frattolillo, A., La Barbera, A., Scaramuzzi, F., and Violante, V., Progress report on the research activities on cold fusion at ENEA Frascati, in *Sixth International Conference on Cold Fusion, Progress in New Hydrogen Energy*, Okamoto, M. New Energy and Industrial Technology Development Organization, Tokyo Institute of Technology, Tokyo, Japan, Lake Toya, Hokkaido, Japan, 1996, pp. 145.

[135.] Cellucci, F., Cignini, P. L., Gigli, G., Gozzi, D., Tomellini, M., Cisbani, E., Frullani, S., Garibaldi, F., Jodice, M., and Urciuoli, G. M., X-ray, heat excess and ^{4}He in the electrochemical confinement of deuterium in palladium, in *Sixth International Conference on Cold Fusion, Progress in New Hydrogen Energy*, Okamoto, M. New Energy and Industrial Technology Development Organization, Tokyo Institute of Technology, Tokyo, Japan, Lake Toya, Hokkaido, Japan, 1996, pp. 3.

[136.] Arata, Y. and Zhang, Y.-C., Achievement of solid-state plasma fusion ("cold fusion"), in *Sixth International Conference on Cold Fusion, Progress in New Hydrogen Energy*, Okamoto, M. New Energy and Industrial Technology Development Organization, Tokyo Institute of Technology, Tokyo, Japan, Lake Toya, Hokkaido, Japan, 1996, pp. 129.

[137.] Arata, Y. and Zhang, Y. C., Achievement of solid-state plasma fusion ("cold fusion"), *Koon Gakkaishi* 21 ((6)), 303 (in Japanese), 1995.

[138.] Zhang, Z., Sun, X., Zhou, W., Zhang, L., Li, B., Wang, M., Yan, B., and Tan, F., Precision calorimetric studies of H_2O electrolysis, *J. Thermal Anal.* 45, 99, 1995.

[139.] Takahashi, R., Synthesis of substance and generation of heat in charcoal cathode in electrolysis of H_2O and D_2O using various alkalihydrooxides, in *5th International Conference on Cold Fusion*, Pons, S. IMRA Europe, Sophia Antipolis Cedex, France, Monte-Carlo, Monaco, 1995, pp. 619.

[140.] Takahashi, A., Inokuchi, T., Chimi, Y., Ikegawa, T., Kaji, N., Nitta, Y., Kobayashi, K., and Taniguchi, M., Experimental correlation between excess heat and nuclear products, in *5th International Conference on Cold Fusion*, Pons, S. IMRA Europe, Sophia Antipolis Cedex, France, Monte-Carlo, Monaco, 1995, pp. 69.

[141.] Samgin, A. L., Finodeyev, O., Tsvetkov, S. A., Andreev, V. S., Khokhlov, V. A., Filatov, E. S., Murygin, I. V., Gorelov, V. P., and Vakarin, S. V., Cold fusion and anomalous effects in deuteron conductors during non-stationary high-temperature electrolysis, in *5th International Conference on Cold Fusion*, Pons, S. IMRA Europe, Sophia Antipolis Cedex, France, Monte-Carlo, Monaco, 1995, pp. 201.

[142.] Ota, K., Yamaki, K., Tanabe, S., Yoshitake, H., and Kamiya, N., Effect of boron for the heat production at the heavy water electrolysis using palladium cathodes, in *5th International Conference on Cold Fusion*, Pons, S. IMRA Europe, Sophia Antipolis Cedex, France, Monte-Carlo, Monaco, 1995, pp. 132.

[143.] Ogawa, H., Yoshida, S., Yoshinaga, Y., Aida, M., and Okamoto, M., Correlation of excess heat and neutron emission in Pd-Li-D electrolysis, in *5th International Conference on Cold Fusion*, Pons, S. IMRA Europe, Sophia Antipolis Cedex, France, Monte-Carlo, Monaco, 1995, pp. 116.

[144.] Noble, G., Dash, J., and McNasser, L., Electrolysis of heavy water with a palladium and sulfate composite, in *5th International Conference on Cold Fusion*, Pons, S. IMRA Europe, Sophia Antipolis Cedex, France, Monte-Carlo, Monaco, 1995, pp. 136.

[145.] Miles, M. H., The extraction of information from an integrating open calorimeter in Fleischmann-Pons effect experiments, in *5th International Conference on Cold Fusion*, Pons, S. IMRA Europe, Sophia Antipolis Cedex, France, Monte-Carlo, Monaco, 1995, pp. 97.

[146.] Karabut, A. B., Kucherov, Y. R., and Savvatimova, I. B., Excess heat measurements in glow discharge using flow "calorimeter-2", in *5th International Conference on Cold Fusion*, Pons, S. IMRA Europe, Sophia Antipolis Cedex, France, Monte-Carlo, Monaco, 1995, pp. 223.

[147.] Isagawa, S., Kanda, Y., and Suzuki, T., Heat production and trial to detect nuclear products from palladium-deuterium electrolysis cells, in *5th International Conference on Cold Fusion*, Pons, S. IMRA Europe, Sophia Antipolis Cedex, France, Monte-Carlo, Monaco, 1995, pp. 124.

[148.] Hasegawa, N., Sumi, M., Takahashi, M., Senjuh, T., Asami, N., Sakai, T., and Shigemitsu, T., Electrolytic deuterium absorption by Pd cathode and a consideration for high D/Pd ratio, in *5th International Conference on Cold Fusion*, Pons, S. IMRA Europe, Sophia Antipolis Cedex, France, Monte-Carlo, Monaco, 1995, pp. 449.

[149.] Gozzi, D., Caputo, R., Cignini, P. L., Tomellini, M., Gigli, G., Balducci, G., Cisbani, E., Frullani, S., Garibaldi, F., Jodice, M., and Urciuoli, G. M., Calorimetric and nuclear byproduct measurements in electrochemical confinement of deuterium in palladium, *J. Electroanal. Chem.* 380, 91, 1995.

[150.] Dufour, J., Foos, J., and Millot, J. P., Excess energy in the system palladium/hydrogen isotopes. Measurement of excess energy per atom hydrogen, in *5th International Conference on Cold Fusion*, Pons, S. IMRA Europe, Sophia Antipolis Cedex, France, Monte-Carlo, Monaco, 1995, pp. 495.

[151.] Cravens, D., A report on testing the Patterson power cell, *Infinite Energy* 1 (1), 21, 1995.

[152.] Cravens, D., Flowing electrolyte calorimetry, in *5th International Conference on Cold Fusion*, Pons, S. IMRA Europe, Sophia Antipolis Cedex, France, Monte-Carlo, Monaco, 1995, pp. 79.

[153.] Celani, F., Spallone, A., Tripodi, P., Petrocch, A., Di Gioacchino, D., Marini, P., Di Stefano, V., Pace, S., and Mancini, A., High power µs pulsed electrolysis using palladium wires: evidence for a possible "phase" transition under deuterium overloaded conditions and related excess heat, in *5th International Conference on Cold Fusion*, Pons, S. IMRA Europe, Sophia Antipolis Cedex, France, Monte-Carlo, Monaco, 1995, pp. 57.

[154.] Biberian, J.-P., Excess heat measurements in AlLaO$_3$ doped with deuterium, in *5th International Conference on Cold Fusion*, Pons, S. IMRA Europe, Sophia Antipolis Cedex, France, Monte-Carlo, Monaco, 1995, pp. 49.

[155.] Bertalot, L., De Ninno, A., De Marco, F., La Barbera, A., Scaramuzzi, F., and Violante, V., Power excess production in electrolysis experiments at ENEA Frascati, in *5th International Conference on Cold Fusion*, Pons, S. IMRA Europe, Sophia Antipolis Cedex, France, Monte-Carlo, Monaco, 1995, pp. 34.

[156.] Storms, E., Some characteristics of heat production using the "cold fusion" effect, *Trans. Fusion Technol.* 26 (4T), 96, 1994.

[157.] Storms, E., Some characteristics of heat production using the "cold fusion" effect, in *Fourth International Conference on Cold Fusion* Electric Power Research Institute 3412 Hillview Ave., Palo Alto, CA 94304, Lahaina, Maui, 1993, pp. 4.

[158.] Notoya, R., Noya, Y., and Ohnishi, T., Tritium generation and large excess heat evolution by electrolysis in light and heavy water-potassium carbonate solutions with nickel electrodes, *Fusion Technol.* 26, 179, 1994.

[159.] Miles, M., Bush, B. F., and Lagowski, J. J., Anomalous effects involving excess power, radiation, and helium production during D_2O electrolysis using palladium cathodes, *Fusion Technol.* 25, 478, 1994.

[160.] Miles, M. H. and Bush, B. F., Heat and helium measurements in deuterated palladium, *Trans. Fusion Technol.* 26 (#4T), 156, 1994.

[161.] McKubre, M. C. H., Crouch-Baker, S., Rocha-Filho, R. C., Smedley, S. I., Tanzella, F. L., Passell, T. O., and Santucci, J., Isothermal flow calorimetric investigations of the D/Pd and H/Pd systems, *J. Electroanal. Chem.* 368, 55, 1994.

[162.] Focardi, S., Habel, R., and Piantelli, F., Anomalous heat production in Ni-H systems, *Nuovo Cimento* 107A, 163, 1994.

[163.] Bush, R. T. and Eagleton, R. D., Evidence for electrolytically induced transmutation and radioactivity correlated with excess heat in electrolytic cells with light water rubidium salt electrolytes, *Trans. Fusion Technol.* 26 (4T), 344, 1994.

[164.] Bockris, J. O. M., Sundaresan, R., Letts, D., and Minevski, Z., Triggering of heat and sub-surface changes in Pd-D systems, in *Fourth International Conference on Cold Fusion* Electric Power Research Institute 3412 Hillview Ave., Palo Alto, CA 94304, Lahaina, Maui, 1993, pp. 1.

[165.] Bockris, J. O. M., Sundaresan, R., Minevski, Z., and Letts, D., Triggering of heat and sub-surface changes in Pd-D systems, *Trans. Fusion Technol.* 26 (#4T), 267, 1994.

[166.] Arata, Y. and Zhang, Y.-C., A new energy caused by "spillover-deuterium", *Proc. Japan. Acad.* 70 ser. B, 106, 1994.

[167.] Zhang, Q. Q., Gou, Z., Zhu, J., Lou, F., Liu, J. S., Miao, B., Ye, A., and Cheng, S., The excess heat experiments on cold fusion in a titanium lattice, in *Fourth International Conference on Cold Fusion* Electric Power Research Institute 3412 Hillview Ave., Palo Alto, CA 94304, Lahaina, Maui, 1993, pp. 17.

[168.] Storms, E., Measurements of excess heat from a Pons-Fleischmann-type electrolytic cell using palladium sheet, *Fusion Technol.* 23, 230, 1993.

[169.] Storms, E., Measurement of excess heat from a Pons-Fleischmann type electrolytic cell, in *Third International Conference on Cold Fusion, "Frontiers of Cold Fusion"*, Ikegami, H. Universal Academy Press, Inc., Tokyo, Japan, Nagoya Japan, 1992, pp. 21.

[170.] Ramamurthy, H., Srinivasan, M., Mukherjee, U. K., and Adi Babu, P., Further studies on excess heat generation in $Ni-H_2O$ electrolytic cells, in *Fourth International Conference on Cold Fusion* Electric Power Research Institute 3412 Hillview Ave., Palo Alto, CA 94304, Lahaina, Maui, 1993, pp. 15.

[171.] Pons, S. and Fleischmann, M., Heat after death, in *Fourth International Conference on Cold Fusion* Electric Power Research Institute 3412 Hillview Ave., Palo Alto, CA 94304, Lahaina, Maui, 1993, pp. 8.

[172.] Pons, S. and Fleischmann, M., Heat after death, *Trans. Fusion Technol.* 26 (4T), 97, 1994.

[173.] Ota, K., Yoshitake, H., Yamazaki, O., Kuratsuka, M., Yamaki, K., Ando, K., Iida, Y., and Kamiya, N., Heat measurement of water electrolysis using Pd cathode and the electrochemistry, in *Fourth International Conference on Cold Fusion* Electric Power Research Institute 3412 Hillview Ave., Palo Alto, CA 94304, Lahaina, Maui, 1993, pp. 5.

[174.] Okamoto, M., Yoshinaga, Y., Aida, M., and Kusunoki, T., Excess heat generation, voltage deviation, and neutron emission in D_2O-LiOD systems, in *Fourth International Conference on Cold Fusion* Electric Power Research Institute 3412 Hillview Ave., Palo Alto, CA 94304, Lahaina, Maui, 1993, pp. 3.

[175.] Okamoto, M., Yoshinaga, Y., Aida, M., and Kusunoki, T., Excess heat generation, voltage deviation, and neutron emission in D_2O-LiOD systems, *Trans. Fusion Technol.* 26 (4T), 176, 1994.

[176.] Ohmori, T. and Enyo, M., Excess heat evolution during electrolysis of H_2O with nickel, gold, silver, and tin cathodes, *Fusion Technol.* 24, 293, 1993.

[177.] Ohmori, T. and Enyo, M., Excess heat production during electrolysis of H_2O on Ni, Au, Ag and Sn electrodes in alkaline media, in *Third International Conference on Cold Fusion, "Frontiers of*

Cold Fusion", Ikegami, H. Universal Academy Press, Inc., Tokyo, Japan, Nagoya Japan, 1992, pp. 427.

[178.] Mizuno, T., Enyo, M., Akimoto, T., and Azumi, K., Anomalous heat evolution from $SrCeO_3$-type proton conductors during absorption/desorption in alternate electric field, in *Fourth International Conference on Cold Fusion* Electric Power Research Institute 3412 Hillview Ave., Palo Alto, CA 94304, Lahaina, Maui, 1993, pp. 14.

[179.] Miles, M. H. and Bush, B. F., Heat and helium measurements in deuterated palladium, in *Fourth International Conference on Cold Fusion* Electric Power Research Institute 3412 Hillview Ave., Palo Alto, CA 94304, Lahaina, Maui, 1993, pp. 6.

[180.] Miles, M. H., Hollins, R. A., Bush, B. F., Lagowski, J. J., and Miles, R. E., Correlation of excess power and helium production during D_2O and H_2O electrolysis using palladium cathodes, *J. Electroanal. Chem.* 346, 99, 1993.

[181.] Hugo, M., A home cold fusion experiment, in *Fourth International Conference on Cold Fusion* Electric Power Research Institute 3412 Hillview Ave., Palo Alto, CA 94304, Lahaina, Maui, 1993, pp. 22.

[182.] Hasegawa, N., Hayakawa, N., Tsuchida, Y., Yamamoto, Y., and Kunimatsu, K., Observation of excess heat during electrolysis of 1 M LiOD in a fuel cell type closed cell, in *Fourth International Conference on Cold Fusion* Electric Power Research Institute 3412 Hillview Ave., Palo Alto, CA 94304, Lahaina, Maui, 1993, pp. 3.

[183.] Gozzi, D., Caputo, R., Cignini, P. L., Tomellini, M., Gigli, G., Balducci, G., Cisbani, E., Frullani, S., Garibaldi, F., Jodice, M., and Urciuoli, G. M., Excess heat and nuclear product measurements in cold fusion electrochemical cells, in *Fourth International Conference on Cold Fusion* Electric Power Research Institute 3412 Hillview Ave., Palo Alto, CA 94304, Lahaina, Maui, 1993, pp. 2.

[184.] Fleischmann, M. and Pons, S., Calorimetry of the $Pd-D_2O$ system: from simplicity via complications to simplicity, *Phys. Lett. A* 176, 118, 1993.

[185.] Dufour, J., Foos, J., and Millot, J. P., Cold fusion by sparking in hydrogen isotopes. Energy balances and search for fusion by-products. A strategy to prove the reality of cold fusion, in *Fourth International Conference on Cold Fusion* Electric Power Research Institute 3412 Hillview Ave., Palo Alto, CA 94304, Lahaina, Maui, 1993, pp. 9.

[186.] Dufour, J., Cold fusion by sparking in hydrogen isotopes, *Fusion Technol.* 24, 205, 1993.

[187.] Dufour, J., Foos, J., and Millot, J. P., Cold fusion by sparking in hydrogen isotopes. Energy balances and search for fusion by-products, *Trans. Fusion Technol.* 26 (#4T), 375, 1994.

[188.] Criddle, E. E., Evidence of agglomeration and syneresis in regular and excess heat cells in water, in *Fourth International Conference on Cold Fusion* Electric Power Research Institute 3412 Hillview Ave., Palo Alto, CA 94304, Lahaina, Maui, 1993, pp. 32.

[189.] Celani, F., Spallone, A., Tripodi, P., Nuvoli, A., Petrocchi, A., Di Gioacchino, D., Boutet, M., Marini, P., and Di Stefano, V., High power μs pulsed electrolysis for large deuterium loading on Pd plates, in *Fourth International Conference on Cold Fusion* Electric Power Research Institute 3412 Hillview Ave., Palo Alto, CA 94304, Lahaina, Maui, 1993, pp. 22.

[190.] Celani, F., Spallone, A., Tripodi, P., Nuvoli, A., Petrocchi, A., Di Gioacchino, D., Boutet, M., Marini, P., and Di Stefano, V., High power μs pulsed electrolysis for large deuterium loading in Pd plates, *Trans. Fusion Technol.* 26 (#4T), 127, 1994.

[191.] Bush, R. T. and Eagleton, R. D., Calorimetric studies for several light water electrolytic cells with nickel fibrex cathodes and electrolytes with alkali salts of potassium, rubidium, and cesium, in *Fourth International Conference on Cold Fusion* Electric Power Research Institute 3412 Hillview Ave., Palo Alto, CA 94304, Lahaina, Maui, 1993, pp. 13.

[192.] Bertalot, L., De Marco, F., De Ninno, A., La Barbera, A., Scaramuzzi, F., Violante, V., and Zeppa, P., Study of deuterium charging in palladium by the electrolysis of heavy water: heat excess production, *Nuovo Cimento Soc. Ital. Fis. D* 15, 1435, 1993.

[193.] Bertalot, L., De Marco, F., De Ninno, A., La Barbera, A., Scaramuzzi, F., Violante, V., and Zeppa, P., Study of deuterium charging in palladium by the electrolysis of heavy water: Search for heat excess and nuclear ashes, in *Third International Conference on Cold Fusion, "Frontiers of Cold Fusion"*, Ikegami, H. Universal Academy Press, Inc., Tokyo, Japan, Nagoya Japan, 1992, pp. 365.

[194.] Bertalot, L., Bettinali, L., De Marco, F., Violante, V., De Logu, P., Dikonimos, T., and La Barbera, A., Analysis of tritium and heat excess in electrochemical cells with Pd cathodes, in *Second Annual Conference on Cold Fusion, "The Science of Cold Fusion"*, Bressani, T., Del Giudice, E., andPreparata, G. Societa Italiana di Fisica, Bologna, Italy, Como, Italy, 1991, pp. 3.

[195.] Bazhutov, Y. N., Chertov, Y. P., Krivoshein, A. A., Skuratnik, Y. B., and Khokhlov, N. I., Excess heat observation during electrolysis of Cs_2CO_3 solution in light water, in *Fourth International Conference on Cold Fusion* Electric Power Research Institute 3412 Hillview Ave., Palo Alto, CA 94304, Lahaina, Maui, 1993, pp. 24.

[196.] Aoki, T., Kurata, Y., and Ebihara, H., Study of concentrations of helium and tritium in electrolytic cells with excess heat generation, in *Fourth International Conference on Cold Fusion* Electric Power Research Institute 3412 Hillview Ave., Palo Alto, CA 94304, Lahaina, Maui, 1993, pp. 23.

[197.] Aoki, T., Kurata, Y., Ebihara, H., and Yoshikawa, N., Helium and tritium concentrations in electrolytic cells, *Trans. Fusion Technol.* 26 (4T), 214, 1994.

[198.] Yuan, L. J., Wan, C. M., Liang, C. Y., and Chen, S. K., Neutron monitoring on cold-fusion experiments, in *Third International Conference on Cold Fusion, "Frontiers of Cold Fusion"*, Ikegami, H. Universal Academy Press, Inc., Tokyo, Japan, Nagoya Japan, 1992, pp. 461.

[199.] Wan, C. M., Chen, S. K., Liang, C. Y., Linn, C. J., Chu, S. B., and Wan, C. C., Anomalous heat generation/absorption in $Pd/Pd/LiOD/D_2O/Pd$ electrolysis system, in *Third International Conference on Cold Fusion, "Frontiers of Cold Fusion"*, Ikegami, H. Universal Academy Press, Inc., Tokyo, Japan, Nagoya Japan, 1992, pp. 389.

[200.] Takahashi, A., Mega, A., Takeuchi, T., Miyamaru, H., and Iida, T., Anomalous excess heat by D_2O/Pd cell under L-H mode electrolysis, in *Third International Conference on Cold Fusion, "Frontiers of Cold Fusion"*, Ikegami, H. Universal Academy Press, Inc., Tokyo, Japan, Nagoya Japan, 1992, pp. 79.

[201.] Takahashi, A., Iida, T., Takeuchi, T., and Mega, A., Excess heat and nuclear products by D_2O/Pd electrolysis and multibody fusion, *Int. J. Appl. Electromagn. Mater.* 3, 221, 1992.

[202.] Srinivasan, M., Shyam, A., Sankaranarayanan, T. K., Bajpai, M. B., Ramamurthy, H., Mukherjee, U. K., Krishnan, M. S., Nayar, M. G., and Naik, Y. P., Tritium and excess heat generation during electrolysis of aqueous solutions of alkali salts with nickel cathode, in *Third International Conference on Cold Fusion, "Frontiers of Cold Fusion"*, Ikegami, H. Universal Academy Press, Inc., Tokyo, Japan, Nagoya Japan, 1992, pp. 123.

[203.] Ray, M. K. S., Saini, R. D., Das, D., Chattopadhyay, G., Parthasarathy, R., Garg, S. P., Venkataramani, R., Sen, B. K., Iyengar, T. S., Kutty, K. K., Wagh, D. N., Bajpai, H. N., and Iyer, C. S. P., The Fleischmann-Pons phenomenon - a different perspective, *Fusion Technol.* 22, 395, 1992.

[204.] Oyama, N., Terashima, T., Kasahara, S., Hatozaki, O., Ohsaka, T., and Tatsuma, T., Electrochemical calorimetry of D_2O electrolysis using a palladium cathode in a closed cell system, in *Third International Conference on Cold Fusion, "Frontiers of Cold Fusion"*, Ikegami, H. Universal Academy Press, Inc., Tokyo, Japan, Nagoya Japan, 1992, pp. 67.

[205.] Ota, K., Kuratsuka, M., Ando, K., Iida, Y., Yoshitake, H., and Kamiya, N., Heat production at the heavy water electrolysis using mechanically treated cathode, in *Third International Conference on Cold Fusion, "Frontiers of Cold Fusion"*, Ikegami, H. Universal Academy Press, Inc., Tokyo, Japan, Nagoya Japan, 1992, pp. 71.

[206.] Notoya, R. and Enyo, M., Excess heat production in electrolysis of potassium carbonate solution with nickel electrodes, in *Third International Conference on Cold Fusion, "Frontiers of Cold Fusion"*, Ikegami, H. Universal Academy Press, Inc., Tokyo, Japan, Nagoya Japan, 1992, pp. 421.

[207.] Noninski, V. C., Excess heat during the electrolysis of a light water solution of K_2CO_3 with a nickel cathode, *Fusion Technol.* 21, 163, 1992.

208. Mizuno, T., Akimoto, T., Azumi, K., and Enyo, M., Cold fusion reaction products and behavior of deuterium absorption in Pd electrode, in *Third International Conference on Cold Fusion, "Frontiers of Cold Fusion"*, Ikegami, H. Universal Academy Press, Inc., Tokyo, Japan, Nagoya Japan, 1992, pp. 373.

209. Miyamaru, H. and Takahashi, A., Periodically current-controlled electrolysis of D_2O/Pd system for excess heat production, in *Third International Conference on Cold Fusion, "Frontiers of Cold Fusion"*, Ikegami, H. Universal Academy Press, Inc., Tokyo, Japan, Nagoya Japan, 1992, pp. 393.

210. McKubre, M. C. H., Crouch-Baker, S., Riley, A. M., Smedley, S. I., and Tanzella, F. L., Excess power observations in electrochemical studies of the D/Pd system; the influence of loading, in *Third International Conference on Cold Fusion, "Frontiers of Cold Fusion"*, Ikegami, H. Universal Academy Press, Inc., Tokyo, Japan, Nagoya Japan, 1992, pp. 5.

211. Kunimatsu, K., Hasegawa, N., Kubota, A., Imai, N., Ishikawa, M., Akita, H., and Tsuchida, Y., Deuterium loading ratio and excess heat generation during electrolysis of heavy water by palladium cathode in a closed cell using a partially immersed fuel cell anode, in *Third International Conference on Cold Fusion, "Frontiers of Cold Fusion"*, Ikegami, H. Universal Academy Press, Inc., Tokyo, Japan, Nagoya Japan, 1992, pp. 31.

212. Hasegawa, N., Kunimatsu, K., Ohi, T., and Terasawa, T., Observation of excess heat during electrolysis of 1 M LiOD in a fuel cell type closed cell, in *Third International Conference on Cold Fusion, "Frontiers of Cold Fusion"*, Ikegami, H. Universal Academy Press, Inc., Tokyo, Japan, Nagoya Japan, 1992, pp. 377.

213. Kobayashi, M., Imai, N., Hasegawa, N., Kubota, A., and Kunimatsu, K., Measurements of D/Pd and excess heat during electrolysis of LiOD in a fuel-cell type closed cell using a palladium sheet cathode, in *Third International Conference on Cold Fusion, "Frontiers of Cold Fusion"*, Ikegami, H. Universal Academy Press, Inc., Tokyo, Japan, Nagoya Japan, 1992, pp. 385.

214. Karabut, A. B., Kucherov, Y. R., and Savvatimova, I. B., Nuclear product ratio for glow discharge in deuterium, *Phys. Lett. A* 170, 265, 1992.

215. Karabut, A. B., Kucherov, Y. R., and Savvatimova, I. B., The investigation of deuterium nuclei fusion at glow discharge cathode, *Fusion Technol.* 20, 924, 1991.

216. Isagawa, S., Kanada, Y., and Suzuki, T., Search for excess heat, neutron emission and tritium yield from electrochemically charged palladium in D_2O, in *Third International Conference on Cold Fusion, "Frontiers of Cold Fusion"*, Ikegami, H. Universal Academy Press, Inc., Tokyo, Japan, Nagoya Japan, 1992, pp. 477.

217. Gozzi, D., Cignini, P. L., Caputo, R., Tomellini, M., Balducci, G., Gigli, G., Cisbani, E., Frullani, S., Garibaldi, F., Jodice, M., and Urciuoli, G. M., Experiments with global detection of cold fusion byproducts, in *Third International Conference on Cold Fusion, "Frontiers of Cold Fusion"*, Ikegami, H. Universal Academy Press, Inc., Tokyo, Japan, Nagoya Japan, 1992, pp. 155.

218. Celani, F., Spallone, A., Tripodi, P., and Nuvoli, A., Measurement of excess heat and tritium during self-biased pulsed electrolysis of Pd-D_2O, in *Third International Conference on Cold Fusion, "Frontiers of Cold Fusion"*, Ikegami, H. Universal Academy Press, Inc., Tokyo, Japan, Nagoya Japan, 1992, pp. 93.

219. Bush, R. T., A light water excess heat reaction suggests that 'cold fusion' may be 'alkali-hydrogen fusion', *Fusion Technol.* 22, 301, 1992.

220. Bush, R. T. and Eagleton, R. D., Experimental studies supporting the transmission resonance model for cold fusion in light water: II. Correlation of X-Ray emission with excess power, in *Third International Conference on Cold Fusion, "Frontiers of Cold Fusion"*, Ikegami, H. Universal Academy Press, Inc., Tokyo, Japan, Nagoya Japan, 1992, pp. 409.

221. Yun, K.-S., Ju, J.-B., Cho, B.-W., Cho, W.-I., and Park, S.-Y., Calorimetric observation of heat production during electrolysis of 0.1 M LiOD + D_2O solution, *J. Electroanal. Chem.* 306, 279, 1991.

222. Will, F. G., Cedzynska, K., Yang, M. C., Peterson, J. R., Bergeson, H. E., Barrowes, S. C., West, W. J., and Linton, D. C., Studies of electrolytic and gas phase loading of palladium with deuterium, in *Second Annual Conference on Cold Fusion, "The Science of Cold Fusion"*, Bressani,

T., Giudice, E. D., andPreparata, G. Societa Italiana di Fisica, Bologna, Italy, Como, Italy, 1991, pp. 373.

[223.] Bush, B. F., Lagowski, J. J., Miles, M. H., and Ostrom, G. S., Helium production during the electrolysis of D_2O in cold fusion experiments, *J. Electroanal. Chem.* 304, 271, 1991.

[224.] Szpak, S., Mosier-Boss, P. A., and Smith, J. J., On the behavior of Pd deposited in the presence of evolving deuterium, *J. Electroanal. Chem.* 302, 255, 1991.

[225.] Noninski, V. C. and Noninski, C. I., Determination of the excess energy obtained during the electrolysis of heavy water, *Fusion Technol.* 19, 364, 1991.

[226.] Eagleton, R. D. and Bush, R. T., Calorimetric experiments supporting the transmission resonance model for cold fusion, *Fusion Technol.* 20, 239, 1991.

[227.] Mills, R. L. and Kneizys, P., Excess heat production by the electrolysis of an aqueous potassium carbonate electrolyte and the implications for cold fusion, *Fusion Technol.* 20, 65, 1991.

[228.] McKubre, M. C. H., Rocha-Filho, R. C., Smedley, S. I., Tanzella, F. L., Crouch-Baker, S., Passell, T. O., and Santucci, J., Isothermal flow calorimetric investigations of the D/Pd system, in *Second Annual Conference on Cold Fusion, "The Science of Cold Fusion"*, Bressani, T., Del Giudice, E., andPreparata, G. Societa Italiana di Fisica, Bologna, Italy, Como, Italy, 1991, pp. 419.

[229.] Liaw, B. Y., Tao, P.-L., Turner, P., and Liebert, B. E., Elevated-temperature excess heat production in a Pd + D system, *J. Electroanal. Chem.* 319, 161, 1991.

[230.] Liaw, B. Y., Tao, P.-L., Turner, P., and Liebert, B. E., Elevated temperature excess heat production using molten-salt electrochemical techniques, in *8th World Hydrogen Energy Conf* Hawaii Natural Energy Institute, 2540 Dole St., Holmes Hall 246, Honolulu, HI 96822, Honolulu, HI, 1990, pp. 49.

[231.] Zhang, Z. L., Yan, B. Z., Wang, M. G., Gu, J., and Tan, F., Calorimetric observation combined with the detection of particle emissions during the electrolysis of heavy water, in *Anomalous Nuclear Effects in Deuterium/Solid Systems, "AIP Conference Proceedings 228"*, Jones, S., Scaramuzzi, F., andWorledge, D. American Institute of Physics, New York, Brigham Young Univ., Provo, UT, 1990, pp. 572.

[232.] Scott, C. D., Greenbaum, E., Michaels, G. E., Mrochek, J. E., Newman, E., Petek, M., and Scott, T. C., Preliminary investigation of possible low-temperature fusion, *J. Fusion Energy* 9 (2), 115, 1990.

[233.] Scott, C. D., Mrochek, J. E., Scott, T. C., Michaels, G. E., Newman, E., and Petek, M., The initiation of excess power and possible products of nuclear interactions during the electrolysis of heavy water, in *The First Annual Conference on Cold Fusion*, Will, F. National Cold Fusion Institute, University of Utah Research Park, Salt Lake City, Utah, 1990, pp. 164.

[234.] Scott, C. D., Mrochek, J. E., Scott, T. C., Michaels, G. E., Newman, E., and Petek, M., Measurement of excess heat and apparent coincident increases in the neutron and gamma-ray count rates during the electrolysis of heavy water, *Fusion Technol.* 18, 103, 1990.

[235.] Schreiber, M., Gür, T. M., Lucier, G., Ferrante, J. A., Chao, J., and Huggins, R. A., Recent measurements of excess energy production in electrochemical cells containing heavy water and palladium, in *The First Annual Conference on Cold Fusion*, Will, F. National Cold Fusion Institute, University of Utah Research Park, Salt Lake City, Utah, 1990, pp. 44.

[236.] Gür, T. M., Schreiber, M., Lucier, G., Ferrante, J. A., and Huggins, R. A., Experimental Considerations Involved in the Generation of Excess Power as a Result of the Electrochemical Insertion of Hydrogen and Deuterium in Palladium, in *8th World Hydrogen Energy Conf.* Hawaii Natural Energy Insitute, 2540 Dole St., Holmes Hall 246, Honolulu, HI 96822, Honolulu, HI, 1990, pp. 31.

[237.] Pons, S. and Fleischmann, M., Calorimetry of the palladium-deuterium system, in *The First Annual Conference on Cold Fusion*, Will, F. National Cold Fusion Institute, University of Utah Research Park, Salt Lake City, Utah, 1990, pp. 1.

[238.] Fleischmann, M., Pons, S., Anderson, M. W., Li, L. J., and Hawkins, M., Calorimetry of the palladium-deuterium-heavy water system, *J. Electroanal. Chem.* 287, 293, 1990.

[239.] Yang, C.-S., Liang, C.-Y., Perng, T. P., Yuan, L.-J., Wang, C.-M., and Wang, C.-C., Observation of excess heat and tritium on electrolysis of D₂O, in *8th World Hydrogen Energy Conf.* Hawaii Natural Energy Insitute, 2540 Dole St., Holmes Hall 246, Honolulu, HI 96822, Honolulu, HI, 1990, pp. 95.

[240.] Guruswamy, S. and Wadsworth, M. E., Metallurgical Aspects in Cold Fusion Experiments, in *The First Annual Conference on Cold Fusion*, Will, F. National Cold Fusion Institute, University of Utah Research Park, Salt Lake City, Utah, 1990, pp. 314.

[241.] Lewis, D. and Sköld, K., A phenomenological study of the Fleischmann-Pons effect, *J. Electroanal. Chem.* 294, 275, 1990.

[242.] Oriani, R. A., Nelson, J. C., Lee, S.-K., and Broadhurst, J. H., Calorimetric measurements of excess power output during the cathodic charging of deuterium into palladium, *Fusion Technol.* 18, 652, 1990.

[243.] Miles, M. H., Park, K. H., and Stilwell, D. E., Electrochemical calorimetric studies of the cold fusion effect, in *The First Annual Conference on Cold Fusion*, Will, F. National Cold Fusion Institute, University of Utah Research Park, Salt Lake City, Utah, 1990, pp. 328.

[244.] Miles, M. H., Park, K. H., and Stilwell, D. E., Electrochemical calorimetric evidence for cold fusion in the palladium-deuterium system, *J. Electroanal. Chem.* 296, 241, 1990.

[245.] Hutchinson, D. P., Bennet, C. A., Richards, R. K., Bullock, J., and Powell, G. L., Report No. ORNL/TM-11356, 1990.

[246.] Powell, G. L., Bullock, I. J. S., Hallman, R. L., Horton, P. J., and Hutchinson, D. P., The preparation of palladium for cold fusion experiments, *J. Fusion Energy* 9 (3), 355, 1990.

[247.] Bullock, J. S., Powell, G. L., and Hutchinson, D. P., Electrochemical factors in cold fusion experiments, *J. Fusion Energy* 9, 275, 1990.

[248.] McKubre, M. C. H., Rocha-Filho, R. C., Smedley, S., Tanzella, F. L., Chao, J., Chexal, B., Passell, T. O., and Santucci, J., Calorimetry and electrochemistry in the D/Pd system, in *The First Annual Conference on Cold Fusion*, Will, F. National Cold Fusion Institute, University of Utah Research Park, Salt Lake City, Utah, 1990, pp. 20.

[249.] Appleby, A. J., Kim, Y. J., Murphy, O. J., and Srinivasan, S., Anomalous calorimetric results during long-term evolution of deuterium on palladium from alkaline deuteroxide electrolyte, in *The First Annual Conference on Cold Fusion*, Will, F. National Cold Fusion Institute, University of Utah Research Park, Salt Lake City, Utah, 1990, pp. 32.

[250.] Appleby, A. J., Srinivasan, S., Kim, Y. J., Murphy, O. J., and Martin, C. R., Evidence for excess heat generation rates during electrolysis of D₂O in LiOD using a palladium cathode-A microcalorimetric study, in *Workshop on Cold Fusion Phenomena*, Santa Fe, NM, 1989.

[251.] Kainthla, R. C., Velev, O. A., Kaba, L., Lin, G. H., Packham, N. J. C., Szklarczyk, M., Wass, J., and Bockris, J. O. M., Sporadic observation of the Fleischmann-Pons heat effect, *Electrochim. Acta* 34, 1315, 1989.

[252.] Belzner, A., Bischler, U., Crouch-Baker, S., Guer, T. M., Lucier, G., Schreiber, M., and Huggins, R. A., Two fast mixed-conductor systems: deuterium and hydrogen in palladium - thermal measurements and experimental considerations, *J. Fusion Energy* 9 (2), 219, 1990.

[253.] Belzner, A., Bischler, U., Crouch-Baker, S., Guer, T. M., Lucier, G., Schreiber, M., and Huggins, R. A., Recent results on mixed conductors containing hydrogen or deuterium, *Solid State Ionics* 40/41, 519, 1990.

[254.] Santhanam, K. S. V., Ragarajan, J., Braganza, O. N., Haram, S. K., Limaye, N. M., and Mandal, K. C., Electrochemically initiated cold fusion of deuterium, *Indian J. Technol.* 27, 175, 1989.

[255.] Fleischmann, M., Pons, S., and Hawkins, M., Electrochemically Induced Nuclear Fusion of Deuterium, *J. Electroanal. Chem.* 261, 301 and errata in Vol. 263, 1989.

[256.] Jow, T. R., Plichta, E., Walker, C., Slane, S., and Gilman, S., Calorimetric studies of deuterated Pd electrodes, *J. Electrochem. Soc.* 137 (8), 2473, 1990.

[257.] Armstrong, R. D., Charles, E. A., Fells, I., Molyneux, L., and Todd, M., Some aspects of thermal energy generation during the electrolysis of D₂O using a palladium cathode, *Electrochim. Acta* 34, 1319, 1989.

[258.] Blaser, J. P., Haas, O., Ptitjean, C., Barbero, C., Bertl, W., Lou, K., Mathias, M., Baumann, P., Daniel, H., Hartmann, J., Hechtl, E., Ackerbauer, P., Kammel, P., Scrinzi, A., Zmeskal, H., Kozlowski, T., Kipfer, R., Baur, H., Signr, P., and Wieler, R., Experimental investigation of cold fusion phenomena in palladium, *Chimia* 43, 262, 1989.

[259.] Chemla, M., Chevalet, J., and Bury, R., Heat evolution involved with the electrochemical discharge of hydrogen and deuterium on palladium, *C. R. Acad. Sci., Ser. 2* 309, 987 (in French), 1989.

[260.] Chu, C. W., Xue, Y. Y., Meng, R. L., Hor, P. H., Huang, Z. J., and Gao, L., Search for the proposed cold fusion of D in Pd, *Mod. Phys. Lett. B* 3, 753, 1989.

[261.] Divisek, J., Fuerst, L., and Balej, J., Energy balance of D_2O electrolysis with a palladium cathode. Part II. Experimental results, *J. Electroanal. Chem.* 278, 99, 1989.

[262.] Gillespie, D. J., Kamm, G. N., Ehrlich, A. C., and Mart, P. L., A search for anomalies in the palladium-deuterium system, *Fusion Technol.* 16, 526, 1989.

[263.] Shapovalov, V. L., Test for additional heat evolution in electrolysis of heavy water with palladium cathode, *JETP* 50, 117, 1989.

[264.] Alessandrello, A., Bellotti, E., Cattadori, C., Antonione, C., Bianchi, G., Rondinini, S., Torchio, S., Fiorini, E., Guiliani, A., Ragazzi, S., Zanotti, L., and Gatti, C., Search for cold fusion induced by electrolysis in palladium, *Il Nuovo Cimento* A103, 1617, 1990.

[265.] Chemla, M., Chevalet, J., Bury, R., and Perie, M., Experimental investigation of thermal and radiation effects induced by deuterium discharge at the palladium electrode, *J. Electroanal. Chem.* 277, 93, 1990.

[266.] Gottesfeld, S., Anderson, R. E., Baker, D. A., Bolton, R. D., Butterfield, K. B., Garzon, F. H., Goulding, C. A., Johnson, M. W., Leonard, E. M., Springer, T. E., and Zawodzinski, T., Experiments and nuclear measurements in search of cold fusion processes, *J. Fusion Energy* 9 (3), 287, 1990.

[267.] Godshall, N. A., Roth, E. P., Kelly, M. J., Guilinger, T. R., and Ewing, R. I., Calorimetric and thermodynamic analysis of palladium-deuterium electrochemical cells, *J. Fusion Energy* 9, 229, 1990.

[268.] Hayden, M. E., Närger, U., Booth, J. L., Whitehead, L. A., Hardy, W. N., Carolan, J. F., Wishnow, E. H., Balzarini, D. A., Brewer, J. H., and Blake, C. C., High precision calorimetric search for evidence of cold fusion using in situ catalytic recombination of evolved gases, *J. Fusion Energy* 9 (2), 161, 1990.

[269.] Ilic, R., Rant, J., Sutej, T., Dobersek, M., Kristov, E., Skvarc, J., and Kozelj, M., Investigation of the deuterium-deuterium fusion reaction in cast, annealed, and cold-rolled palladium, *Fusion Technol.* 18, 505, 1990.

[270.] Storms, E., What conditions are required to initiate the LENR effect?, in *Tenth International Conference on Cold Fusion*, Hagelstein, P. L. andChubb, S. R. World Scientific Publishing Co., Cambridge, MA, 2003, pp. 285.

[271.] Szpak, S., Mosier-Boss, P. A., Dea, J., and Gordon, F., Polarized D+/Pd-D_2O system: hot spots and "mini-explosions", in *Tenth International Conference on Cold Fusion*, Hagelstein, P. L. andChubb, S. R. World Scientific Publishing Co., Cambridge, MA, 2003, pp. 13.

[272.] Szpak, S., Mosier-Boss, P. A., Young, C., and Gordon, F., Evidence of nuclear reactions in the Pd lattice, *Naturwiss.* 92, 394, 2005.

[273.] Shirai, O., Kihara, S., Sohrin, Y., and Matsui, M., Some experimental results relating to cold nuclear fusion, *Bull. Inst. Chem. Res., Kyoto Univ.* 69, 550, 1991.

[274.] Zhang, Z., Liu, F., Liu, M., Wang, Z., Zhong, F., and Wu, F., Calorimetric studies on the electrorefining process of copper, *J. Thermal Anal.* 50, 89, 1997.

[275.] Biberian, J. P., Excess heat during co-deposition of palladium and deuterium, in *ASTI-5* www.iscmns.org/, Asti, Italy, 2004.

[276.] Liaw, B. Y., Tao, P.-L., and Liebert, B. E., Helium analysis of palladium electrodes after molten salt electrolysis, *Fusion Technol.* 23, 92, 1993.

277. Okamoto, H. and Nezu, S., Measurements of hydrogen loading ratio of Pd anodes polarized in LiH-LiCl-KCl molten salt systems, *Trans. Fusion Technol.* 26 (#4T), 59, 1994.

278. Karabut, A. B., Kucherov, Y. R., and Savvatimova, I. B., Cold fusion observation at gas-discharge device cathode, in *Anniversary Specialist Conf. on Nucl. Power Eng. in Space*, Obninsk, Russia, 1990.

279. Karabut, A. B., Kucherov, Y. R., and Savvatimova, I. B., Nuclear reactions at the cathode in a gas discharge, *Sov. Tech. Phys. Lett.* 16 (12), 46353, 1990.

280. Savvatimova, I., Kucherov, Y., and Karabut, A., Cathode material change after deuterium glow discharge experiments, in *Fourth International Conference on Cold Fusion* Electric Power Research Institute 3412 Hillview Ave., Palo Alto, CA 94304, Lahaina, Maui, 1993, pp. 16.

281. Savvatimova, I., Kucherov, Y., and Karabut, A., Cathode material change after deuterium glow discharge experiments, *Trans. Fusion Technol.* 26 (4T), 389, 1994.

282. Savvatimova, I. and Karabut, A. B., Radioactivity of the cathode samples after gow discharge, in *5th International Conference on Cold Fusion*, Pons, S. IMRA Europe, Sophia Antipolis Cedex, France, Monte-Carlo, Monaco, 1995, pp. 209.

283. Savvatimova, I. and Karabut, A., Nuclear reaction products registration on the cathode after glow discharge., in *5th International Conference on Cold Fusion*, Pons, S. IMRA Europe, Sophia Antipolis Cedex, France, Monte-Carlo, Monaco, 1995, pp. 213.

284. Karabut, A. B., Kolomeychenko, S. A., and Savvatimova, I. B., High energy phenomena in glow discharge experiments, in *5th International Conference on Cold Fusion*, Pons, S. IMRA Europe, Sophia Antipolis Cedex, France, Monte-Carlo, Monaco, 1995, pp. 241.

285. Savvatimova, I. B. and Karabut, A. B., Nuclear reaction products detected at the cathode after a glow discharge in deuterium, *Poverkhnost(Surface)* (1), 63 (in Russian), 1996.

286. Savvatimova, I. B. and Karabut, A. B., Radioactivity of palladium cathodes after irradiation in a glow discharge, *Poverkhnost (Surface)* (1), 76 (in Russian), 1996.

287. Karabut, A., Kucherov, Y., and Savvatimova, I., Possible nuclear reactions mechanisms at glow discharge in deuterium, *J. New Energy* 1 (1), 20, 1996.

288. Karabut, A., Analysis of experimental results on excess heat power production, impurity nuclides yield in the cathode material and penetrating radiation in experiments with high-current glow discharge, in *8th International Conference on Cold Fusion*, Scaramuzzi, F. Italian Physical Society, Bologna, Italy, Lerici (La Spezia), Italy, 2000, pp. 329.

289. Karabut, A. B., X-ray emission in the high-current glow discharge experiments, in *The 9th International Conference on Cold Fusion, Condensed Matter Nuclear Science*, Li, X. Z. Tsinghua Univ. Press, Tsinghua Univ., Beijing, China, 2002, pp. 155.

290. Lipson, A. G., Karabut, A. B., and Roussetsky, A. S., Anomalous enhancement of DD-reaction, alpha emission and X-ray generation in the high current pulsing deuterium glow-discharge with Ti-cathode at the voltages ranging from 0.8-2.5 kV., in *The 9th International Conference on Cold Fusion, Condensed Matter Nuclear Science*, Li, X. Z. Tsinghua Univ. Press, Tsinghua Univ., Beijing, China, 2002, pp. 208.

291. Karabut, A., Research into powerful solid X-ray laser (wave length is 0.8-1.2 nm) with excitation of high current glow discharge ions, in *11th International Conf. on Emerging Nuclear Energy Syst.*, Albuquerque, NM, 2002, pp. 374.

292. Karabut, A., Generation of heat, long-living atomic levels in the solid medium (1 to 3 keV) and accumulation of nuclear reaction products in a cathode of a glow discharge chamber, in *11th International Conf. on Emerging Nuclear Energy Syst.*, Albuquerque, NM, 2002.

293. Karabut, A. B. and Kolomeychenko, S. A., Experiments characterizing the X-ray emission from solid-state cathode using a high-current glow discharge, in *Tenth International Conference on Cold Fusion*, Hagelstein, P. L. andChubb, S. R. World Scientific Publishing Co., Cambridge, MA, 2003, pp. 585.

294. Karabut, A. B., Experimental research Into secondary penetrating radiation when X-Ray beams of solid laser interact with various materials targets, in *Tenth International Conference on Cold*

Fusion, Hagelstein, P. L. andChubb, S. R. World Scientific Publishing Co., Cambridge, MA, 2003, pp. 597.

[295.] Lipson, A. G., Roussetski, A. S., Karabut, A. B., and Miley, G. H., Strong enhancement of DD-reaction accompanied by X-ray generation in a pulsed low voltage high-current deuterium glow discharge with a titanuim cathode, in *Tenth International Conference on Cold Fusion*, Hagelstein, P. L. andChubb, S. R. World Scientific Publishing Co., Cambridge, MA, 2003, pp. 635.

[296.] Karabut, A., Research into characteristics of X-ray emission laser beams from solid-state cathode medium of high-current glow dischange, in *11th International Conference on Cold Fusion*, Biberian, J.-P. World Scientific Co., Marseilles, France, 2004, pp. 253.

[297.] Nakamura, K., Kawase, T., and Ogura, I., Possibility of element transmutation by arcing in water, *Kinki Daigaku Genshiryoku Kenkyusho Nenpo* 33, 25 (in Japanese), 1996.

[298.] Mizuno, T., Ohmori, T., and Akimoto, T., Probability of neutron and heat emission from Pt electrode induced by discharge in alkaline solution, in *The Seventh International Conference on Cold Fusion*, Jaeger, F. ENECO, Inc., Salt Lake City, UT, Vancouver, Canada, 1998, pp. 247.

[299.] Ohmori, T., Recent development in solid state nuclear transmutation occurring by the electrolysis, *Curr. Topics Electrochem.* 7, 101, 2000.

[300.] Cirillo, D. and Iorio, V., Transmutation of metal at low energy in a confined plasma in water, in *11th International Conference on Cold Fusion*, Biberian, J.-P. World Scientific Co., Marseilles, France, 2004, pp. 492.

[301.] McKubre, M. C. H., Bush, B. F., Crouch-Baker, S., Hauser, A., Jevtic, N., Smedley, S., Srinivasan, M., Tanzella, F. L., Williams, M., and Wing, S., Loading, calorimetric and nuclear investigation of the D/Pd system, in *Fourth International Conference on Cold Fusion* Electric Power Research Institute 3412 Hillview Ave., Palo Alto, CA 94304, Lahaina, Maui, 1993, pp. 5.

[302.] McKubre, M. C. H., Crouch-Baker, S., Hauser, A. K., Smedley, S. I., Tanzella, F. L., Williams, M. S., and Wing, S. S., Concerning reproducibility of excess power production, in *5th International Conference on Cold Fusion*, Pons, S. IMRA Europe, Sophia Antipolis Cedex, France, Monte-Carlo, Monaco, 1995, pp. 17.

[303.] McKubre, M. C. H., Crouch-Baker, S., Tanzella, F. L., Williams, M., and Wing, S., New hydrogen energy research at SRI, in *Sixth International Conference on Cold Fusion, Progress in New Hydrogen Energy*, Okamoto, M. New Energy and Industrial Technology Development Organization, Tokyo Institute of Technology, Tokyo, Japan, Lake Toya, Hokkaido, Japan, 1996, pp. 75.

[304.] McKubre, M. C. H. and Tanzella, F. L., Materials issues of loading deuterium into palladium and the association with excess heat production, in *The Seventh International Conference on Cold Fusion*, Jaeger, F. ENECO, Inc., Salt Lake City, UT, Vancouver, Canada, 1998, pp. 230.

[305.] McKubre, M. C. H., Tanzella, F. L., Tripodi, P., Di Gioacchino, D., and Violante, V., Finite element modeling of the transient colorimetric behavior of the MATRIX experimental apparatus: ^{4}He and excess of power production correlation through numerical results, in *8th International Conference on Cold Fusion*, Scaramuzzi, F. Italian Physical Society, Bologna, Italy, Lerici (La Spezia), Italy, 2000, pp. 23.

[306.] McKubre, M. C., Tanzella, F., and Tripodi, P., Evidence of d-d fusion products in experiments conducted with palladium at near ambient temperatures, *Trans. Am. Nucl. Soc.* 83, 367, 2000.

[307.] McKubre, M. C. H., Tanzella, F., Tripodi, P., and Violante, V., Progress towards replication, in *The 9th International Conference on Cold Fusion, Condensed Matter Nuclear Science*, Li, X. Z. Tsinghua Univ. Press, Tsinghua Univ., Beijing, China, 2002, pp. 241.

[308.] McKubre, M. C. H., Review of experimental measurements involving dd reactions, PowerPoint slides, in *Tenth International Conference on Cold Fusion*, Hagelstein, P. L. andChubb, S. R. World Scientific Publishing Co., Cambridge, MA, 2003.

[309.] McKubre, M. C., Tanzella, F., Hagelstein, P. L., Mullican, K., and Trevithick, M., The need for triggering in cold fusion reactions, in *Tenth International Conference on Cold Fusion*, Hagelstein, P. L. andChubb, S. R. World Scientific Publishing Co., Cambridge, MA, 2003, pp. 199.

[310.] Arata, Y. and Zhang, Y. C., Achievement of intense 'cold fusion' reaction, *Kaku Yugo Kenkyu* 62, 398 (In Japanese), 1989.

[311.] Arata, Y. and Zhang, Y.-C., Achievement of an intense cold fusion reaction, *Fusion Technol.* 18, 95, 1990.

[312.] Arata, Y. and Zhang, Y. C., Achievement of intense 'cold' fusion reaction, *Proc. Jpn. Acad., Ser. B* 66, 1, 1990.

[313.] Arata, Y. and Zhang, Y. C., 'Cold' fusion in deuterated complex cathode, *Kaku Yugo Kenkyu* 67 ((5)), 432 (in Japanese), 1992.

[314.] Arata, Y. and Zhang, Y.-C., Excess heat in a double structure deuterated cathode, *Kakuyuogo Kenkyo* 69 ((8)), 963 (in Japanese), 1993.

[315.] Arata, Y. and Zhang, Y.-C., A new energy generated in DS-cathode with 'Pd-black', *Koon Gakkaishi* 20 (4), 148 (in Japanese), 1994.

[316.] Arata, Y. and Zhang, Y.-C., Achievement of solid-state plasma fusion ("cold fusion"), *Proc. Japan Acad.* 71 (Ser B), 304, 1995.

[317.] Arata, Y. and Zhang, Y.-C., Peculiar relation between hot plasma fusion and solid-state plasma fusion ("cold fusion"), *Koon Gakkaishi* 21, 130 (in Japanese), 1995.

[318.] Arata, Y. and Zhang, Y.-C., Cold fusion caused by 'lattice quake', *Koon Gakkaishi* 21, 43 (in Japanese), 1995.

[319.] Arata, Y. and Zhang, Y. C., Excess heat and mechanism in cold fusion reaction, in *5th International Conference on Cold Fusion*, Pons, S. IMRA Europe, Sophia Antipolis Cedex, France, Monte-Carlo, Monaco, 1995, pp. 483.

[320.] Arata, Y. and Zhang, Y. C., Cold fusion caused by 'lattice quake', *Koon Gakkaishi* 21, 43 (in Japanese), 1995.

[321.] Arata, Y. and Zhang, Y. C., Peculiar relation between hot plasma fusion and solid-state plasma fusion ("cold fusion"), *Koon Gakkaishi* 21, 130 (in Japanese), 1995.

[322.] Arata, Y. and Zhang, Y. C., Helium (^{4}He, ^{3}He) within deuterated Pd-black, *Proc. Jpn. Acad., Ser. B* 73, 1, 1997.

[323.] Arata, Y. and Zhang, Y.-C., Deuterium nuclear reaction process within solid, *J. New Energy* 2 (1), 27, 1997.

[324.] Arata, Y., Zhang, Y.-C., Fujita, H., and Inoue, A., Discovery of solid deuterium nuclear fusion of pycnodeuterium-lumps solidified locally within nano-Pd particles, *Koon Gakkaishi* 29, 68, 2003.

[325.] Arata, Y. and Zhang, Y.-C., The basics of nuclear fusion reactor using solid pycnodeuterium as nuclear fuel, *High Temp. Soc, Japan* 29, 1, 2003.

[326.] Clarke, B. W., Oliver, B. M., McKubre, M. C. H., Tanzella, F. L., and Tripodi, P., Search for ^{3}He and ^{4}He in Arata-style palladium cathodes II: Evidence for tritium production, *Fusion Sci. & Technol.* 40, 152, 2001.

[327.] Arata, Y. and Zhang, Y.-C., Development of "DS-reactor" as a practical reactor of "cold fusion" based on the "DS-cell" with "DS-cathode", in *12th International Conference on Cold Fusion*, Yokohama, Japan, 2005.

[328.] Dardik, I., Zilov, T., Branover, H., El-Boher, A., Greenspan, E., Khachaturov, B., Krakov, V., Lesin, S., and Tsirlin, M., Excess heat in electrolysis experiments at Energetics Technologies, in *6th International Workship on Anomalies in Hydrogen/Deuterium Loaded Metals*, Siena, Italy, 2005.

[329.] Mills, R. L., Reply to 'Comments on "Excess heat production by the electrolysis of an aqueous potassium carbonate electrolyte and the implications for cold fusion"', *Fusion Technol.* 21, 96, 1992.

[330.] Notoya, R., Nuclear products of cold Fusion caused by electrolysis in alkali metallic ions solutions, in *5th International Conference on Cold Fusion*, Pons, S. IMRA Europe, Sophia Antipolis Cedex, France, Monte-Carlo, Monaco, 1995, pp. 531.

[331.] Notoya, R., Cold fusion by electrolysis in a light water-potassium carbonate solution with a nickel electrode, *Fusion Technol.* 24, 202, 1993.

[332.] Srinivasan, M. and McKubre, M. C. H., 1994.

[333.] Srinivasan, M., Babu, P. A., Bajpai, M. B., Gupta, D. S., Mukherjee, U. K., Ramamurthy, H., Sankarnarainan, T. K., Sinha, A., and Shyam, A., 1994.

[334.] Notoya, R., Low temperature nuclear change of alkali metallic ions caused by electrolysis, *J. New Energy* 1 (1), 39, 1996.

[335.] Ohmori, T. and Enyo, M., Iron formation in gold and palladium cathodes, *J. New Energy* 1 (1), 15, 1996.

[336.] Miley, G. H., Possible evidence of anomalous energy effects in H/D-loaded solids-low energy nuclear reactions (LENRS), *J. New Energy* 2 (3/4), 6, 1997.

[337.] S' 'edi, Z., McDonald, R. C., Breen, J. J., Maguire, S. J., and Veranth, J., Calorimetry, excess heat, and Faraday efficiency in Ni-H_2O electrolytic cells, *Fusion Technol.* 28, 1720, 1995.

[338.] Shkedi, Z., Response to "Comments on 'Calorimetry, excess heat, and Faraday efficiency in Ni-H_2O electrolytic cells'". *Fusion Technol.* 30, 133, 1996.

[339.] Patterson, J. A., US Patent #5,494,559, 1996, System for electrolysis. Feb. 27, 1996.

[340.] Patterson, J. A., US Patent # 5,318,675, 1994, Method for electrolysis of water to form metal hydride. June 7, 1994.

[341.] Rothwell, J., CETI's 1 kilowatt cold fusion device denonstrated, *Infinite Energy* 1 (5&6), 18, 1996.

[342.] Focardi, S., Gabbani, V., Montalbano, V., Piantelli, F., and Veronesi, F., Evidence of electromagnetic radiation from Ni-H systems, in *11th International Conference on Cold Fusion*, Biberian, J.-P. World Scientific Co., Marseilles, France, 2004, pp. 70.

[343.] Campari, E. G., Focardi, S., Gabbani, V., Montalbano, V., Piantelli, F., and Stanghini, C., Nuclear reactions in Ni-H systems, in *6th International Workship on Anomalies in Hydrogen/Deuterium Loaded Metals*, Siena, Italy, 2005.

[344.] Baranowskyi, B., *Hydrogen in metals II* Springer, Berlin, 1978.

[345.] Storms, E., Review of experimental observations about the cold fusion effect, *Fusion Technol.* 20, 433, 1991.

[346.] Romodanov, V. A., Tritium generation during the interaction of plasma glow discharge with metals and a magnetic field, in *Tenth International Conference on Cold Fusion*, Hagelstein, P. L. andChubb, S. R. World Scientific Publishing Co., Cambridge, MA, 2003, pp. 325.

[347.] Violante, V., Tripodi, P., Di Gioacchino, D., Borelli, R., Bettinali, L., Santoro, E., Rosada, A., Sarto, F., Pizzuto, A., McKubre, M. C. H., and Tanzella, F., X-ray emission during electrolysis of light water on palladium and nickel thin films, in *The 9th International Conference on Cold Fusion, Condensed Matter Nuclear Science*, Li, X. Z. Tsinghua Univ. Press, Tsinghua Univ., Beijing, China, 2002, pp. 376.

[348.] Celani, F., Spallone, A., Marini, P., di Stefano, V., Nakamura, M., Righi, E., Trenta, G., Quercia, P., Mancini, A., D'Agostaro, G., Catena, C., Sandri, S., Nobili, C., and Andressi, V., Evidence of anomalous tritium excess in D/Pd overloading experiments, in *The 9th International Conference on Cold Fusion, Condensed Matter Nuclear Science*, Li, X. Z. Tsinghua Univ. Press, Tsinghua Univ., Beijing, China, 2002, pp. 36.

[349.] Yamada, H., Narita, S., Inamura, I., Nakai, M., Iwasaki, K., and Baba, M., Tritium production in palladium deuteride/hydride in evacuated chamber, in *8th International Conference on Cold Fusion*, Scaramuzzi, F. Italian Physical Society, Bologna, Italy, Lerici (La Spezia), Italy, 2000, pp. 241.

[350.] Romodanov, V. A., Skuratnik , Y. B., and Pokrovsky, A. K., Generation of tritium for deuterium interaction with metals, in *8th International Conference on Cold Fusion*, Scaramuzzi, F. Italian Physical Society, Bologna, Italy, Lerici (La Spezia), Italy, 2000, pp. 265.

[351.] Szpak, S., Mosier-Boss, P. A., Boss, R. D., and Smith, J. J., On the behavior of the Pd/D system: evidence for tritium production, *Fusion Technol.* 33, 38, 1998.

[352.] Romodanov, V. A., Savin, V., Skuratnik, Y., and Yuriev, M., Nuclear reactions in condensed media and X-ray, in *The Seventh International Conference on Cold Fusion*, Jaeger, F. ENECO, Inc., Salt Lake City, UT., The Seventh International Conference on Cold Fusion, 1998, pp. 330.

[353.] Romodanov, V. A., Savin, V., Skuratnik, Y., and Yuriev, M., High-temperature nuclear reactions in condensed media, in *The Seventh International Conference on Cold Fusion*, Jaeger, F. ENECO, Inc., Salt Lake City, UT., Vancouver, Canada, 1998, pp. 325.

[354.] Romodanov, V. A., Savin, V., Skuratnik, Y., and Majorov, V., Tritium generations in metals at thermal activation, in *The Seventh International Conference on Cold Fusion*, Jaeger, F. ENECO, Inc., Salt Lake City, UT, Vancouver, Canada, 1998, pp. 319.

[355.] Claytor, T. N., Schwab, M. J., Thoma, D. J., Teter, D. F., and Tuggle, D. G., Tritium production from palladium alloys, in *The Seventh International Conference on Cold Fusion*, Jaeger, F. ENECO, Inc., Salt Lake City, UT., Vancouver, Canada, 1998, pp. 88.

[356.] Sankaranarayanan, T. K., Srinivasan, M., Bajpai, M. B., and Gupta, D. S., Investigation of low-level tritium generation in Ni-H2O electrolytic cells, *Fusion Technol.* 30, 349, 1996.

[357.] Romodanov, V. A., Savin, V. I., Skuratnik, Y. B., and Majorov, V. N., The nuclear reactions in condensed media for interaction of charge particles in energy region is forming by maximum elastic losses, in *Sixth International Conference on Cold Fusion, Progress in New Hydrogen Energy*, Okamoto, M. New Energy and Industrial Technology Development Organization, Tokyo Institute of Technology, Tokyo, Japan, Lake Toya, Hokkaido, Japan, 1996, pp. 340.

[358.] Itoh, T., Iwamura, Y., Gotoh, N., and Toyoda, I., Observation of nuclear products in gas release experiments with electrochemically deuterated palladium, in *Sixth International Conference on Cold Fusion, Progress in New Hydrogen Energy*, Okamoto, M. New Energy and Industrial Technology Development Organization, Tokyo Institute of Technology, Tokyo, Japan, Lake Toya, Hokkaido, Japan, 1996, pp. 410.

[359.] Itoh, T., Iwamura, Y., Gotoh, N., and Toyoda, I., Observation of nuclear products under vacuum conditions from deuterated palladium with high loading ratio, in *5th International Conference on Cold Fusion*, Pons, S. IMRA Europe, Sophia Antipolis Cedex, France, Monte-Carlo, Monaco, 1995, pp. 189.

[360.] Claytor, T. N., Jackson, D. D., and Tuggle, D. G., Tritium production from low voltage deuterium discharge on palladium and other metals, *Infinite Energy* 2 (7), 39, 1996.

[361.] Tuggle, D. G., Claytor, T. N., and Taylor, S. F., Tritium evolution from various morphologies of palladium, *Trans. Fusion Technol.* 26 (#4T), 221, 1994.

[362.] Sankaranarayanan, T. K., Sprinivasan, M., Bajpai, M. B., and Gupta, D. S., Evidence for tritium generation in self-heated nickel wires subjected to hydrogen gas absorption/desorption cycles., in *5th International Conference on Cold Fusion*, Pons, S. IMRA Europe, Sophia Antipolis Cedex, France, Monte-Carlo, Monaco, 1995, pp. 173.

[363.] Sankaranarayanan, M., Srinivasan, M., Bajpai, M., and Gupta, D., Investigation of low level tritium generation in Ni-H_2O electrolytic cells, in *Fourth International Conference on Cold Fusion* Electric Power Research Institute 3412 Hillview Ave., Palo Alto, CA 94304, Lahaina, Maui, 1993, pp. 3.

[364.] Lipson, A. G., Sakov, D. M., Lyakhov, B. F., Saunin, E. I., and Deryagin, B. V., Generation of the products of DD nuclear fusion in high-temperature superconductors YBa2Cu3O7-Dy near the superconducting phase transition, *Tech. Phys.* 40, 839, 1995.

[365.] Notoya, R., Alkali-hydrogen cold fusion accompanied by tritium production on nickel, *Trans. Fusion Technol.* 26 (#4T), 205, 1994.

[366.] Will, F. G., Cedzynska, K., and Linton, D. C., Reproducible tritium generation in electrochemical cells employing palladium cathodes with high deuterium loading, *J. Electroanal. Chem.* 360, 161, 1993.

[367.] Will, F. G., Cedzynska, K., and Linton, D. C., Tritium generation in palladium cathodes with high deuterium loading, in *Fourth International Conference on Cold Fusion* Electric Power Research Institute 3412 Hillview Ave., Palo Alto, CA 94304, Lahaina, Maui, 1993, pp. 8.

[368.] Stukan, P. A., Rumyantsev, Y. M., and Shishkov, A. V., Generation of hard radiation and accumulation of tritium during electrolysis of heavy water, *High Energy Chem.* 27, 461, 1993.

[369.] Chien, C.-C. and Huang, T. C., Tritium production by electrolysis of heavy water, *Fusion Technol.* 22, 391, 1992.

[370.] Notoya, R., Alkali-hydrogen cold fusion accompanied by tritium production on nickel, in *Fourth International Conference on Cold Fusion* Electric Power Research Institute 3412 Hillview Ave., Palo Alto, CA 94304, Lahaina, Maui, 1993, pp. 1.

[371.] Mengoli, G., Fabrizio, M., Manduchi, C., Zannoni, G., Riccardi, L., and Buffa, A., Tritium and neutron emission in D_2O electrolysis at Pd and Ti cathodes, *J. Electroanal. Chem.* 322, 107, 1992.

[372.] Mengoli, G., Fabrizio, M., Manduchi, C., Zannoni, G., Riccardi, L., Veronesi, F., and Buffa, A., The observation of tritium in the electrolysis of D_2O at palladium sheet electrodes, *J. Electroanal. Chem.* 304, 279, 1991.

[373.] Stella, B., Alessio, M., Carradi, M., Croce, F., Ferrarotto, F., Improta, S., Iucci, N., Milone, V., Villoresi, G., Celani, F., and Spallone, A., The FERMI apparatus and a measurement of tritium production in an electrolytic experiment, in *Third International Conference on Cold Fusion, "Frontiers of Cold Fusion"*, Ikegami, H. Universal Academy Press, Inc., Tokyo, Japan, Nagoya Japan, 1992, pp. 503.

[374.] Sevilla, J., Escarpizo, B., Fernandez, F., Cuevas, F., and Sanchez, C., Time-evolution of tritium concentration in the electrolyte of prolonged cold fusion experiments and its relation to Ti cathode surface treatment, in *Third International Conference on Cold Fusion, "Frontiers of Cold Fusion"*, Ikegami, H. Universal Academy Press, Inc., Tokyo, Japan, Nagoya Japan, 1992, pp. 507.

[375.] Matsumoto, O., Kimura, K., Saito, Y., Uyama, H., Yaita, T., Yamaguchi, A., and Suenaga, O., Detection of neutron and tritium during electrolysis of D_2SO_4-D_2O solution, in *Third International Conference on Cold Fusion, "Frontiers of Cold Fusion"*, Ikegami, H. Universal Academy Press, Inc., Tokyo, Japan, Nagoya Japan, 1992, pp. 495.

[376.] Lee, K. H. and Kim, Y. M., The change of tritium concentration during the electrolysis of D_2O in various electrolytic cells, in *Third International Conference on Cold Fusion, "Frontiers of Cold Fusion"*, Ikegami, H. Universal Academy Press, Inc., Tokyo, Japan, Nagoya Japan, 1992, pp. 511.

[377.] Gozzi, D., Cignini, P. L., Tomellini, M., Frullani, S., Garibaldi, F., Ghio, F., Jodice, M., and Urciuoli, G. M., Neutron and tritium evidence in the electrolytic reduction of deuterium on palladium electrodes, *Fusion Technol.* 21, 60, 1992.

[378.] Gozzi, D., Cignini, P. L., Petrucci, L., Tomellini, M., Frullani, S., Garibaldi, F., Ghio, F., Jodice, M., and Urciuoli, G. M., First results from a ten electrolytic cells experiment, in *Anomalous Nuclear Effects in Deuterium/Solid Systems, "AIP Conference Proceedings 228"*, Jones, S., Scaramuzzi, F., andWorledge, D. American Institute of Physics, New York, Brigham Young Univ., Provo, UT, 1990, pp. 481.

[379.] Gozzi, D., Cignini, P. L., Petrucci, L., Tomellini, M., De Maria, G., Frullani, S., Garibaldi, F., Ghio, F., Jodice, M., and Tabet, E., Nuclear and thermal effects during electrolytic reduction of deuterium at palladium cathode, *J. Fusion Energy* 9 (3), 241, 1990.

[380.] Claytor, T. N., Tuggle, D. G., and Taylor, S. F., Evolution of tritium from deuterided palladium subject to high electrical currents, in *Third International Conference on Cold Fusion, "Frontiers of Cold Fusion"*, Ikegami, H. Universal Academy Press, Inc., Tokyo, Japan, Nagoya Japan, 1992, pp. 217.

[381.] Clarke, B. W. and Clarke, R. M., Search for (3)H, (3)He, and (4)He in D_2-loaded titanium, *Fusion Technol.* 21, 170, 1992.

[382.] Chien, C.-C., Hodko, D., Minevski, Z., and Bockris, J. O. M., On an electrode producing massive quantities of tritium and helium, *J. Electroanal. Chem.* 338, 189, 1992.

[383.] Bockris, J., Chien, C., Hodko, D., and Minevski, Z., Tritium and helium production in palladium electrodes and the fugacity of deuterium therein, in *Third International Conference on Cold Fusion, "Frontiers of Cold Fusion"*, Ikegami, H. Universal Academy Press, Inc., Tokyo, Japan, Nagoya Japan, 1992, pp. 231.

[384.] Lin, G. H., Kainthla, R. C., Packham, N. J. C., and Bockris, J. O. M., Electrochemical fusion: a mechanism speculation, *J. Electroanal. Chem.* 280, 207, 1990.

[385.] Szpak, S., Mosier-Boss, P. A., and Smith, J. J., Reliable procedure for the initiation of the Fleischmann-Pons effect, in *Second Annual Conference on Cold Fusion, "The Science of Cold*

Fusion", Bressani, T., Del Giudice, E., andPreparata, G. Societa Italiana di Fisica, Bologna, Italy, Como, Italy, 1991, pp. 87.

[386.] Mengoli, G., Fabrizio, M., Manduchi, C., Zannoni, G., Riccardi, L., and Buffa, A., Tritium and neutron emission in conventional and contact glow discharge electrolysis of D_2O at Pd and Ti cathodes, in *Second Annual Conference on Cold Fusion, "The Science of Cold Fusion"*, Bressani, T., Del Giudice, E., andPreparata, G. Societa Italiana di Fisica, Bologna, Italy, Como, Italy, 1991, pp. 65.

[387.] Lanza, F., Bertolini, G., Vocino, V., Parnisari, E., and Ronsecco, C., Tritium production resulting from deuteration of different metals and alloys, in *Second Annual Conference on Cold Fusion, "The Science of Cold Fusion"*, Bressani, T., Del Giudice, E., andPreparata, G. Societa Italiana di Fisica, Bologna, Italy, Como, Italy, 1991, pp. 151.

[388.] Kochubey, D. I., Babenko, V. P., Vargaftik, M. N., and I., M. I., Enrichment of deuterium with tritium in the presence of a palladium-561 giant cluster, *J. Molec. Catal.* 66, 99, 1991.

[389.] De Ninno, A., Scaramuzzi, F., Frattolillo, A., Migliori, S., Lanza, F., Scaglione, S., Zeppa, P., and Pontorieri, C., The production of neutrons and tritium in the deuterium gas-titanium interaction, in *Second Annual Conference on Cold Fusion, "The Science of Cold Fusion"*, Bressani, T., Del Giudice, E., andPreparata, G. Societa Italiana di Fisica, Bologna, Italy, Como, Italy, 1991, pp. 129.

[390.] Claytor, T. N., Tuggle, D. G., and Menlove, H. O., Tritium generation and neutron measurements in Pd-Si under high deuterium gas pressure, in *Second Annual Conference on Cold Fusion, "The Science of Cold Fusion"*, Bressani, T., Del Giudice, E., andPreparata, G. Societa Italiana di Fisica, Bologna, Italy, Como, Ita, 1991, pp. 395.

[391.] Wolf, K. L., Packham, N. J. C., Lawson, D., Shoemaker, J., Cheng, F., and Wass, J. C., Neutron emission and the tritium content associated with deuterium-loaded palladium and titanium metals, *J. Fusion Energy* 9 (2), 105, 1990.

[392.] Storms, E. and Talcott, C. L., Electrolytic tritium production, *Fusion Technol.* 17, 680, 1990.

[393.] Storms, E. and Talcott, C. L., A study of electrolytic tritium production, in *The First Annual Conference on Cold Fusion*, Will, F. National Cold Fusion Institute, University of Utah Research Park, Salt Lake City, Utah, 1990, pp. 149.

[394.] Srinivasan, M., Shyam, A., Kaushik, T. C., Rout, R. K., Kulkarni, L. V., Krishnan, M. S., Malhotra, S. K., Nagvenkar, V. G., and Iyengar, P. K., Observation of tritium in gas/plasma loaded titanium samples, in *Anomalous Nuclear Effects in Deuterium/Solid Systems, "AIP Conference Proceedings 228"*, Jones, S., Scaramuzzi, F., andWorledge, D. American Institute of Physics, New York, Brigham Young Univ., Provo, UT, 1990, pp. 514.

[395.] Kaushik, T. C., Shyam, A., Srinivasan, M., Rout, R. K., Kulkarni, L. V., Krishnan, M. S., Malhotra, S. K., and Nagvenkar, V. B., Preliminary report on direct measurement of tritium in liquid nitrogen treated TiDx chips, *Indian J. Technol.* 28, 667, 1990.

[396.] Sona, P. G., Parmigiani, F., Barberis, F., Battaglia, A., Berti, R., Buzzanca, G., Capelli, A., Capra, D., and Ferrari, M., Preliminary tests on tritium and neutrons in cold nuclear fusion within palladium cathodes, *Fusion Technol.* 17, 713, 1990.

[397.] Matsumoto, O., Kimura, K., Saito, Y., Uyama, H., and Yaita, T., Tritium production rate, in *Anomalous Nuclear Effects in Deuterium/Solid Systems, "AIP Conference Proceedings 228"*, Jones, S., Scaramuzzi, F., andWorledge, D. American Institute of Physics, New York, Brigham Young Univ., Provo, UT, 1990, pp. 494.

[398.] Fernández, F., Sevilla, J., Escarpizo, B., and Sánchez, C., Nuclear effects in electrolytically deuterated Ti and Pd electrodes, in *Anomalous Nuclear Effects in Deuterium/Solid Systems, "AIP Conference Proceedings 228"*, Jones, S., Scaramuzzi, F., andWorledge, D. American Institute of Physics, New York, Brigham Young Univ., Provo, UT, 1990, pp. 130.

[399.] Claytor, T. N., Tuggle, D. G., Menlove, H. O., Seeger, P. A., Doty, W. R., and Rohwer, R. K., Tritium and neutron measurements from deuterated Pd-Si, in *Anomalous Nuclear Effects in Deuterium/Solid Systems, "AIP Conference Proceedings 228"*, Jones, S., Scaramuzzi, F., andWorledge, D. American Institute of Physics, New York, Brigham Young Univ., Provo, UT, 1990, pp. 467.

[400.] Chêne, J. and Brass, A. M., Tritium production during the cathodic discharge of deuterium on palladium, *J. Electroanal. Chem.* 280, 199, 1990.

[401.] Bockris, J. O. M., Lin, G. H., Kainthla, R. C., Packham, N. J. C., and Velev, O., Does tritium form at electrodes by nuclear reactions?, in *The First Annual Conference on Cold Fusion*, Will, F. National Cold Fusion Institute, University of Utah Research Park, Salt Lake City, Utah, 1990, pp. 137.

[402.] Packham, N. J. C., Wolf, K. L., Wass, J. C., Kainthla, R. C., and Bockris, J. O. M., Production of tritium from D_2O electrolysis at a palladium cathode, *J. Electroanal. Chem.* 270, 451, 1989.

[403.] Lin, G. H., Kainthla, R. C., Packham, N. J. C., Velev, O. A., and Bockris, J., On electrochemical tritium production, *Int. J. Hydrogen Energy* 15 (8), 537, 1990.

[404.] Iyengar, P. K. and Srinivasan, M., Overview of BARC Studies in Cold Fusion, in *The First Annual Conference on Cold Fusion*, Will, F. National Cold Fusion Institute, University of Utah Research Park, Salt Lake City, Utah, 1990, pp. 62.

[405.] Iyengar, P. K. and Srinivasan, M., Report No. B.A.R.C.1500, 1989.

[406.] Adzic, R. G., Bae, I., Cahan, B., and Yeager, E., Tritium measurements and deuterium loading in D_2O electrolysis with a palladium cathode, in *The First Annual Conference on Cold Fusion*, Will, F. National Cold Fusion Institute, University of Utah Research Park, Salt Lake City, Utah, 1990, pp. 261.

[407.] Sánchez, C., Sevilla, J., Escarpizo, B., Fernández, F. J., and Canizares, J., Nuclear products detection during electrolysis of heavy water with titanium and platinum electrodes, *Solid State Commun.* 71, 1039, 1989.

[408.] Anderson, D. M. and Bockris, J. O. M., Cold fusion at Texas A&M, *Science* 249, 463, 1990.

[409.] Anderson, J., Bockris, J. O. M., Worledge, D. H., and Taubes, G., Letters and response about cold fusion at Texas A&M, *Science* 249, 463-465, 1990.

[410.] Bockris, J. O. M., Cold fusion II: The story continues, *New Scientist* 19, 50, 1991.

[411.] Bockris, J. O. M., Accountability and academic freedom: The battle concerning research on cold fusion at Texas A&M University, *Accountability in Research* 8, 103, 2000.

[412.] Iyengar, P. K., Srinivasan, M., Sikka, S. K., Shyam, A., Chitra, V., Kulkarni, L. V., Rout, R. K., Krishnan, M. S., Malhotra, S. K., Gaonkar, D. G., Sadhukhan, H. K., Nagvenkar, V. B., Nayar, M. G., Mitra, S. K., Raghunathan, P., Degwekar, S. B., Radhakrishnan, T. P., Sundaresan, R., Arunachalam, J., Raju, V. S., Kalyanaraman, R., Gangadharan, S., Venkateswaran, G., Moorthy, P. N., Venkateswarlu, K. S., Yuvaraju, B., Kishore, K., Guha, S. N., Panajkar, M. S., Rao, K. A., Raj, P., Suryanarayana, P., Sathyamoorthy, A., Datta, T., Bose, H., Prabhu, L. H., Sankaranarayanan, S., Shetiya, R. S., Veeraraghavan, N., Murthy, T. S., Sen, B. K., Joshi, P. V., Sharma, K. G. B., Joseph, T. B., Iyengar, T. S., Shrikhande, V. K., Mittal, K. C., Misra, S. C., Lal, M., and Rao, P. S., Bhabha Atomic Research Centre studies on cold fusion, *Fusion Technol.* 18, 32., 1990.

[413.] Li, X. Z., A Chinese view on the summary of the condensed matter nuclear science, *J. Fusion Energy* 23 (3), 217, 2004.

[414.] Cedzynska, K., Barrowes, S. C., Bergeson, H. E., Knight, L. C., and Will, F. G., Tritium analysis in palladium with an open system analytical procedure, *Fusion Technol.* 20, 108, 1991.

[415.] Cedzynska, K. and Will, F. G., Closed-system analysis of tritium in palladium, *Fusion Technol.* 22, 156, 1992.

[416.] Aiello, S., De Filippo, E., Lanzano, G., Lo Nigro, S., and Pagano, A., Nuclear fusion experiment in palladium charged by deuterium gas, *Fusion Technol.* 18, 115, 1990.

[417.] Claytor, T. N., Seeger, P., Rohwer, R. K., Tuggle, D. G., and Doty, W. R., Tritium and neutron measurements of a solid state cell, in *NSF/EPRI Workshop on Anomalous Effects in Deuterated Materials* LA-UR-89-39-46, Washington, DC, 1989.

[418.] Claytor, T. N., Jackson, D. D., and Tuggle, D. G., Tritium production from a low voltage deuterium discharge of palladium and other metals, *J. New Energy* 1 (1), 111, 1996.

419. Romodanov, V. A., Savin, V., Elksnin, V., and Skuratnik, Y., Reproducibility of tritium generation from nuclear reactions in condensed matter, in *Fourth International Conference on Cold Fusion* Electric Power Research Institute 3412 Hillview Ave., Palo Alto, CA 94304, Lahaina, Maui, 1993, pp. 15.

420. Alekseev, V. A., Vasil'ev, V. I., Romodanov, V. A., Ryshkov, Y. F., Rylov, S. V., Savin, V. I., Skuratnik, Y. B., and Strunnikov, V. M., Tritium production in the interaction of dense streams of deuterium plasma with metal surfaces, *Tech. Phys. Lett.* 21, 231, 1995.

421. Lipson, A. G., Klyuev, V. A., Deryagin, B. V., Toporov, Y. P., and Sakov, D. M., Anomalous beta activity of products of mechanical working of a titanium- deuterated material, *Sov. Tech. Phys. Lett.* 15 (10), 783, 1989.

422. Shioe, Y., Mondal, N. N., Chiba, M., Hirose, T., Fujii, M., Nakahara, H., Sueki, K., Shirakawa, T., and M., U., Measurement of neutron production rate regarding the quantity of $LiNbO_3$ in the fracturing process under D2 atmosphere, *Nuovo Cimento Soc. Ital. Fis. A* 112 A, 1059, 1999.

423. Lipson, A. G., Kuznetsov, V. A., Ivanova, T. S., Saunin, E. I., and Ushakov, S. I., Possibility of mechanically stimulated transmutation of carbon nuclei in ultradisperse deuterium-containing media, *Tech. Phys.* 42, 676, 1997.

424. Shirakawa, T., Chiba, M., Fujii, M., Sueki, K., Miyamoto, S., Nakamitu, Y., Toriumi, H., Uehara, T., Miura, H., Watanabe, T., Fukushima, K., Hirose, T., Seimiya, T., and Nakahara, H., A neutron emission from lithium niobate fracture, *Chem. Lett.*, 897, 1993.

425. Ochiai, K., Maruta, K., Miyamaru, H., and Takahashi, A., Measurement of high-energetic particles from titanium sheets implanted with deuterium, in *The Seventh International Conference on Cold Fusion*, Jaeger, F. ENECO, Inc., Salt Lake City, UT., Vancouver, Canada, 1998, pp. 274.

426. Miles, M., Bush, B. F., Ostrom, G. S., and Lagowski, J. J., Heat and helium production in cold fusion experiments, in *Second Annual Conference on Cold Fusion, "The Science of Cold Fusion"*, Bressani, T., Giudice, E. D., andPreparata, G. Societa Italiana di Fisica, Bologna, Italy, Como, Italy, 1991, pp. 363.

427. Miles, M. and Bush, B. F., Search for anomalous effects involving excess power and helium during D_2O electrolysis using palladium cathodes, in *Third International Conference on Cold Fusion, "Frontiers of Cold Fusion"*, Ikegami, H. Universal Academy Press, Inc., Tokyo, Japan, Nagoya Japan, 1992, pp. 189.

428. Miles, M., Johnson, K. B., and Imam, M. A., Heat and helium measurements using palladium and palladium alloys in heavy water, in *Sixth International Conference on Cold Fusion, Progress in New Hydrogen Energy*, Okamoto, M. New Energy and Industrial Technology Development Organization, Tokyo Institute of Technology, Tokyo, Japan, Lake Toya, Hokkaido, Japan, 1996, pp. 20.

429. Miles, M., Correlation of excess enthalpy and helium-4 production: A review, in *Tenth International Conference on Cold Fusion*, Hagelstein, P. L. andChubb, S. R. World Scientific Publishing Co., Cambridge, MA, 2003, pp. 123.

430. Miles, M. and Jones, C. P., Cold fusion experimenter Miles responds to critic, *21st Century Sci. & Technol.* Spring, 75, 1992.

431. Miles, M. H., Reply to 'An assessment of claims of excess heat in cold fusion calorimetry', *J. Phys. Chem. B* 102, 3648, 1998.

432. Miles, M. H., Reply to 'Examination of claims of Miles *et al.* in Pons-Fleischmann-type cold fusion experiments', *J. Phys. Chem. B* 102, 3642, 1998.

433. Jones, S. E. and Hansen, L. D., Examination of claims of Miles et al in Pons-Fleischmann-Type cold fusion experiments, *J. Phys. Chem.* 99, 6966, 1995.

434. Jones, S. E., Hansen, L. D., and Shelton, D. S., An assessment of claims of excess heat in cold fusion calorimetry, *J. Phys. Chem. B* 102, 3647, 1998.

435. Gozzi, D., Caputo, R., Cignini, P. L., Tomellini, M., Gigli, G., Balducci, G., Cisbani, E., Frullani, S., Garibaldi, F., Jodice, M., and Urciuoli, G. M., Helium-4 quantitative measurements in the gas phase of cold fusion electrochemical cells, in *Fourth International Conference on Cold*

Fusion Electric Power Research Institute 3412 Hillview Ave., Palo Alto, CA 94304, Lahaina, Maui, 1993, pp. 6.

[436.] Gozzi, D., Caputo, R., Cignini, P. L., Tomellini, M., Gigli, G., Balducci, G., Cisbani, E., Frullani, S., Garibaldi, F., Jodice, M., and Urciuoli, G. M., Quantitative measurements of helium-4 in the gas phase of Pd + D2O electrolysis, *J. Electroanal. Chem.* 380, 109, 1995.

[437.] Case, L. C., Catalytic fusion of deuterium into helium-4, in *The Seventh International Conference on Cold Fusion*, Jaeger, F. ENECO, Inc., Salt Lake City, UT, Vancouver, Canada, 1998, pp. 48.

[438.] Hagelstein, P. L., McKubre, M. C., Nagel, D. J., Chubb, T., and Hekman, R., New physical effects in metal deuterides. Report of the review on low energy nuclear reactions, in "Review of Low Energy Nuclear Reactions", DoE, Office of Sci., Washington, DC, 2004, in *11th International Conference on Cold Fusion*, Biberian, J.-P. World Scientific Co., Marseilles, France, 2004, pp. 23.

[439.] Morrey, J. R., Caffee, M. W., Farrar IV, H., Hoffman, N. J., Hudson, G. B., Jones, R. H., Kurz, M. D., Lupton, J., Oliver, B. M., Ruiz, B. V., Wacker, J. F., and van Veen, A., Measurements of helium in electrolyzed palladium, *Fusion Technol.* 18, 659, 1990.

[440.] Yamaguchi, E. and Nishioka, T., Helium-4 production and its correlation with heat evolution, *Oyo Butsuri* 62 (7), 712 (in Japanese), 1993.

[441.] Ueda, S., Yasuda, K., and Takahashi, A., Study of excess heat and nuclear products with closed electrolysis system and quadrupole mass spectrometer, in *The Seventh International Conference on Cold Fusion*, Jaeger, F. ENECO, Inc., Salt Lake City, UT, Vancouver, Canada, 1998, pp. 398.

[442.] George, R., Production of He-4 from deuterium during contact with nano-particle palladium on carbon at 200° C and 3 atmosphere deuterium pressure, www.hooked.net/~rgeorge/catalystpaper1.htm, 1999.

[443.] Matsunaka, M., Isobe, Y., Ueda, S., Yabuta, K., Ohishi, T., Mori, H., and Takahashi, A., Studies of coherent deuteron fusion and related nuclear reactions in solid., in *The 9th International Conference on Cold Fusion, Condensed Matter Nuclear Science*, Li, X. Z. Tsinghua Univ., Beijing, China, Tsinghua Univ., Beijing, China, 2002, pp. 237.

[444.] Zhang, Q. F., Gou, Q. Q., Zhu, Z. H., Xio, B. L., Lou, J. M., Liu, F. S., S., J. X., Ning, Y. G., Xie, H., and Wang, Z. G., The detection of 4-He in Ti-cathode on cold fusion, in *Third International Conference on Cold Fusion, "Frontiers of Cold Fusion"*, Ikegami, H. Universal Academy Press, Inc., Tokyo, Japan, Nagoya Japan, 1992, pp. 531.

[445.] Clarke, W. B., Search for ^{3}He and ^{4}He in Arata-style palladium cathodes I: A negative result, *Fusion Sci. & Technol.* 40, 2001.

[446.] Clarke, W. B. and Oliver, B. M., Response to "Comments on 'Search for 3He and 4He in Arata-Style Palladium Cathodes II: Evidence for Tritium Production" (Lett. to Ed;), *Fusion Sci. Technol.* 41, 153, 2002.

[447.] Clarke, W. B., Jenkins, W. J., and Top, Z., Determination of tritium by mass spectrometric measurement of ^{3}He, *Int. J. Appl. Radia. Isot.* 27, 515, 1976.

[448.] Mamyrin, B. A., Khabarin, L. V., and Yudenich, V. S., Anomalously high isotope ratio in helium in technical-grade metals and semiconductors, *Sov. Phys. Dokl.* 23, 581, 1978.

[449.] DoE, Report No. www.LENR-CANR.org, 2004.

[450.] DoE, Report of the review of low energy nuclear reactions, in *Review of Low Energy Nuclear Reactions* Department of Energy, Office of Science, Washington, DC, 2004.

[451.] Storms, E., Report No. http://lenr-canr.org/acrobat/StormsEaresponset.pdf, 2005.

[452.] Iwamura, Y., Itoh, T., Sakano, M., Yamazaki, N., Kuribayashi, S., Terada, Y., Ishikawa, T., and Kasagi, J., Observation of nuclear transmutation reactions induced by D2 gas permeation through Pd complexes, in *ICCF-11, International Conference on Condensed Matter Nuclear Science*, Biberian, J. P. World Scientific, Marseilles, France, 2004, pp. 339.

[453.] Sundaresan, R. and Bockris, J. O. M., Anomalous reactions during arcing between carbon rods in water, *Fusion Technol.* 26, 261, 1994.

[454.] Hanawa, T., X-ray spectroscropic analysis of carbon arc products in water, in *8th International Conference on Cold Fusion*, Scaramuzzi, F. Italian Physical Society, Bologna, Italy, Lerici (La Spezia), Italy, 2000, pp. 147.

[455.] Singh, M., Saksena, M. D., Dixit, V. S., and Kartha, V. B., Verification of the George Oshawa experiment for anomalous production of iron from carbon arc in water, *Fusion Technol.* 26, 266, 1994.

[456.] Jiang, X. L., Han, L. J., and Kang, W., Anomalous element production induced by carbon arcing under water, in *The Seventh International Conference on Cold Fusion*, Jaeger, F. ENECO, Inc., Salt Lake City, UT., Vancouver, Canada, 1998, pp. 172.

[457.] Ransford, H. E., Non-Stellar nucleosynthesis: Transition metal production by DC plasma-discharge electrolysis using carbon electrodes in a non-metallic cell, *Infinite Energy* 4 (23), 16, 1999.

[458.] Miley, G., Characteristics of reaction product patterns in thin metallic films experiments, in *Asti Workshop on Anomalies in Hydrogen/Deuterium Loaded Metals*, Collis, W. J. M. F. Italian Phys. Soc., Villa Riccardi, Rocca d'Arazzo, Italy, 1997, pp. 77.

[459.] Collis, W. J. M. F., Cold fusion or cold fission?, in *Asti Workshop on Anomalies in Hydrogen/Deuterium Loaded Metals*, Collis, W. J. M. F. Italian Phys. Soc., Villa Riccardi, Rocca d'Arazzo, Italy, 1997.

[460.] Takahashi, A., Production of stable isotopes by selective channel photofission of Pd, *Jpn. J. Appl. Phys. A* 40 (12), 7031-7046, 2001.

[461.] Miley, G. H., Narne, G., Williams, M. J., Patterson, J. A., Nix, J., Cravens, D., and Hora, H., Quantitative observations of transmutation products occuring in thin-film coated microspheres during electrolysis, in *Sixth International Conference on Cold Fusion, Progress in New Hydrogen Energy*, Okamoto, M. New Energy and Industrial Technology Development Organization, Tokyo Institute of Technology, Tokyo, Japan, Lake Toya, Hokkaido, Japan, 1996, pp. 629.

[462.] Mizuno, T., Ohmori, T., and Enyo, M., Isotopic changes of the reaction products induced by cathodic electrolysis in Pd, *J. New Energy* 1 (3), 31, 1996.

[463.] Violante, V., Castagna, E., Sibilia, C., Paoloni, S., and Sarto, F., Analysis of Ni-hydride thin film after surface plasmons generation by laser technique, in *Tenth International Conference on Cold Fusion*, Hagelstein, P. L. andChubb, S. R. World Scientific Publishing Co., Cambridge, MA, 2003, pp. 421.

[464.] Ohmori, T., Yamada, H., Narita, S., Mizuno, T., and Aoki, Y., Enrichment of ^{41}K isotope in potassium formed on and in a rhenium electrode during plasma electrolysis in K_2CO_3/H_2O and K_2CO_3/D_2O solutions., *J. Appl. Electrochem.* 33, 643, 2003.

[465.] Ohmori, T., Mizuno, T., Nodasaka, Y., and Enyo, M., Transmutation in a gold-light water electrolysis system, *Fusion Technol.* 33, 367, 1998.

[466.] Donohue, D. L. and Petek, M., Isotopic measurements of palladium metal containing protium and deuterium by glow discharge mass spectrometry, *Anal. Chem.* 63, 740, 1991.

[467.] Mizuno, T., Ohmori, T., and Enyo, M., Anomalous isotopic distribution in palladium cathode after electrolysis, *J. New Energy* 1 (2), 37, 1996.

[468.] Savvatimova, I. B., Senchukova, A. D., and Chernov, I. P., Transmutation phenomena in a palladium cathode after ions irradiation at glow discharge, in *The Sixth International Conference on Cold Fusion*, Okamoto, M. The Institute of Applied Energyy, Lake Toya, Japan, 1996, pp. 575.

[469.] Violante, V., Apicella, M., Capobianco, L., Sarto, F., Roada, A., Santoro, E., McKubre, M. C. H., Tanzella, F. L., and Sibilia, C., Search for nuclear ashes In electrochemical experiments, in *Tenth International Conference on Cold Fusion*, Hagelstein, P. L. andChubb, S. R. World Scientific Publishing Co., Cambridge, MA, 2003, pp. 405.

[470.] Kim, Y. E. and Passell, T. O., Alternative interpretations of low-energy nuclear reaction processes with deuterated metals based on the Bose-Einstein condensation mechanism, in *11th International Conference on Cold Fusion*, Biberian, J.-P. World Scientific Co., Marseilles, France, 2004, pp. 718.

[471.] Qiao, G. S., Han, X. L., Kong, L. C., Zheng, S. X., Huang, H. F., Yan, Y. J., Wu, Q. L., Deng, Y., Lei, S. L., and Li, X. Z., Nuclear products in a gas-loading D/Pd and H/Pd system, in *The Seventh International Conference on Cold Fusion*, Jaeger, F., Vancouver, Canada, 1998, pp. 314.

[472.] Li, X. Z., Yue, W. Z., Huang, G. S., Shi, H., Gao, L., Liu, M. L., and Bu, F. S., "Excess heat" measurement in gas-loading D/Pd system, in *Sixth International Conference on Cold Fusion, Progress in New Hydrogen Energy*, Okamoto, M. New Energy and Industrial Technology Development Organization, Tokyo Institute of Technology, Tokyo, Japan, Lake Toya, Hokkaido, Japan, 1996, pp. 455.

[473.] Oates, W. A. and Flanagan, T. B., Formation of nearly stoichiometric palladium-hydrogen systems, *Nature Phys. Sci.* 231, 19, 1971.

[474.] Bush, R. T. and Eagleton, R. D., Evidence for electrolytically induced transmutation and radioactivity correlated with excess heat in electrolytic cells with light water rubidium salt electrolytes, in *Fourth International Conference on Cold Fusion* Electric Power Research Institute 3412 Hillview Ave., Palo Alto, CA 94304, Lahaina, Maui, 1993, pp. 2.

[475.] Iwamura, Y., Sakano, M., and Itoh, T., Elemental analysis of Pd complexes: effects of D_2 gas permeation, *Jpn. J. Appl. Phys. A* 41 (7), 4642, 2002.

[476.] Iwamura, Y., Itoh, T., Sakano, M., Sakai, S., and Kuribayashi, S., Low energy nuclear transmutation in condensed matter induced by D_2 gas permeation through Pd complexes: correlation between deuterium flux and nuclear products, in *Tenth International Conference on Cold Fusion*, Hagelstein, P. L. andChubb, S. R. World Scientific Publishing Co., Cambridge, MA, 2003, pp. 435.

[477.] Iwamura, Y., Itoh, T., Sakano, M., Yamazaki, N., Kuribayashi, S., Terada, Y., and Ishikawa, T., Observation of surface distribution of products by X-ray fluorescence spectrometry during D_2 gas permeation through Pd complexes, in *12th International Conference on Condensed Matter Nuclear Science*, Takahashi, A., Yokohama, Japan, 2005.

[478.] Iwamura, Y., Itoh, T., and Sakano, M., U.S.A. US 2002/0080903 A1 and EP 1 202 290 A2, 2002, Nuclide transmutation device and nuclide transmutation method. Jun. 27, 2002.

[479.] Kitamura, A., Nishio, R., Iwai, H., Satoh, R., Taniike, A., and Furuyama, Y., In-situ accelerator analysis of palladium complex under deuterium permeation, in *12th International Conference on Condensed Matter Nuclear Science*, Takahashi, A., Yokohama, Japan, 2005.

[480.] Higashiyama, Y., Sakano, M., Miyamaru, H., and Takahashi, A., Replication of MHI transmutation experiment by D_2 gas permeation through Pd complex, in *Tenth International Conference on Cold Fusion*, Hagelstein, P. L. andChubb, S. R. World Scientific Publishing Co., Cambridge, MA, 2003, pp. 447.

[481.] Wang, Q. and Dash, J., Effect of an additive on thermal output during electrolysis of heavy water with a palladium cathode, in *12th International Conference on Condensed Matter Nuclear Science*, Takahashi, A., Yokohama, Japan, 2005.

[482.] Szpak, S., Mosier-Boss, P. A., and Gordon, F., Precursors and the fusion reactions in polarized Pd/D-D$_2$O systems: Effect of an external electric field, in *11th International Conference on Cold Fusion*, Biberian, J.-P. World Scientific Co., Marseilles, France, 2004, pp. 359.

[483.] Lochak, G. and Urutskoev, L., Low-energy nuclear reactions and the leptonic monopole, in *11th International Conference on Cold Fusion*, Biberian, J.-P. World Scientific Co., Marseilles, France, 2004, pp. 421.

[484.] Celani, F., Spallone, A., Righi, E., Trenta, G., Catena, C., D'Agostaro, G., Quercia, P., Andreassi, V., Marini, P., Di Stefano, V., Nakamura, M., Mancini, A., Sona, P. G., Fontana, F., Gamberale, L., Garbelli, D., Celia, E., Falcioni, F., Marchesini, M., Novaro, E., and Mastromatteo, U., Innovative procedure for the, in situ, measurement of the resistive thermal coefficient of H(D)/Pd during electrolysis; cross-comparison of new elements detected in the Th-Hg-Pd-D(H) electroytic cells, in *11th International Conference on Cold Fusion*, Biberian, J.-P. World Scientific Co., Marseilles, France, 2004, pp. 108.

485. Yamada, H., Narita, S., Onodera, H., Suzuki, N., Tanaka, N., and Nyui, T., Analysis by time-of-flight secondary ion mass spectroscopy for nuclear products In hydrogen penetration through palladium, in *Tenth International Conference on Cold Fusion*, Hagelstein, P. L. andChubb, S. R. World Scientific Publishing Co., Cambridge, MA, 2003, pp. 455.

486. Passell, T. O., Pd110/Pd108 ratios and trace element changes in particulate palladium exposed to deuterium gas, in *Tenth International Conference on Cold Fusion*, Hagelstein, P. L. andChubb, S. R. World Scientific Publishing Co., Cambridge, MA, 2003, pp. 399.

487. Yamada, H., Narita, S., Fujii, Y., Sato, T., Sasaki, S., and Omori, T., Production of Ba and several anomalous elements in Pd under light water electrolysis, in *The 9th International Conference on Cold Fusion, Condensed Matter Nuclear Science*, Li, X. Z. Tsinghua Univ. Press, Tsinghua Univ., Beijing, China, 2002, pp. 420.

488. Vysotskii, V. I., Kornilova, A. A., Samoylenko, I. I., and Zykov, G. A., Catalytic influence of caesium on the effectiveness of nuclear transmutation on intermediate and heavy mass isotopes in growing biological cultures, in *The 9th International Conference on Cold Fusion, Condensed Matter Nuclear Science*, Li, X. Z. Tsinghua Univ. Press, Tsinghua Univ., Beijing, China, 2002, pp. 391.

489. Iwamura, Y., Itoh, T., Sakano, M., and Sakai, S., Observation of low energy nuclear reactions induced by D_2 gas permeation through Pd complexes, in *The Ninth International Conference on Cold Fusion (ICCF9)*, Li, X. Z. Tsinghua University, Beijing, China, 2002, pp. 141.

490. Goryachev, I. V., Registration of synthesis of $_{45}Rh^{102}$ in media of excited nuclei of $_{28}Ni^{58}$., in *The 9th International Conference on Cold Fusion, Condensed Matter Nuclear Science*, Li, X. Z. Tsinghua Univ. Press, Tsinghua Univ., Beijing, China, 2002, pp. 109.

491. Di Giulio, M., Filippo, E., Manno, D., and Nassisi, V., Analysis of nuclear transmutations observed in D- and H-loaded films, *J. Hydrogen Eng.* 27, 527, 2002.

492. Arapi, A., Ito, R., Sato, N., Itagaki, M., Narita, S., and Yamada, H., Experimental observation of the new elements production in the deuterated and/or hydride palladium electrodes, exposed to low energy DC glow discharge, in *The 9th International Conference on Cold Fusion, Condensed Matter Nuclear Science*, Li, X. Z. Tsinghua Univ. Press, Tsinghua Univ., Beijing, China, 2002, pp. 1.

493. Yamada, H., Uchiyama, K., Kawata, N., Kurisawa, Y., and Nakamura, M., Producing a radioactive source in a deuterated palladium electrode under direct-current glow discharge, *Fusion Technol.* 39, 253, 2001.

494. Wang, T., Zhu, Y., Wang, Z., Li, S., and Zheng, S., Nuclear phemonena in P+Ti_2H_x experiments, in *8th International Conference on Cold Fusion*, Scaramuzzi, F. Italian Physical Society, Bologna, Italy, Lerici (La Spezia), Italy, 2000, pp. 317.

495. Vysotskii, V., Kornilova, A. A., Samoylenko, I. I., and Zykov, G. A., Experimental observation and study of controlled transmutation of intermediate mass isotopes in growing biological cultures, in *8th International Conference on Cold Fusion*, Scaramuzzi, F. Italian Physical Society, Bologna, Italy, Lerici (La Spezia), Italy, 2000, pp. 135.

496. Passell, T. O. and George, R., Trace elements added to palladium by exposure to gaseous deuterium, in *8th International Conference on Cold Fusion*, Scaramuzzi, F. Italian Physical Society, Bologna, Italy, Lerici (La Spezia), Italy, 2000, pp. 129.

497. Nassisi, V. and Longo, M. L., Experimental results of transmutation of elements observed in etched palladium samples by an excimer laser, *Fusion Technol.* 37 (May), 247, 2000.

498. Li, X. Z., Yan, Y. J., Tian, J., Mei, M. Y., Deng, Y., Yu, W. Z., Tang, G. Y., and Cao, D. X., Nuclear transmutation in Pd deuteride, in *8th International Conference on Cold Fusion*, Scaramuzzi, F. Italian Physical Society, Bologna, Italy, Lerici (La Spezia), Italy, 2000, pp. 123.

499. Iwamura, Y., Itoh, T., and Sakano, M., Nuclear products and their time dependence induced by continuous diffusion of deuterium through multi-layer palladium containing low work function material, in *8th International Conference on Cold Fusion*, Scaramuzzi, F. Italian Physical Society, Bologna, Italy, Lerici (La Spezia), Italy, 2000, pp. 141.

[500.] Castellano, Di Giulio, M., Dinescu, M., Nassisi, V., Conte, A., and Pompa, P. P., Nuclear transmutation in deutered Pd films irradiated by an UV laser, in *8th International Conference on Cold Fusion*, Scaramuzzi, F. Italian Physical Society, Bologna, Italy, Lerici (La Spezia), Italy, 2000, pp. 287.

[501.] Ohmori, T. and Mizuno, T., Nuclear transmutation reaction caused by light water electrolysis on tungsten cathode under incandescent conditions, *Infinite Energy* 5 (27), 34, 1999.

[502.] Klopfenstein, M. F. and Dash, J., Thermal imaging during electrolysis of heavy water with a Ti cathode, in *The Seventh International Conference on Cold Fusion*, Jaeger, F. Vancouver, Canada, Vancouver, Canada, 1998, pp. 98.

[503.] Kong, L. C., Han, X. L., Zheng, S. X., Huang, H. F., Yan, Y. J., Wu, Q. L., Deng, Y., Lei, S. L., Li, C. X., and Li, X. Z., Nuclear products and transmutation in a gas-loading D/Pd and H/Pd system, *J. New Energy* 3 (1), 20, 1998.

[504.] Notoya, R., Ohnishi, T., and Noya, Y., Products of nuclear processes caused by electrolysis on nickel and platinum electrodes in solutions of alkali-metallic ions, in *The Seventh International Conference on Cold Fusion*, Jaeger, F. ENECO, Inc., Salt Lake City, UT., Vancouver, Canada, 1998, pp. 269.

[505.] Nassisi, V., Transmutation of elements in saturated palladium hydrides by an XeCl excimer laser, *Fusion Technol.* 33, 468, 1998.

[506.] Nassisi, V., Incandescent Pd and Anomalous Distribution of Elements in Deuterated Samples Processed by an Excimer Laser, *J. New Energy* 2 (3/4), 14, 1997.

[507.] Jiang, X.-L., Chen, C.-Y., Fu, D.-F., Han, L.-J., and Kang, W., Tip effect and nuclear-active sites, in *The Seventh International Conference on Cold Fusion*, Jaeger, F. ENECO, Inc., Salt Lake City, UT., Vancouver, Canada, 1998, pp. 175.

[508.] Nakamura, K., Kishimoto, Y., and Ogura, I., Element conversion by arcing in aqueous solution, *J. New Energy* 2 (2), 53, 1997.

[509.] Qiao, G. S., Han, X. M., Kong, L. C., and Li, X. Z., Nuclear transmutation in a gas-loading system, *J. New Energy* 2 (2), 48, 1997.

[510.] Dash, J., Kopecek, R., and Miguet, S., Excess heat and unexpected elements from aqueous electrolysis with titanium and palladium cathodes, in *32nd Intersociety Energy Conversion Engineering Conference*, 1997, pp. 1350-1355.

[511.] Yamada, H., Nonaka, H., Dohi, A., Hirahara, H., Fujiwara, T., Li, X.-Z., and Chiba, A., Carbon production on palladium point electrode with neutron burst under DC glow discharge in pressurized deuterium gas, *J. New Energy* 1 (4), 55, 1996.

[512.] Miley, G. H. and Patterson, J. A., Nuclear transmutations in thin-film nickel coatings undergoing electrolysis, *J. New Energy* 1 (3), 5, 1996.

[513.] Notoya, R., Low temperature nuclear change of alkali metallic ions caused by electrolysis, *J. New Energy* 1, 39, 1996.

[514.] Mizuno, T., Inoda, K., Akimoto, T., Azumi, K., Kitaichi, M., Kurokawa, K., Ohmori, T., and Enyo, M., Formation of ^{197}Pt radioisotopes in solid state electrolyte treated by high temperature electrolysis in D_2 gas., *Infinite Energy* 1 (4), 9, 1995.

[515.] Mizuno, T., Akimoto, T., Azumi, K., Kitaichi, M., Kuroiwa, K., and Enyo, M., Excess heat evolution and analysis of elements for solid state electrolyte in deuterium atmosphere during applied electric field, *J. New Energy* 1 (1), 79, 1996.

[516.] Dash, J. and Miguet, S., Microanalysis of Pd cathodes after electrolysis in aqueous acids, *J. New Energy* 1 (1), 23, 1996.

[517.] Miguet, S. and Dash, J., Microanalysis of palladium after electrolysis in heavy water, *J. New Energy* 1 (1), 23, 1996.

[518.] Matsumoto, T., Experiments of underwater spark discharge with pinched electrodes, *J. New Energy* 1 (4), 79, 1996.

[519.] Bush, R. T., Electrolytic stimulated cold nuclear synthesis of strontium from rubidium, *J. New Energy* 1, 28, 1996.

[520.] Komaki, H., An Approach to the Probable Mechanism of the Non-Radioactive Biological Cold Fusion or So-Called Kervran Effect (Part 2), in *Fourth International Conference on Cold Fusion* Electric Power Research Institute 3412 Hillview Ave., Palo Alto, CA 94304, Lahaina, Maui, 1993, pp. 44.

[521.] Dillon, C. T. and Kennedy, B. J., The electrochemically formed palladium-deuterium system. I. Surface composition and morphology, *Aust. J. Chem.* 46, 663, 1993.

[522.] Bush, R. T. and Eagleton, R. D., Experimental studies supporting the transmission resonance model for cold fusion in light water: I. Correlation of isotopic and elemental evidence with excess energy., in *Third International Conference on Cold Fusion, "Frontiers of Cold Fusion"*, Ikegami, H. Universal Academy Press, Inc., Tokyo, Japan, Nagoya Japan, 1992, pp. 405.

[523.] Bush, R. T., Electrolytically simulated cold nuclear synthesis of strontium from rubidium, *J. New Energy* 1 (1), 28, 1996.

[524.] Rolison, D. R. and O'Grady, W. E., Observation of elemental anomalies at the surface of palladium after electrochemical loading of deuterium or hydrogen, *Anal. Chem.* 63, 1697, 1991.

[525.] Greber, T., Fischer, A., Rheme, C., Drissi, S., Osterwalder, J., Kern, J., and Schlapbach, L., Cold fusion experiments in Fribourg, in *Understanding Cold Fusion Phenomena*, Ricci, R. A., Sindoni, E., andDeMarco, F., 1989.

[526.] Melich, M. E. and Hansen, W. N., Some lessons from 3 Years of electrochemical calorimetry, in *Third International Conference on Cold Fusion, "Frontiers of Cold Fusion"*, Ikegami, H. Universal Academy Press, Inc., Tokyo, Japan, Nagoya Japan, 1992, pp. 397.

[527.] Miles, M. H. and Bush, B. F., Calorimetric principles and problems in Pd-D$_2$O electrolysis, in *Third International Conference on Cold Fusion, "Frontiers of Cold Fusion"*, Ikegami, H. Universal Academy Press, Inc., Tokyo, Japan, Nagoya Japan, 1992, pp. 113.

[528.] Fleischmann, M., The experimenters' regress, in *5th International Conference on Cold Fusion*, Pons, S. IMRA Europe, Sophia Antipolis Cedex, France, Monte-Carlo, Monaco, 1995, pp. 152.

[529.] Perez-Pariente, J., Evidence on the occurrence of LENR-type processes in alchemist transmutation, in *11th International Conference on Cold Fusion*, Biberian, J.-P. World Scientific Co., Marseilles, France, 2004, pp. 554.

[530.] Cau, A., Natural nuclear synthesis of superheavy elements (SHE), *J. New Energy* 1 (3), 155, 1996.

[531.] Monti, R. A., Low energy nuclear reactions: Experimental evidence for the alpha extended model of the atom, *J. New Energy* 1 (3), 131, 1996.

[532.] Monti, R. A., Nuclear transmutation processes of lead, silver, thorium, uranium, in *The Seventh International Conference on Cold Fusion*, Jaeger, F. ENECO, Inc., Salt Lake City, UT., Vancouver, Canada, 1998, pp. 264.

[533.] Champion, J., *Producing precious metals at home* Discovery Publishing, Westboro, WI, 1994.

[534.] Lipson, A. G., Miley, G., Lyakhov, B. F., and Roussetski, A. S., Energetic charged particles emission from hydrogen-loaded Pd and Ti cathodes and its enhancement by He-4 implantation, in *11th International Conference on Cold Fusion*, Biberian, J.-P. World Scientific Co, Marseilles, France, 2004, pp. 324.

[535.] Kowalski, L., Jones, S., Letts, D., and Cravens, D., Charged particles from Ti and Pd foils, in *11th International Conference on Cold Fusion*, Biberian, J.-P. World Scientific Co., Marseille, France, 2004, pp. 269.

[536.] Takahashi, A., Progress in condensed matter nuclear science, in *12 th International Conference on Cold Fusion*, Takahashi, A. World Scientific Co., Yokohama, Japan, 2006.

[537.] Oriani, R. A. and Fisher, J. C., Detection of energetic charged particles during electrolysis, in *Tenth International Conference on Cold Fusion*, Hagelstein, P. L. andChubb, S. R. World Scientific Publishing Co., Cambridge, MA, 2003, pp. 577.

[538.] Oriani, R. A. and Fisher, J. C., Energetic charged particles produced during electrolysis, in *Tenth International Conference on Cold Fusion*, Hagelstein, P. L. andChubb, S. R. World Scientific Publishing Co., Cambridge, MA, 2003, pp. 567.

[539.] Oriani, R. A. and Fisher, J. C., Generation of nuclear tracks during electrolysis, *Jpn. J. Appl. Phys. A* 41, 6180-6183, 2002.

[540.] Keeney, F., Jones, S. E., Johnson, A., Buehler, D. B., Cecil, F. E., Hubler, G. K., Hagelstein, P. L., Scott, M., and Ellsworth, J., Charged-particle emissions from metal deuterides, in *Tenth International Conference on Cold Fusion*, Hagelstein, P. L. andChubb, S. R. World Scientific Publishing Co., Cambridge, MA, 2003, pp. 509.

[541.] Lipson, A. G., Roussetski, A. S., Miley, G. H., and Saunin, E. I., Phenomenon of an energetic charged particle emission from hydrogen/deuterium loaded metals, in *Tenth International Conference on Cold Fusion*, Hagelstein, P. L. andChubb, S. R. World Scientific Publishing Co., Cambridge, MA, 2003, pp. 539.

[542.] Cecil, E., Liu, H., and Galovich, C. S., Energetic charged particles from deuterium metal systems, in *Tenth International Conference on Cold Fusion*, Chubb, P. I. H. a. S. R. World Scientific Publishing Co., Cambridge, MA, 2003, pp. 535.

[543.] Afonichev, D., High-frequency radiation and tritium channel, in *Tenth International Conference on Cold Fusion*, Hagelstein, P. L. andChubb, S. R. World Scientific Publishing Co., Cambridge, MA, 2003, pp. 353.

[544.] Afonichev, D. and Murzinova, M., Indicator of the process of cold fusion, *Int. J. Hydrogen Energy* 28, 1005-1010, 2003.

[545.] Lipson, A. G., Roussetsky, A. S., Miley, G. H., and Castano, C. H., In-situ charged particles and X-ray detection in Pd thin film-cathodes during electrolysis in Li_2SO_4/H_2O., in *The 9th International Conference on Cold Fusion, Condensed Matter Nuclear Science*, Li, X. Z. Tsinghua Univ. Press, Tsinghua Univ., Beijing, China, 2002, pp. 218.

[546.] Miley, G. H., Hora, H., Lipson, A. G., Kim, S. O., Luo, N., Castano, C. H., and Woo, T., Progress in thin-film LENR research at the University of Illinois., in *The 9th International Conference on Cold Fusion, Condensed Matter Nuclear Science*, Li, X. Z. Tsinghua Univ. Press, Tsinghua Univ., Beijing, China, 2002, pp. 255.

[547.] Lipson, A. G., Lyakhov, B. F., Roussetski, A. S., Akimoto, T., Mizuno, T., Asami, N., Shimada, R., Miyashita, S., and Takahashi, A., Evidence for low-intensity D-D reaction as a result of exothermic deuterium desorption from Au/Pd/PdO:D heterostructure, *Fusion Technol.* 38, 238, 2000.

[548.] Lipson, A. G., Lyakhov, B. F., Rousstesky, A. S., and Asami, N., Evidence for DD-reaction and a long-range alpha emission in Au/Pd/PdO:D heterstructure as a result of exothermic deuterium deposition, in *8th International Conference on Cold Fusion*, Scaramuzzi, F. Italian Physical Society, Bologna, Italy, Lerici (La Spezia), Italy, 2000, pp. 231.

[549.] Roussetski, A. S., Application of CR-39 plastic track detector for detection of DD and DT-reaction products in cold fusion experiments, in *8th International Conference on Cold Fusion*, Scaramuzzi, F. Italian Physical Society, Bologna, Italy, Lerici (La Spezia), Italy, 2000, pp. 253.

[550.] Roussetski, A. S., Investigation of nuclear emissions in the process of D(H) escaping from deuterized (hydrogenized) PdO-Pd-PdO and PdO-Ag samples, in *Sixth International Conference on Cold Fusion, Progress in New Hydrogen Energy*, Okamoto, M. New Energy and Industrial Technology Development Organization, Tokyo Institute of Technology, Tokyo, Japan, Lake Toya, Hokkaido, Japan, 1996, pp. 345.

[551.] Savvatimova, I., Reproducibility of experiments in glow discharge and processes accompanying deuterium ions bombardment, in *8th International Conference on Cold Fusion*, Scaramuzzi, F. Italian Physical Society, Bologna, Italy, Lerici (La Spezia), Italy, 2000, pp. 277.

[552.] Szpak, S. and Mosier-Boss, P. A., Nuclear and thermal events associated with Pd + D co-deposition, *J. New Energy* 1 (3), 54, 1996.

[553.] Lin, G. H. and Bockris, J. O. M., Anomalous radioactivity and unexpected elements as a result of heating inorganic mixtures, *J. New Energy* 1 (3), 100, 1996.

[554.] Notoya, R., Ohnishi, T., and Noya, Y., Nuclear reactions caused by electrolysis in light and heavy water solutions, *J. New Energy* 1 (4), 40, 1996.

555. Rout, R. K., Shyam, A., Srinivasan, M., Garg, A. B., and Shrikhande, V. K., Reproducible, anomalous emissions from palladium deuteride/hydride, *Fusion Technol.* 30, 273, 1996.

556. Rout, R. K., Shyam, A., Srinivasan, M., and Garg, A. B., Phenomenon of low energy emission from hydrogen/deuterium loaded palladium, in *Third International Conference on Cold Fusion, "Frontiers of Cold Fusion"*, Ikegami, H. Universal Academy Press, Inc., Tokyo, Japan, Nagoya Japan, 1992, pp. 547.

557. Matsumoto, T., Cold fusion experiments using sparking discharges in water, in *5th International Conference on Cold Fusion*, Pons, S. IMRA Europe, Sophia Antipolis Cedex, France, Monte-Carlo, Monaco, 1995, pp. 583.

558. Matsumoto, T., Extraordinary traces produced during pulsed discharge in water, *Bull. Faculty of Eng., Hokkaido Univ,* 175, 1995.

559. Manduchi, C., Zannoni, G., Milli, E., Riccardi, L., Mengoli, G., Fabrizio, M., and Buffa, A., Anomalous effects during the interaction of subatmospheric $D_2(H_2)$ with Pd from 900° C to room temperature, *Nuovo Cimento Soc. Ital. Fis. A* 107 A (2), 171, 1994.

560. Koval'chuk, E. P., Yanchuk, O. M., and Reshetnyak, O. V., Electromagnetic radiation during electrolysis of heavy water, *Phys. Lett. A* 189, 15, 1994.

561. Taniguchi, R., Characteristic peak structures on charged particle spectra during electrolysis experiment, *Trans. Fusion Technol.* 26, 186, 1994.

562. Taniguchi, R. and Yamamoto, T., Fine structure of the charged particle bursts induced by D_2O electrolysis, in *Third International Conference on Cold Fusion, "Frontiers of Cold Fusion"*, Ikegami, H. Universal Academy Press, Inc., Tokyo, Japan, Nagoya Japan, 1992, pp. 519.

563. Matsumoto, T., Cold fusion experiments with ordinary water and thin nickel foil, *Fusion Technol.* 24, 296, 1993.

564. Matsumoto, T., Observation of gravity decays of multiple-neutron nuclei during cold fusion, *Fusion Technol.* 22, 164, 1992.

565. Mo, D. W., Zhang, L., Chen, B. X., Liu, Y. S., Doing, S. Y., Yao, M. Y., Zhou, L. Y., Huang, H. G., Li, X. Z., Shen, X. D., Wang, S. C., Kang, T. S., and Huang, N. Z., Real time measurements of the energetic charged particles and the loading ratio (D/Pd), in *Third International Conference on Cold Fusion, "Frontiers of Cold Fusion"*, Ikegami, H. Universal Academy Press, Inc., Tokyo, Japan, Nagoya Japan, 1992, pp. 535.

566. Mo, W., Liu, Y. S., Zhou, L. Y., Dong, S. Y., Wang, K. L., Wang, S. C., and Li, X. Z., Search for precursor and charged particles in "cold fusion", in *Second Annual Conference on Cold Fusion, "The Science of Cold Fusion"*, Bressani, T., Del Giudice, E., andPreparata, G. Societa Italiana di Fisica, Bologna, Italy, Como, Italy, 1991, pp. 123.

567. Long, H. Q., Sun, S. H., Liu, H. Q., Xie, R. S., Zhang, X. W., and Zhang, W. S., Anomalous effects in deuterium/metal systems, in *Third International Conference on Cold Fusion, "Frontiers of Cold Fusion"*, Ikegami, H. Universal Academy Press, Inc., Tokyo, Japan, Nagoya Japan, 1992, pp. 447.

568. Uchida, H., Hamada, Y., Matsumura, Y., and Hayashi, T., Detection of radioactive emissions in the electrolytic deuteriding-dedeuteriding reactions of Pd and Ti, in *Third International Conference on Cold Fusion, "Frontiers of Cold Fusion"*, Ikegami, H. Universal Academy Press, Inc., Tokyo, Japan, Nagoya Japan, 1992, pp. 539.

569. Wang, K. L., Li, X. Z., Dong, S. Y., Wang, S. C., Mo, D. W., Luo, C. M., Lin, Q. R., Wu, X. D., Li, W. Z., Zhu, Y. F., Zhou, P. L., and Chang, L., Search for better material for cold fusion experiment using CR-39 detector, in *Second Annual Conference on Cold Fusion, "The Science of Cold Fusion"*, Bressani, T., Del Giudice, E., andPreparata, G. Societa Italiana di Fisica, Bologna, Italy, Como, Italy, 1991, pp. 163.

570. Wang, C., Kang, T. S., Wang, K. L., Dong, S. Y., Feng, Y. Y., Mo, D. W., and Li, X. Z., Identification of the energetic charged particles in gas-loading experiment of "cold fusion" using CR-39 plastic track detector., in *Second Annual Conference on Cold Fusion, "The Science of Cold Fusion"*, Bressani, T., Del Giudice, E., andPreparata, G. Societa Italiana di Fisica, Bologna, Italy, Como, Italy, 1991, pp. 169.

[571.] Jin, S., Zhang, F., Yao, D., and Wu, B., Anomalous nuclear events in deuterium palladium systems, in *Second Annual Conference on Cold Fusion, "The Science of Cold Fusion"*, Bressani, T., Del Giudice, E., andPreparata, G. Societa Italiana di Fisica, Bologna, Italy, Como, Italy, 1991, pp. 145.

[572.] Dong, S. Y., Wang, K. L., Feng, Y. Y., Chang, L., Luo, C. M., Hu, R. Y., Zhou, P. L., Mo, D. W., Zhu, Y. F., Song, C. L., Chen, Y. T., Yao, M. Y., Ren, C., Chen, Q. K., and Li, X. Z., Precursors to 'cold fusion' phenomenon and the detection of energetic charged particles in deuterium/solid systems, *Fusion Technol.* 20, 330, 1991.

[573.] Li, X., Dong, S. Y., Wang, K. L., Feng, Y. Y., Luo, C., Hu, R., Zhou, P., Mo, D., Zhu, Y., Song, C., Chen, Y., Yao, M., Ren, C., and Chen, Q., The precursor of "cold fusion" phenomenon in deuterium/solid systems, in *Anomalous Nuclear Effects in Deuterium/Solid Systems, "AIP Conference Proceedings 228"*, Jones, S., Scaramuzzi, F., andWorledge, D. American Institute of Physics, New York, Brigham Young Univ., Provo, UT, 1990, pp. 419.

[574.] Taniguchi, R. and Yamamoto, T., High sensitivity measurement of charged particles emitted during pulsed electrolysis of D_2O, in *Anomalous Nuclear Effects in Deuterium/Solid Systems, "AIP Conference Proceedings 228"*, Jones, S., Scaramuzzi, F., andWorledge, D. American Institute of Physics, New York, Brigham Young Univ., Provo, UT, 1990, pp. 445.

[575.] Taniguchi, R., Yamamoto, T., and Irie, S., Detection of charged particles emitted by electrolytically induced cold nuclear fusion, *Jpn. J. Appl. Phys.* 28 (11), L2021, 1989.

[576.] Matsumoto, T., Progress of NATTOH model and new particles emitted during cold fusion, in *Anomalous Nuclear Effects in Deuterium/Solid Systems, "AIP Conference Proceedings 228"*, Jones, S., Scaramuzzi, F., andWorledge, D. American Institute of Physics, New York, Brigham Young Univ., Provo, UT, 1990, pp. 827.

[577.] Matsumoto, T., Prediction of new particle emission on cold fusion, *Fusion Technol.* 18, 647, 1990.

[578.] Jones, S. E., Bartlett, T. K., Buehler, D. B., Czirr, J. B., Jensen, G. L., and Wang, J. C., Preliminary results from the BYU charged-particle spectrometer, in *Anomalous Nuclear Effects in Deuterium/Solid Systems, "AIP Conference Proceedings 228"*, Jones, S., Scaramuzzi, F., andWorledge, D. American Institute of Physics, New York, Brigham Young Univ., Provo, UT, 1990, pp. 397.

[579.] Cecil, F. E., Liu, H., Beddingfield, D. H., and Galovich, C. S., Observation of charged-particle bursts from deuterium-loaded thin-titanium foils, in *Anomalous Nuclear Effects in Deuterium/Solid Systems, "AIP Conference Proceedings 228"*, Jones, S., Scaramuzzi, F., andWorledge, D. American Institute of Physics, New York, Brigham Young Univ., Provo, UT, 1990, pp. 375.

[580.] Celani, F., DiStefano, V., Pace, S., and Bianco, S., Report No. LNF-89/048(P), 1989.

[581.] Bennington, S. M., Sokhi, R. S., Stonadge, P. R., Ross, D. K., Benham, M. J., Beynon, T. D., Whithey, P., Harris, I. R., and Farr, J. P. G., A search for the emission of x-rays from electrolytically charged palladium-deuterium, *Electrochim. Acta* 34, 1323, 1989.

[582.] Deakin, M. R., Fox, J. D., Kemper, K. W., Myers, E. G., Shelton, W. N., and Skofronick, J. G., Search for cold fusion using x-ray detection, *Phys. Rev. C: Nucl. Phys* 40 (5), R1851, 1989.

[583.] Briand, J. P., Ban, G., Froment, M., Keddam, M., and Abel, F., Cold fusion rates in titanium foils, *Phys. Lett.* A145 (4), 187, 1990.

[584.] Brudanin, V. B., Bystritskii, V. M., Egorov, V. G., Shamsutdinov, S. G., Shyshkin, A. L., Stolupin, V. A., and Yutlandov, I. A., Does cold nuclear fusion exist?, *Phys. Lett. A* 146, 347, 1990.

[585.] Antanasijevic, R. D., Konjevic, D. J., Maric, Z., Sevic, D. M., Vigier, J. P., and Zaric, A. J., "Cold fusion" in terms of new quantum chemistry: The role of magnetic interactions in dense physica media, in *5th International Conference on Cold Fusion*, Pons, S. IMRA Europe, Sophia Antipolis Cedex, France, Monte-Carlo, Monaco, 1995, pp. 505.

[586.] Iwamura, Y., Gotoh, N., Itoh, T., and Toyoda, I., Characteristic X-ray and neutron emissions from electrochemically deuterated palladium, in *5th International Conference on Cold Fusion*, Pons, S. IMRA Europe, Sophia Antipolis Cedex, France, Monte-Carlo, Monaco, 1995, pp. 197.

[587.] Abriola, D., Achterberg, E., Davidson, M., Debray, M., Etchegoyen, M. C., Fazzini, N., Niello, J. F., Ferrero, A. M. J., Filevich, A., Galia, M. C., Garavaglia, R., G., G. B., Gettar, R. T., Gil, S., Grahmann, H., Huck, H., Jech , A., Kreiner, A. J., Macchiavelli, A. O., Magallanes, J. F., Maqueda, E., Marti, G., Pacheco, A. J., Percz, M. L., Pomar, C., Ramirez, M., and Scassera, M., Examination of nuclear measurement conditions in cold fusion experiments, *J. Electroanal. Chem.* 265, 355, 1989.

[588.] Gai, M., Rugari, S. L., France, R. H., Lund, B. J., Zhao, Z., Davenport, A. J., Isaacs, H. S., and Lynn, K. G., Upper limits on neutron and gamma-ray emission from cold fusion, *Nature (London)* 340, 29, 1989.

[589.] Petrasso, R. D., Chen, X., Wenzel, K. W., Parker, R. R., Li, C. K., and Fiore, C., Measurement of g-rays from cold fusion, *Nature (London)* 339, 667, 1989.

[590.] Savvatimova, I. and Dash, J., Emission registration on films during glow discharge experiments, in *The 9th International Conference on Cold Fusion, Condensed Matter Nuclear Science*, Li, X. Z. Tsinghua Univ. Press, Tsinghua Univ., Beijing, China, 2002, pp. 312.

[591.] Narita, S., Yamada, H., Arapi, A., Sato, N., Kato, D., Yamamura, M., and Itagaki, M., Gamma ray detection and surface analysis on palladium electrode in DC glow-like discharge experiment, in *Tenth International Conference on Cold Fusion*, Hagelstein, P. L. andChubb, S. R. World Scientific Publishing Co., Cambridge, MA, 2003, pp. 603.

[592.] Roussetski, A. S., CR-39 track detectors in cold fusion experiments: Review and perspectives, in *11th International Conference on Cold Fusion*, Biberian, J.-P. World Scientific Co, Marseilles, France, 2004, pp. 274.

[593.] Price, P. B., Barwick, S. W., Williams, W. T., and Porter, J. D., Search for energetic-charged-particle emission from deuterated Ti and Pd foils, *Phys. Rev. Lett.* 63 (18), 1926, 1989.

[594.] Baranowski, B., Filipek, S. M., Szustakowski, M., Farny, J., and Woryna, W., Search for 'cold fusion' in some Me-D systems at high pressures of gaseous deuterium, *J. Less-Common Met.* 158, 347, 1990.

[595.] Barwick, S. W., Price, P. B., Williams, W. T., and Porter, J. D., Search for 0.8 MeV (3)He nuclei emitted from Pd and Ti exposed to high pressure D_2, *J. Fusion Energy* 9 (3), 273, 1990.

[596.] Brudanin, V. B., Bystritsky, V. M., Egorov, V. G., Stetsenko, S. G., and Yutlandov, I. A., Search for the cold fusion d(d,^{4}He) in electrolysis of D_2O, *Phys. Lett. A* 151 (9), 543, 1990.

[597.] Miljanic, S., Jevtic, N., Pesic, S., Ninkovic, M., Nikolic, D., Josipovic, M., Petkovska, L. J., and Bacic, S., An attempt to replicate cold fusion claims, *Fusion Technol.* 18, 340, 1990.

[598.] Jin, S., Zhang, F., Yao, D., Wang, Q., Wu, B., Feng, Y., and Chen, M., Anomalous nuclear effects in palladium-deuterium systems during the gas discharge process, *Gaojishu Tongxun* 1 (5), 25 (In Chinese), 1991.

[599.] Oriani, R. A. and Fisher, J. C., Energetic particle showers in the vapor from electrolysis, in *11th International Conference on Cold Fusion*, Biberian, J.-P. World Scientific Co., Marseilles, France, 2004, pp. 281.

[600.] Oriani, R. A. and Fisher, J. C., Nuclear reactions produced in an operating electrolytic cell, in *11th International Conference on Cold Fusion*, Biberian, J.-P. World Scientific Co., Marseilles, France, 2004, pp. 295.

[601.] Minari, T., Nishio, R., Taniike, A., Furuyama, Y., and Kitamura, A., Experiments on condensed matter nuclear events in Kobe University, in *11th International Conference on Cold Fusion*, Biberian, J.-P. World Scientific Co., Marseilles, France, 2004, pp. 218.

[602.] Huke, A., Czerski, K., Dorsch, T., and Heide, P., Evidence for a target-material dependence of the neutron-proton branching ratio in d+d reactions for deuterium energies below 20 kev, in *11th International Conference on Cold Fusion*, Biberian, J.-P. World Scientific Co., Marseilles, Franc, 2004, pp. 210.

[603.] Huke, A., Czerski, K., and Heide, P., Accelerator experiments and theoretical models for the electron screening effect in metallic environments, in *11th International Converence on Cold Fusion*, Biberian, J.-P. World Scientific Co., Marseilles, France, 2004, pp. 194.

[604.] Czerski, K., Heide, P., and Huke, A., Electron screening constraints for the cold fusion, in *11th International Conference on Cold Fusion*, Biberian, J.-P. World Scientific Co., Marseilles, France, 2004, pp. 228.

[605.] Takahashi, A., Miyadera, H., Ochiai, K., Katayama, Y., Hayashi, T., and Dairaku, T., Studies on 3D fusion reactions in TiDx under Ion beam implantation, in *Tenth International Conference on Cold Fusion*, Hagelstein, P. L. andChubb, S. R. World Scientific Publishing Co., Cambridge, MA, 2003, pp. 657.

[606.] Dairaku, T., Katayama, Y., Hayashi, T., Isobe, Y., and Takahashi, A., Sudies of nuclear-reactions-in-solid in titanium deuteride under ion implantation., in *The 9th International Conference on Cold Fusion, Condensed Matter Nuclear Science*, Li, X. Z. Tsinghua Univ. Press, Tsinghua Univ., Beijing, China, 2002, pp. 73.

[607.] Takahashi, A., Maruta, K., Ochiai, K., and Miyamaru, H., Detection of three-body deuteron fusion in titanium deuteride under the stimulation by a deuteron beam, *Phys. Lett. A* 255, 89, 1999.

[608.] Kitamura, A., Awa, Y., Minari, T., Kubota, N., Taniike, A., and Furuyama, Y., D(d,p)t reaction rate enhancement in a mixed layer of Au and Pd, in *Tenth International Conference on Cold Fusion*, Hagelstein, P. L. andChubb, S. R. World Scientific Publishing Co., Cambridge, MA, 2003, pp. 623.

[609.] Miyamoto, M., Awa, Y., Kubota, N., Tamiike, A., Furuyama, Y., and Kitamura, A., Deuterium ion beam irradiation of palladium under in situ control of deuterium density., in *The 9th International Conference on Cold Fusion, Condensed Matter Nuclear Science*, Li, X. Z. Tsinghua Univ. Press, Tsinghua Univ., Beijing, China, 2002, pp. 261.

[610.] Wang, T., Wang, Z., Chen, J., Jin, G., and Piao, Y., Investigating the unknown nuclear reaction in a low-energy (E<300 keV) p + T_2H_x experiment, *Fusion Technol.* 37, 146, 2000.

[611.] Kubota, A., Taniike, A., and Kitamura, A., Production of high energy charged particles during deuteron implantation of titanium deuterides, in *8th International Conference on Cold Fusion*, Scaramuzzi, F. Italian Physical Society, Bologna, Italy, Lerici (La Spezia), Italy, 2000, pp. 311.

[612.] Kamada, K., Katano, Y., Ookubo, N., and Yoshizawa, I., Anomalous heat evolution of deuteron implanted Al upon electron bombardment. IV Trial to observe the nuclear reactions, in *8th International Conference on Cold Fusion*, Scaramuzzi, F. Italian Physical Society, Bologna, Italy, Lerici (La Spezia), Italy, 2000, pp. 341.

[613.] Kamada, K., Kinoshita, H., and Takahashi, H., Anomalous heat evolution of deuteron implanted Al on electron bombardment, in *5th International Conference on Cold Fusion*, Pons, S. IMRA Europe, Sophia Antipolis Cedex, France, Monte-Carlo, Monaco, 1995, pp. 41.

[614.] Kamada, K., Electron impact H-H and D-D fusions in molecules embedded in Al., in *Third International Conference on Cold Fusion, "Frontiers of Cold Fusion"*, Ikegami, H. Universal Academy Press, Inc., Tokyo, Japan, Nagoya Japan, 1992, pp. 551.

[615.] Kamada, K., Electron impact H-H and D-D fusions in molecules embedded in Al. 1. Experimental results, *Jpn. J. Appl. Phys. A* 31 (9), L1287, 1992.

[616.] Wang, T., Ochiai, K., Wang, Z., Jing, G., Iida, T., and Takahashi, A., Anomalous radiation induced by 1-300 keV deuteron ion beam implantation on palladium and titanium, in *The Seventh International Conference on Cold Fusion*, Jaeger, F. ENECO, Inc., Salt Lake City, UT., Vancouver, Canada, 1998, pp. 490.

[617.] Wang, T., Ochiai, K., Maruta, K., Iida, T., and Takahashi, A., Study of possible indirect fusion reaction in solids, in *The Seventh International Conference on Cold Fusion*, Jaeger, F. ENECO, Inc., Salt Lake City, UT., Vancouver, Canada, 1998, pp. 485.

[618.] Wang, T., Ochiai, K., Maruta, K., Datenichi, J., Sugimoto, H., Iida, T., Takahashi, A., and Piao, Y., Nuclear and atomic cluster effect of deuterium molecular ion (D_3+), in *The Seventh International Conference on Cold Fusion*, Jaeger, F. ENECO, Inc., Salt Lake City, UT., Vancouver, Canada, 1998, pp. 480.

[619.] Takahashi, A., Maruta, K., Ochiai, K., Miyamaru, H., and Iida, T., Anomalous enhancement of three-body deuteron fusion in titanium-deuteride with low-energy D+ beam implantation, *Fusion Technol.* 34, 256, 1998.

[620.] Ochiai, K., Iida, T., Beppu, N., Maruta, K., Miyamaru, H., and Takahashi, A., Deuteron fusion experiments in metal foils implanted with deutron beams, in *Sixth International Conference on Cold Fusion, Progress in New Hydrogen Energy*, Okamoto, M. New Energy and Industrial Technology Development Organization, Tokyo Institute of Technology, Tokyo, Japan, Lake Toya, Hokkaido, Japan, 1996, pp. 377.

[621.] Kasagi, J., Yuki, H., Baba, T., Noda, T., Taguchi, J., and Galster, W., Energetic protons and alpha particles emitted in 150-keV deuteron bombardment on deuterated Ti, *J. Phys. Soc. Japan* 64, 777-783, 1998.

[622.] Kasagi, J., Yuki, H., Itoh, T., Kasajima, N., Ohtsuki, T., and Lipson, A. G., Anomalously enhanced D(d,p)T reaction in Pd and PdO observed at very low bombarding energies, in *The Seventh International Conference on Cold Fusion*, Jaeger, F. ENECO, Inc., Salt Lake City, UT., Vancouver, Canada, 1998, pp. 180.

[623.] Wang, T., Piao, Y., Hao, J., Wang, X., Jin, G., and Niu, Z., Anomalous phenomena in E<18 KeV hydrogen ion beam implantation experiments on Pd and Ti, in *Sixth International Conference on Cold Fusion, Progress in New Hydrogen Energy*, Okamoto, M. New Energy and Industrial Technology Development Organization, Tokyo Institute of Technology, Tokyo, Japan, Lake Toya, Hokkaido, Japan, 1996, pp. 401.

[624.] Shinojima, H., Hishioka, T., Shikano, K., and Kanbe, H., Studies of d-d reactions in deuterated palladium by using low-energy deuterium ion bombardment, in *5th International Conference on Cold Fusion*, Pons, S. IMRA Europe, Sophia Antipolis Cedex, France, Monte-Carlo, Monaco, 1995, pp. 255.

[625.] Kitamura, A., Saitoh, T., and Itoh, T., In-situ ERD analysis of hydrogen isotopes during deuterium implantation of Pd, in *5th International Conference on Cold Fusion*, Pons, S. IMRA Europe, Sophia Antipolis Cedex, France, Monte-Carlo, Monaco, 1995, pp. 579.

[626.] Chindarkar, A. R., Paithankar, A. S., Bhagwat, A. M., Naik, G. R., Iyyengar, S. K., and Srinivasan, M., Observation of anomalous emissions of high energy ($\approx$1 MeV) charged particles when 5 keV protons impinge on palladium and titanium foils, *Trans. Fusion Technol.* 26 (4T), 197, 1994.

[627.] Baranov, D., Bazhutov, Y., Khokhov, N., Koretsky, V. P., Kuznetsov, A. B., Skuratnik, Y., and Sukovatkin, N., Experimental testing of the Erzion model by reacting of electron flux on the target, in *Fourth International Conference on Cold Fusion* Electric Power Research Institute 3412 Hillview Ave., Palo Alto, CA 94304, Lahaina, Maui, 1993, pp. 8.

[628.] Kasagi, J., Ishii, K., Hiraga, M., and Yoshihara, K., Observation of high energy protons emitted in the TiD_x+D reaction at E_d=150 keV and anomalous concentration of ^{3}He, in *Third International Conference on Cold Fusion, "Frontiers of Cold Fusion"*, Ikegami, H. Universal Academy Press, Inc., Tokyo, Japan, Nagoya Japan, 1992, pp. 209.

[629.] Iida, T., Fukuhara, M., Miyazaki, H., Sueyoshi, Y., Sunarno, Datemichi, J., and Takahashi, A., Deuteron fusion experiment with Ti and Pd foils implanted with deuterium beams, in *Third International Conference on Cold Fusion, "Frontiers of Cold Fusion"*, Ikegami, H. Universal Academy Press, Inc., Tokyo, Japan, Nagoya Japan, 1992, pp. 201.

[630.] Rout, R. K., Srinivasan, M., Shyam, A., and Chitra, V., Detection of high tritium activity on the central titanium electrode of a plasma focus device, *Fusion Technol.* 19, 391, 1991.

[631.] Cecil, F. E. and Hale, G. M., Measurement of D-D and D-Li^6 nuclear reactions at very low energies, in *Second Annual Conference on Cold Fusion, "The Science of Cold Fusion"*, Bressani, T., Del Giudice, E., andPreparata, G. Societa Italiana di Fisica, Bologna, Italy, Como, Italy, 1991, pp. 271.

[632.] Beuhler, R. J., Chu, Y. Y., Friedlander, G., Friedman, L., and Kunnmann, W., Deuteron-deuteron fusion by impact of heavy-water clusters on deuterated surfaces, *J. Phys. Chem.* 94, 7665, 1991.

[633.] Beuhler, R. J., Friedlander, G., and Friedman, L., Cluster-impact fusion, *Phys. Rev. Lett.* 63, 1292, 1990.

[634.] Chambers, G. P., Hubler, G. K., and Grabowski, K. S., Search for energetic charged-particle-reaction products during deuterium-charging of metal lattices, in *Anomalous Nuclear Effects in Deuterium/Solid Systems, "AIP Conference Proceedings 228"*, Jones, S., Scaramuzzi, F., andWorledge, D. American Institute of Physics, New York, Brigham Young Univ., Provo, UT, 1990, pp. 383.

[635.] Chambers, G. P., Eridon, J. E., Grabowski, K. S., Sartwell, B. D., and Chrisey, D. B., Charged particle spectra of palladium thin films during low energy deuterium ion implantation, *J. Fusion Energy* 9 (3), 281, 1990.

[636.] Passell, T. O., Personal communication, 1995.

[637.] Bazhutov, Y. N., Kuznetsov, A. B., Surova, T. D., and Chertov, Y. P., Study of the possibility of a cold nuclear fusion reaction by electrolysis of heavy water with a titanium electrode, in *Teo. Eksp. Issled. Vopr. Obshch. Fiz., Min. Obshch. Mashin. SSSR*, 1991, pp. 37 (in Russian).

[638.] Rout, R. K., Shyam, A., Srinivasan, M., and Bansal, A., Copious low energy emissions from palladium loaded with hydrogen or deuterium, *Indian J. Technol.* 29, 571, 1991.

[639.] Garg, A. B., Rout, R. K., Srinivasan, M., Sankarnarayanan, T. K., Shyam, A., and Kulkarni, L. V., Protocol for controlled and rapid loading/unloading of H_2/D_2 gas in self heated Pd wires to trigger nuclear events, in *5th International Conference on Cold Fusion*, Pons, S. IMRA Europe, Sophia Antipolis Cedex, France, Monte-Carlo, Monaco, 1995, pp. 461.

[640.] Rout, R. K., Shyam, A., Srinivasan, M., and Krishnan, M. S., Update on observation of low-energy emissions from deuterated and hydrated palladium, *Indian J. Technol.* 31, 551, 1993.

[641.] Storms, E., A critical evaluation of the Pons-Fleischmann effect: Part 1, *Infinite Energy* 6 (31), 10, 2000.

[642.] Storms, E., A critical evaluation of the Pons-Fleischmann effect: Part 2, *Infinite Energy* 6 (32), 52, 2000.

[643.] Swartz, M. R., Optimal operating point characteristics of nickel light water experiments, in *The Seventh International Conference on Cold Fusion*, Jaeger, F. ENECO, Inc., Salt Lake City, UT, Vancouver, Canada, 1998, pp. 371.

[644.] Huang, N., Effect of light water additions on excess heat generation of palladium deuterium system, in *8th World Hydrogen Energy Conf* Hawaii Natural Energy Insitute, 2540 Dole St., Holmes Hall 246, Honolulu, HI 96822, Honolulu, HI, 1990, pp. 43.

[645.] Storms, E., Cold fusion: An objective assessment, *www.LENR-CANR.org*, 2001.

[646.] Hurtak, J. J. and Bailey, P. G., Cold fusion research: Models and potential benefits, *J. New Energy* 2 (2), 128, 1997.

Chapter 5

[1.] Glück, P., The surfdyne concept: an attempt to solve (or to rename) the puzzles of cold nuclear fusion, *Fusion Technol.* 24, 122, 1993.

[2.] Storms, E., The nature of the energy-active state in Pd-D, *Infinite Energy* (#5 and #6), 77, 1995.

[3.] Farkas, A., On the electrolytic separation of the hydrogen isotopes on a palladium cathode, *Faraday Soc.* 33, 552, 1937.

[4.] Bockris, J. O. M. and Subramanyan, P. K., The equivalent pressure of molecular hydrogen in cavities within metals in terms of the overpotential developed during the evolution of hydrogen, *Electrochim. Acta* 16, 2169, 1971.

[5.] Louthan, J., M. R., Caskey, J., G. R. , Donovan, J. A., and Rawl, J., D. E., Hydrogen embrittlement of metals, *Mater. Sci. and Eng.* 10, 357, 1972.

[6.] Lynch, J. F., Clewley, J. D., and Flanagan, T. B., The formation of voids in palladium metal by the introduction and removal of interstital hydrogen, *Phil. Mag.* 28, 1415, 1973.

[7.] Armacanqui, M. E. and Oriani, R. A., Plastic deformation in B.C.C. alloys induced by hydrogen concentration gradients, *Mater. Sci. and Eng.* 91, 143, 1987.

[8] Bernabei, R., Gannelli, G., Cantelli, R., Cordero, d'Angelo, S., Iucci, N., Picozza, P. G., and Villoresi, G., Neutron monitoring during evolution of deuteride precipitation in Nb, Ta and Ti, *Solid State Commun.* 76, 815, 1990.

[9] Numata, H., Takagi, R., Ohno, I., Kawamura, K., and Haruyama, S., Neutron emission and surface observation during a long-term evolution of deuterium on Pd in 0.1 M LiOD., in *Second Annual Conference on Cold Fusion, "The Science of Cold Fusion"*, Bressani, T., Del Giudice, E., andPreparata, G. Societa Italiana di Fisica, Bologna, Italy, Como, Italy, 1991, pp. 71.

[10] Storms, E., Measurements of excess heat from a Pons-Fleischmann-type electrolytic cell using palladium sheet, *Fusion Technol.* 23, 230, 1993.

[11] Storms, E., A study of those properties of palladium that influence excess energy production by the Pons-Fleischmann effect, *Infinite Energy* 2 (8), 50, 1996.

[12] De Ninno, A., La Barbera, A., and Violante, V., Deformations induced by high loading ratios in palladium-deuterium compounds, *J. Alloys and Compounds* 253-254, 181, 1997.

[13] Storms, E., Formation of b-PdD containing high deuterium concentration using electrolysis of heavy-water, *J. Alloys Comp.* 268, 89, 1998.

[14] Bockris, J. O. M. and Minevski, Z., The mechansim of the evolution of hydrogen on palladium and associated internal damage phenomena, *J. Hydrogen Energy* 25, 747, 2000.

[15] McIntyre, R., Proposal for an experiment designed to seek evidence for cold fusion, in *Tenth International Conference on Cold Fusion*, Hagelstein, P. L. andChubb, S. R. World Scientific Publishing Co., Cambridge, MA, 2003, pp. 611.

[16] Klyuev, V. A., Lipson, A. G., Toporov, Y. P., Deryagin, B. V., Lushohikov, V. I., Streikov, A. V., and Shabalin, E. P., High-energy processes accompanying the fracture of solids, *Sov. Tech. Phys. Lett.* 12, 551, 1986.

[17] Golubnichii, P. I., Kurakin, V. A., Filonenko, A. D., Tsarev, V. A., and Tsarik, A. A., A possible mechanism for cold nuclear fusion, *J Kratk. Soobshch. Fiz.* (6), 56 (In Russian), 1989.

[18] Takeda, T. and Takizuka, T., Fractofusion mechanism, *J. Phys. Soc. Japan* 58 (9), 3073, 1989.

[19] AbuTaha, A. F., Cold fusion - the heat mechanism, *J. Fusion Energy* 9 (3), 345, 1990.

[20] Chechin, V. A., Tsarev, V. A., Golubnichyi, P. I., Philonenko, A. D., and Tsarik, A. A., Fracto-acceleration model of cold nuclear fusion, in *Anomalous Nuclear Effects in Deuterium/Solid Systems, "AIP Conference Proceedings 228"*, Jones, S., Scaramuzzi, F., andWorledge, D. American Institute of Physics, New York, Brigham Young Univ., Provo, UT, 1990, pp. 686.

[21] Mayer, F. J., King, J. S., and Reitz, J. R., Nuclear fusion from crack-generated particle acceleration, *J. Fusion Energy* 9 (3), 269, 1990.

[22] Preparata, G., A New Look at Solid-State Fractures, Particle Emission and 'Cold' Nuclear Fusion, *Nuovo Cimento A* 104, 1259, 1991.

[23] Bagnulo, L. H., Crack-fusion: A plausible explanation of cold fusion, in *Second Annual Conference on Cold Fusion, "The Science of Cold Fusion"*, Bressani, T., Del Giudice, E., andPreparata, G. Societa Italiana di Fisica, Bologna, Italy, Como, Italy, 1991, pp. 267.

[24] Bockris, J. O. M., Hodko, D., and Minevski, Z., Fugacity of hydrogen isotopes in metals: degradation, cracking and cold fusion, *Proc. Electrochem. Soc* 1992, 92, 1992.

[25] Kuehne, R. W. and Sioda, R. E., An extended micro hot fusion model for burst activity in deuterated solids, *Fusion Technol.* 27, 187, 1995.

[26] Bockris, J. O. M., The complex conditions needed to obtain nuclear heat from D-Pd systems., *J. New Energy* 1 (3), 210, 1996.

[27] Frisone, F., Deuteron Interaction Within a Microcrack in a Lattice at Room Temperature, *Fusion Technol.* 39 (2 (March)), 260, 2001.

[28] Frisone, F., Theoretical model of the probability of fusion between deuterons within deformed crystalline lattices with microcracks at room temperature, *Fusion Sci. & Technol.* 40, 139, 2001.

[29] Fulvio, F., Theoretical model on the relationship between low energies in the probability of deuterium nuclei cold fusion, in *The 9th International Conference on Cold Fusion, Condensed Matter Nuclear Science*, Li, X. Z. Tsinghua Univ. Press, Tsinghua Univ., Beijing, China, 2002, pp. 101.

[30.] Frisone, F., Fusion reaction within a microcrack with cubic lattice structure at low energy and study of the nonsemi-classical tunneling, in *Tenth International Conference on Cold Fusion*, Hagelstein, P. L. andChubb, S. R. World Scientific Publishing Co., Cambridge, MA, 2003, pp. 695.

[31.] Vysotskii, V. I. and Kuz'min, R. N., Nonequilibrium Fermi condensate of deuterium atoms in microvoids of crystals and the problem of barrier-free cold nuclear fusion, *Tech. Phys.* 39 (7), 663., 1994.

[32.] Fulvio, F., Theoretical model of the probability of fusion between deuterons within deformed lattices with microcracks at room temperature, in *11th International Conference on Cold Fusion*, Biberian, J.-P. World Scientific Co., Marseilles, France, 2004, pp. 612.

[33.] Shioe, Y., Mondal, N. N., Chiba, M., Hirose, T., Fujii, M., Nakahara, H., Sueki, K., Shirakawa, T., and M., U., Measurement of neutron production rate regarding the quantity of LiNbO3 in the fracturing process under D2 atmosphere, *Nuovo Cimento Soc. Ital. Fis. A* 112 A, 1059, 1999.

[34.] Kaushik, T. C., Kulkarni, L. V., Shyam, A., and Srinivasan, M., Experimental investigations on neutron emission from projectile-impacted deuterated solids, *Physics Lett. A* 232, 384, 1997.

[35.] Jabon, V. D. D., Fedorovich, G. V., and Samsonenko, N. V., Catalitically induced d-d fusion in ferroelectrics, *Braz. J. Phys.* 27, 515, 1997.

[36.] Chiba, M., Shirakawa, T., Fujii, M., Ikebe, T., Yamaoka, S., Sueki, K., Nakahara, H., and Hirose, T., Measurement of neutron emission from LiNbO3 fracture process in D2 and H2 atmosphere, *Nuovo Cimento Soc. Ital. Fis. A* 108, 1277, 1995.

[37.] Shirakawa, T., Chiba, M., Fujii, M., Sueki, K., Miyamoto, S., Nakamitu, Y., Toriumi, H., Uehara, T., Miura, H., Watanabe, T., Fukushima, K., Hirose, T., Seimiya, T., and Nakahara, H., A neutron emission from lithium niobate fracture, *Chem. Lett.*, 897, 1993.

[38.] Yasui, K., Fractofusion mechanism, *Fusion Technol.* 22, 400, 1992.

[39.] Shirakawa, T., Chiba, M., Fujii, M., Sueki, K., Miyamoto, S., Nakamitu, Y., Toriumi, H., Uehara, T., Miura, H., Watanabe, T., Fukushima, K., and Hirose, T., Neutron emission from crushing process of high piezoelectric matter in deuterium gas, in *Third International Conference on Cold Fusion, "Frontiers of Cold Fusion"*, Ikegami, H. Universal Academy Press, Inc., Tokyo, Japan, Nagoya Japan, 1992, pp. 469.

[40.] Lipson, A. G., Kluev, V. A., Mordovin, V. N., Sakov, D. M., Derjaguin, B. V., and Toporov, Y. P., On the initiation of DD reactions in the zirconium-deuterium system, *Phys. Lett. A* 166, 43, 1992.

[41.] Lipson, A. G., Deryagin, B. V., Klyuev, V. A., Toporov, Y. P., Sirotyuk, M. G., Khavroshkin, O. B., and Sakov, D. M., Initiation of nuclear fusion by cavitation action on deuterium-containing media, *Zh. Tekh. Fiz.* 62 (12), 121 (in Russian), 1992.

[42.] Golubnichii, P. I., Kuz'minov, V. V., Merzon, G. I., Pritychenko, B. V., Filonenko, A. D., Tsarev, V. A., and Tsarik, A. A., Correlated neutron and acoustic emission from a deuterium-saturated palladium target, *JETP Lett.* 53, 122, 1991.

[43.] Bittner, M., Meister, A., Seeliger, D., Schwiez, R., and Wüstner, P., Observation of D-D fusion neutrons during degassing of deuterium loaded palladium, in *Second Annual Conference on Cold Fusion, "The Science of Cold Fusion"*, Bressani, T., Del Giudice, E., andPreparata, G. Societa Italiana di Fisica, Bologna, Italy, Como, Italy, 1991, pp. 181.

[44.] Lipson, A. G., Klyuev, V. A., Deryagin, B. V., Toporov, Y. P., Sirotyuk, M. G., Khavroshkin, O. B., and Sakov, D. M., Observation of neutrons from cavitation action on substances containing deuterium, *Pis'ma Zh. Teor. Fiz.* 16 (9), 89 (in Russian), 1990.

[45.] Lipson, A. G., Klyuev, V. A., Toporov, Y. P., and Deryagin, B. V., Neutron generation by mechanical activation of metal surfaces, *Pis'ma Zh. Tekh. Fiz.* 16 (17), 54 (in Russian), 1990.

[46.] Izumida, T., Ozawa, Y., Ozawa, K., Izumi, S., Uchida, S., Miyamoto, T., Yamashita, H., and Miyadera, H., A search for neutron emission from cold nuclear fusion in a titanium-deuterium system, *Fusion Technol.* 18, 641, 1990.

[47.] Govorov, B. V., Gryaznov, V. M., Eremin, N. V., Karavanov, A. N., Roshan, N. R., Tulinov, A. F., and Tyapkin, I. V., Neutron emission from palladium alloys saturated with deuterium, *Russian J. Phys. Chem.* 64 (2), 287, 1990.

[48.] Goldanskii, V. I. and Dalidchik, F. I., On the possibilities of 'cold enhancement' of nuclear fusion, *Phys. Lett. B* 234, 465, 1990.

[49.] Dickinson, J. T., Jensen, L. C., Langford, S. C., Ryan, R. R., and Garcia, E., Fracto-emission from deuterated titanium: Supporting evidence for a fracto-fusion mechanism, *J. Mater. Res.* 5, 109, 1990.

[50.] Lipson, A. G., Sakov, A. G., Klyuev, V. A., Deryagin, B. V., and Toporov, Y. P., Neutron emission during the mechanical treatment of titanium in the presence of deuterated substances, *JETP* 49 (11), 675, 1989.

[51.] Lipson, A. G., Klyuev, V. A., Deryagin, B. V., Toporov, Y. P., and Sakov, D. M., Anomalous beta activity of products of mechanical working of a titanium- deuterated material, *Sov. Tech. Phys. Lett.* 15 (10), 783, 1989.

[52.] Derjaguin, B. V., Lipson, A. G., Kluev, V. A., Sakov, D. M., and Toporov, Y. P., Titanium fracture yields neutrons?, *Nature (London)* 341, 492 (issue 6242, 12. Oct, Scientific Corresp., 1989.

[53.] Deryagin, B. V., Klyuev, V. A., Lipson, A. G., and Toporov, Y. P., Possibility of nuclear reactions during the fracture of solids, *Colloid J. USSR* 48, 8, 1986.

[54.] Davies, J. D. and Cohen, J. S., More on the cold fusion family, *Ettore Majorana Int. Sci. Ser.: Phys. Sci.*, 52, 1990.

[55.] Amato, I., If Not Cold Fusion, Try Fracto-Fusion?, *Science News* 137, 87, 1990.

[56.] Kuehne, R. W., The possible hot nature of cold fusion, *Fusion Technol.* 25, 198, 1994.

[57.] Violante, V., Tripodi, P., Di Gioacchino, D., Borelli, R., Pizzuto, A., McKubre, M. C. H., Tanzella, F., Adrover, A., Giona, M., and Capobianco, L., Metallurgical effects on the dynamic of hydrogen loading in Pd, in *The 9th International Conference on Cold Fusion, Condensed Matter Nuclear Science*, Li, X. Z. Tsinghua Univ. Press, Tsinghua Univ., Beijing, China, 2002, pp. 383.

[58.] Szpak, S., Mosier-Boss, P. A., and Smith, J. J., Deuterium uptake during Pd-D codeposition, *J. Electroanal. Chem.* 379, 121, 1994.

[59.] Swartz, M. R., Codeposition of palladium and deuterium, *Fusion Technol.* 32, 126, 1997.

[60.] Storms, E., Excess power production from platinum cathodes using the Pons-Fleischmann effect, in *8th International Conference on Cold Fusion*, Scaramuzzi, F. Italian Physical Society, Bologna, Italy, Lerici (La Spezia), Italy, 2000, pp. 55.

[61.] Dash, J. and Ambadkar, A., Co-deposition of palladium with hydrogen isotopes, in *11th International Conference on Cold Fusion*, Biberian, J.-P. World Scientific Co., Marseilles, France, 2004, pp. 477.

[62.] Arata, Y. and Zhang, Y.-C., A new energy caused by "spillover-deuterium", *Proc. Japan. Acad.* 70 ser. B, 106, 1994.

[63.] Iwamura, Y., Sakano, M., and Itoh, T., Elemental analysis of Pd complexes: effects of D_2 gas permeation, *Jpn. J. Appl. Phys. A* 41 (7), 4642, 2002.

[64.] Menlove, H. O., Fowler, M. M., Garcia, E., Miller, M. C., Paciotti, M. A., Ryan, R. R., and Jones, S. E., Measurement of neutron emission from Ti and Pd in pressurized D_2 gas and D_2O electrolysis cells, *J. Fusion Energy* 9 (4), 495, 1990.

[65.] Afonichev, D. and Murzinova, M., Indicator of the process of cold fusion, *Int. J. Hydrogen Energy* 28, 1005-1010, 2003.

[66.] Beltyukov, I. L., Bondarenko, N. B., Janelidze, A. A., Gapanov, M. Y., Gribanov, K. G., Kondratov, S. V., Maltsev, A. G., Novikov, P. I., Tsvetkov, S. A., and Zakharov, V. I., Laser-induced cold nuclear fusion in Ti-H_2-D_2-T_2 compositions, *Fusion Technol.* 20, 234, 1991.

[67.] Perfetti, P., Cilloco, F., Felici, R., Capozi, M., and Ippoliti, A., Neutron emission under particular nonequilibrium conditions from palladium and titanium electrolytically charged with deuterium, *Nuovo Cimento Soc. Ital. Fis. D* 11 (6), 921, 1989.

[68.] Filimonov, V. A., Mechanism of cold nuclear fusion, *Pis'ma Zh. Tekh. Fiz.* 16 (20), 29 (in Russian), 1990.

[69.] Barut, A. O., Prediction of new tightly-bound states of H_2+ (D_2+) and 'cold fusion' experiments, *J. Hydrogen Energy* 15, 907, 1990.

[70.] Takahashi, A., Some considerations of multibody fusion in metal-deuterides, *Trans. Fusion Technol.* 26 (4T), 451, 1994.

[71.] Engvild, K. C., Nuclear reaction by three-body recombination between deuterons and the nuclei of lattice-trapped D2 molecules, *Fusion Technol.* 34, 253, 1998.

[72.] Engvild, K. C. and Kowalski, L., Triple deuterium fusion between deuterons and the nuclei of lattice trapped deuterium molecules, in *Tenth International Conference on Cold Fusion*, Hagelstein, P. L. andChubb, S. R. World Scientific Publishing Co., Cambridge, MA, 2003, pp. 825.

[73.] Takahashi, A., Tetrahedral and octahedral resonance fusion under transient condensation of deuterons at lattice focal points, in *ICCF9, Ninth International Conference on Cold Fusion*, Li, X. Z. Tsinghua Univ., China, Beijing, China: Tsinghua University, 2002, pp. 343.

[74.] Takahashi, A., Clean fusion by tetrahedral and octahedral symmetric condensations, in *Proc. JCF5* http://wwwcf.elc.iwate-u.ac.jp/jcf/, Kobe, Japan, 2003, pp. 74-78.

[75.] Hagelstein, P. L., Phonon-exchange models: Some new results, in *11th International Conference on Cold Fusion*, Biberian, J.-P. World Scientific Co., Marseilles, France, 2004, pp. 743.

[76.] Worsham Jr., J. E., Wilkinson, M. K., and Shull, C. G., Neutron-diffraction observations on the palladium-hydrogen and palladium-deuterium systems, *J. Phys. Chem. Solids* 3, 303, 1957.

[77.] Bennington, S. M., Benham, M. J., Stonadge, P. R., Fairclough, J. P. A., and Ross, D. K., In-situ measurements of deuterium uptake into a palladium electrode using time-of-flight neutron diffractometry, *J. Electroanal. Chem.* 281, 323, 1990.

[78.] Dillon, C. T., Kennedy, B. J., and Elcombe, M. M., The electrochemically formed palladium-deuterium system. II. In situ neutron diffraction studies, *Aust. J. Chem.* 46, 681, 1993.

[79.] Wei, S.-H. and Zunger, A., Instability of diatomic deuterium in fcc palladium, *J. Fusion Energy* 9 (4), 367, 1990.

[80.] Christensen, O. B., Ditlevsen, P. D., Jacobsen, K. W., Stoltz, P., Nielsen, O. H., and Norskov, J. K., H-H interactions in Pd, *Phys. Rev. B* 40 (3), 1993, 1993.

[81.] Moizhes, B. Y., Formation of a compact D_2 molecule in interstitial sites - a possible explanation for cold nuclear fusion, *Sov. Tech. Phys. Lett.* 17, 540, 1991.

[82.] Zakowicz, W., Possible resonant mechanism of cold fusion, *Fusion Technol.* 19, 170., 1991.

[83.] Rambaut, M., Experimental evidences for the harmonic oscillator resonance and electron accumulation model of cold fusion, in *5th International Conference on Cold Fusion*, Pons, S. IMRA Europe, Sophia Antipolis Cedex, France, Monte-Carlo, Monac, 1995, pp. 623.

[84.] McNally, J. R., On the possibility of a nuclear mass-energy resonance in deuterium + deuterium reactions at low energy, *Fusion Technol.* 16, 237, 1989.

[85.] Li, X., A new approach towards nuclear fusion without strong nuclear radiation, *Nucl. Fusion Plasma Phys.* 16 (2), 1 (in Chinese), 1996.

[86.] Kim, Y. E., Yoon, J.-H., Zubarev, A. L., and Rabinowitz, M., Reaction barrier transparency for cold fusion with deuterium and hydrogen, *Trans. Fusion Technol.* 26 (4T), 408, 1994.

[87.] Jändel, M., The fusion rate in the transmission resonance model, *Fusion Technol.* 21, 176, 1992.

[88.] Bush, R. T., Cold 'fusion'. The transmission resonance model fits data on excess heat, predicts optimal trigger points, and suggests nuclear reaction scenarios, *Fusion Technol.* 19, 313, 1991.

[89.] Jiang, X. L. and Han, L. J., Dynamic Casimir effect in an electrochemical systems, *J. New Energy* 3 (4), 47, 1999.

[90.] Bockris, J. O. M., Hodko, D., and Minevski, Z., The mechanism of deuterium evolution on palladium: Relation to heat bursts provoked by fluxing deuterium across the interface, in *Second Annual Conference on Cold Fusion, "The Science of Cold Fusion"*, Bressani, T., Del Giudice, E., andPreparata, G. Societa Italiana di Fisica, Bologna, Italy, Como, Italy, 1991, pp. 337.

[91.] Bockris, J. O. M. and Minevski, Z., First experimental establishment of high internal pressure of molecular hydrogen developed in palladium during water electrolysis, *J. Hydrogen Energy* 23 (12), 1079, 1998.

[92.] Letts, D. and Cravens, D., Laser stimulation of deuterated palladium, *Infinite Energy* 9 (50), 10, 2003.

[93.] Letts, D. and Cravens, D., Laser stimulation of deuterated palladium: past and present, in *Tenth International Conference on Cold Fusion*, Hagelstein, P. L. andChubb, S. R. World Scientific Publishing Co., Cambridge, MA, 2003, pp. 159.

[94.] Storms, E., Use of a very sensitive Seebeck calorimeter to study the Pons-Fleischmann and Letts effects, in *Tenth International Conference on Cold Fusion*, Hagelstein, P. L. andChubb, S. R. World Scientific Publishing Co., Cambridge, MA, 2003, pp. 183.

[95.] Violante, V., Castagna, E., Sibilia, C., Paoloni, S., and Sarto, F., Analysis of Ni-hydride thin film after surface plasmons generation by laser technique, in *Tenth International Conference on Cold Fusion*, Hagelstein, P. L. andChubb, S. R. World Scientific Publishing Co., Cambridge, MA, 2003, pp. 421.

[96.] Swartz, M. R., Photo-induced excess heat from laser-irradiated electrically polarized palladium cathodes in D_2O, in *Tenth International Conference on Cold Fusion*, Hagelstein, P. L. andChubb, S. R. World Scientific Publishing Co., Cambridge, MA, 2003, pp. 213.

[97.] Roussetski, A. S., Lipson, A. G., and Andreanov, V. P., Nuclear emissions from titanium hydride/deuteride, induced by powerful picosecond laser beam, in *Tenth International Conference on Cold Fusion*, Chubb, P. I. H. a. S. R. World Scientific Publishing Co., Cambridge, MA, 2003, pp. 559.

[98.] Nassisi, V. and Longo, M. L., Experimental results of transmutation of elements observed in etched palladium samples by an excimer laser, *Fusion Technol.* 37 (May), 247, 2000.

[99.] Castellano, Di Giulio, M., Dinescu, M., Nassisi, V., Conte, A., and Pompa, P. P., Nuclear transmutation in deutered Pd films irradiated by an UV laser, in *8th International Conference on Cold Fusion*, Scaramuzzi, F. Italian Physical Society, Bologna, Italy, Lerici (La Spezia), Italy, 2000, pp. 287.

[100.] Nassisi, V., Transmutation of elements in saturated palladium hydrides by an XeCl excimer laser, *Fusion Technol.* 33, 468, 1998.

[101.] Nassisi, V., Incandescent Pd and Anomalous Distribution of Elements in Deuterated Samples Processed by an Excimer Laser, *J. New Energy* 2 (3/4), 14, 1997.

[102.] Vokhnik, O. M., Goryachev, B. I., Zubrilo, A. A., Kutznetsova, G. P., Popov, Y. V., and Svertilov, S. I., Search for effects related to nuclear fusion in the optical breakdown of heavy water, *Sov. J. Nucl. Phys.* 55 (12), 1772, 1992.

[103.] Karabut, A. B., X-ray emission in the high-current glow discharge experiments, in *The 9th International Conference on Cold Fusion, Condensed Matter Nuclear Science*, Li, X. Z. Tsinghua Univ. Press, Tsinghua Univ., Beijing, China, 2002, pp. 155.

[104.] Arata, Y. and Zhang, Y.-C., Excess heat in a double structure deuterated cathode, *Kakuyuogo Kenkyo* 69 ((8)), 963 (in Japanese), 1993.

[105.] Arata, Y. and Zhang, Y. C., Helium (^4He, ^3He) within deuterated Pd-black, *Proc. Jpn. Acad., Ser. B* 73, 1, 1997.

[106.] Arata, Y. and Zhang, Y. C., Observation of anomalous heat release and helium-4 production from highly deuterated fine particles, *Jpn. J. Appl. Phys. Part 2* 38 (7A), L774, 1999.

[107.] Arata, Y. and Zhang, Y., Development of compact nuclear fusion reactor using solid pycnodeuterium as nuclear fuel, in *Tenth International Conference on Cold Fusion*, Hagelstein, P. L. andChubb, S. R. World Scientific Publishing Co., Cambridge, MA, 2003, pp. 139.

[108.] Clarke, B. W., Oliver, B. M., McKubre, M. C. H., Tanzella, F. L., and Tripodi, P., Search for ^3He and ^4He in Arata-style palladium cathodes II: Evidence for tritium production, *Fusion Sci. & Technol.* 40, 152, 2001.

[109.] McKubre, M. C. H., Tanzella, F. L., Tripodi, P., and Hagelstein, P. L., The emergence of a coherent explanation for anomalies observed in D/Pd and H/Pd system: evidence for ^4He and ^3He production, in *8th International Conference on Cold Fusion*, Scaramuzzi, F. Italian Physical Society, Bologna, Italy, Lerici (La Spezia), Italy, 2000, pp. 3.

[110.] Case, L., There is a Fleischmann-Pons effect. The process is electrolytic, but the effect is catalytic., in *The 9th International Conference on Cold Fusion, Condensed Matter Nuclear Science*, Li, X. Z. Tsinghua Univ. Press, Tsinghua Univ., Beijing, China, 2002, pp. 22.

[111.] Case, L. C., Catalytic fusion of deuterium into helium-4, in *The Seventh International Conference on Cold Fusion*, Jaeger, F. ENECO, Inc., Salt Lake City, UT, Vancouver, Canada, 1998, pp. 48.

[112.] McKubre, M. C. H., Tanzella, F. L., Tripodi, P., Di Gioacchino, D., and Violante, V., Finite element modeling of the transient colorimetric behavior of the MATRIX experimental apparatus: ^{4}He and excess of power production correlation through numerical results, in *8th International Conference on Cold Fusion*, Scaramuzzi, F. Italian Physical Society, Bologna, Italy, Lerici (La Spezia), Italy, 2000, pp. 23.

[113.] George, R., Production of He-4 from deuterium during contact with nano-particle palladium on carbon at 200° C and 3 atmosphere deuterium pressure, www.hooked.net/~rgeorge/catalystpaper1.htm, 1999.

[114.] Clarke, W. B., Production of ^{4}He in D$_2$-loaded palladium-carbon catalyst. I, *Fusion Sci. & Technol.* 43, 122, 2003.

[115.] Nag, N. K., A study on the formation of palladium hydride in a carbon-supported palladium catalyst, *J. Phys. Chem. B* 105, 5945, 2001.

[116.] Arata, Y. and Zhang, Y. C., Formation of condensed metallic deuterium lattice and nuclear fusion, *Proc. Jpn. Acad., Ser. B* 78 (Ser. B), 57, 2002.

[117.] Chen, S.-K., Wan, C.-M., Liu, E.-H., Chu, S.-B., and Liang, C.-Y., The microstructure of electrolytically deuterium-loaded palladium rods, *Fusion Technol.* 29, 302, 1996.

[118.] Dash, J., Noble, G., and Diman, D., Surface morphology and microcomposition of palladium cathodes after electrolysis in acified light and heavy water: correlation with excess heat, *Trans. Fusion Technol.* 26 (4T), 299, 1994.

[119.] Oyama, N., Ozaki, M., Tsukiyama, S., Hatozaki, O., and Kunimatsu, K., In situ interferometric microscopy of Pd electrode surfaces and calorimetry during electrolysis of D$_2$O solution containing sulfur ion, in *Sixth International Conference on Cold Fusion, Progress in New Hydrogen Energy*, Okamoto, M. New Energy and Industrial Technology Development Organization, Tokyo Institute of Technology, Tokyo, Japan, Lake Toya, Hokkaido, Japan, 1996, pp. 234.

[120.] Silver, D. S., Dash, J., and Keefe, P. S., Surface topography of a palladium cathode after electrolysis in heavy water, *Fusion Technol.* 24, 423, 1993.

[121.] Focardi, S., Gabbani, V., Montalbano, V., Piantelli, F., and Veronesi, F., Evidence of electromagnetic radiation from Ni-H systems, in *11th International Conference on Cold Fusion*, Biberian, J.-P. World Scientific Co., Marseilles, France, 2004, pp. 70.

[122.] Focardi, S., Habel, R., and Piantelli, F., Anomalous heat production in Ni-H systems, *Nuovo Cimento* 107A, 163, 1994.

[123.] Cammarota, G., Collis, W. J. M. F., Rizzo, A., and Stremmenos, C., A flow calorimeter study of the Ni/H system, in *Asti Workshop on Anomalies in Hydrogen/Deuterium Loaded Metals*, Collis, W. J. M. F. Italian Phys. Soc., Villa Riccardi, Rocca d'Arazzo, Italy, 1997, pp. 1.

[124.] Mengoli, G., Bernardini, M., Comisso, N., Manduchi, C., and Zannoni, G., The nickel-K$_2$CO$_3$, H$_2$O system: an electrochemical and calorimetric investigation, in *Asti Workshop on Anomalies in Hydrogen/Deuterium Loaded Metals*, Collis, W. J. M. F. Italian Phys. Soc., Riccardi, Rocca d'Arazzo, Italy, 1997, pp. 71.

[125.] Tian, J., Jin, L. H., Weng, Z. K., Song, B., Zhao, X. L., Xiao , Z. J., Chen, G., and Du, B. Q., "Excess heat" during electrolysis in platinum/K$_2$CO$_3$/nickel light water system, in *11th International Conference on Cold Fusion*, Biberian, J.-P. World Scientific Co., Marseilles, France, 2004, pp. 102.

[126.] Notoya, R., Noya, Y., and Ohnishi, T., Tritium generation and large excess heat evolution by electrolysis in light and heavy water-potassium carbonate solutions with nickel electrodes, *Fusion Technol.* 26, 179, 1994.

[127.] Bush, R. T. and Eagleton, R. D., Calorimetric studies for several light water electrolytic cells with nickel fibrex cathodes and electrolytes with alkali salts of potassium, rubidium, and cesium, in *Fourth International Conference on Cold Fusion* Electric Power Research Institute 3412 Hillview Ave., Palo Alto, CA 94304, Lahaina, Maui, 1993, pp. 13.

[128.] Cox, D. M., Fayet, P., Brickman, R., Hahn, M. Y., and Kaldor, A., Abnormally large deuterium uptake on small transition metal clusters, *Catal. Lett.* 4, 271, 1990.

[129.] Everett, D. H. and Sermon, P. A., Crystallite size effects in the palladium/hydrogen system: A smultaneous sorption and X-ray study, *Zeit. Phys. Chem. Neue Folge Bd.* 114 (S), 109, 1979.

[130.] Lynch, J. F. and Flanagan, T. B., An investigation of the dynamic equilibrium between chemisorbed and absorbed hydrogen in the palladium/hydrogen system., *J. Phys. Chem.* 77, 2628, 1973.

[131.] Pons, S. and Fleischmann, M., Heat after death, *Trans. Fusion Technol.* 26 (4T), 97, 1994.

[132.] Packham, N. J. C., Wolf, K. L., Wass, J. C., Kainthla, R. C., and Bockris, J. O. M., Production of tritium from D_2O electrolysis at a palladium cathode, *J. Electroanal. Chem.* 270, 451, 1989.

[133.] Bockris, J. O. M., Lin, G. H., Kainthla, R. C., Packham, N. J. C., and Velev, O., Does tritium form at electrodes by nuclear reactions?, in *The First Annual Conference on Cold Fusion*, Will, F. National Cold Fusion Institute, University of Utah Research Park, Salt Lake City, Utah, 1990, pp. 137.

[134.] Chien, C.-C., Hodko, D., Minevski, Z., and Bockris, J. O. M., On an electrode producing massive quantities of tritium and helium, *J. Electroanal. Chem.* 338, 189, 1992.

[135.] Lin, G. H., Kainthla, R. C., Packham, N. J. C., and Bockris, J. O. M., Electrochemical fusion: a mechanism speculation, *J. Electroanal. Chem.* 280, 207, 1990.

[136.] Yamazaki, O., Yoshitake, H., Kamiya, N., and Ota, K., Hydrogen absorption and Li inclusion in a Pd cathode in LiOH solution, *J. Electroanal. Chem.* 390, 127, 1995.

[137.] Ota, K., Yoshitake, H., Yamazaki, O., Kuratsuka, M., Yamaki, K., Ando, K., Iida, Y., and Kamiya, N., Heat measurement of water electrolysis using Pd cathode and the electrochemistry, in *Fourth International Conference on Cold Fusion* Electric Power Research Institute 3412 Hillview Ave., Palo Alto, CA 94304, Lahaina, Maui, 1993, pp. 5.

[138.] Okamoto, M., Ogawa, H., Yoshinaga, Y., Kusunoki, T., and Odawara, O., Behavior of key elements in Pd for the solid state nuclear phenomena occurred in heavy water electrolysis, in *Fourth International Conference on Cold Fusion* Electric Power Research Institute 3412 Hillview Ave., Palo Alto, CA 94304, Lahaina, Maui, 1993, pp. 14.

[139.] Nakada, M., Kusunoki, T., Okamoto, M., and Odawara, O., A role of lithium for the neutron emission in heavy water electrolysis, in *Third International Conference on Cold Fusion, "Frontiers of Cold Fusion"*, Ikegami, H. Universal Academy Press, Inc., Tokyo, Japan, Nagoya Japan, 1992, pp. 581.

[140.] Brillas, E., Esteve, J., Sardin, G., Casado, J., Domenech, X., and Sanchez-Cabeza, J. A., Product analysis from D_2O electrolysis with Pd and Ti cathodes, *Electrochim. Acta* 37 (2), 215, 1992.

[141.] Astakhov, I. I., Davydov, A. D., Katargin, N. V., Kazarinov, V. E., Kiseleva, I. G., Kriksunov, L. B., Kudryavtsev, D. Y., Lebedev, I. A., Myasoedov, B. F., Shcheglov, O. P., Teplitskaya, G. L., and Tsionskii, V. M., An attempt to detect neutron and gamma radiations in heavy water electrolysis with a palladium cathode, *Electrochim. Acta* 36 (7), 1127, 1991.

[142.] Sona, P. G. and Ferrari, M., The possible negative influence of dissolved O2 in cold nuclear fusion experiments, *Fusion Technol.* 18, 678, 1990.

[143.] Dalard, F., Ulman, M., Augustynski, J., and Selvam, P., Electrochemical incorporation of lithium into palladium from aprotic electrolytes, *J. Electroanal. Chem.* 270, 445, 1989.

[144.] Howald, R. A., Calculation on the palladium-lithium system for cold fusion, *CALPHAD* 14, 1, 1990.

[145.] Kazarinov, V. E., Astakhov, I. I., Teplitskaya, G. L., Kiseleva, I. G., Davydov, A. D., Nekrasova, N. V., Kudryavtsev, D. Y., and Zhukova, T. B., Cathodic behaviour of palladium in electrolytic solutions containing alkali metal ions, *Elektrokhimiya* 27, 9 (in Russian), 1991.

[146.] Kim, Y. E. and Passell, T. O., Alternative interpretations of low-energy nuclear reaction processes with deuterated metals based on the Bose-Einstein condensation mechanism, in *11th International Conference on Cold Fusion*, Biberian, J.-P. World Scientific Co., Marseilles, France, 2004, pp. 718.

[147.] Iwamura, Y., Itoh, T., Gotoh, N., Sakano, M., Toyoda, I., and Sakata, H., Detection of anomalous elements, X-ray and excess heat induced by continous diffusion of deuterium through multi-layer cathode (Pd/CaO/Pd), in *The Seventh International Conference on Cold Fusion*, Jaeger, F. ENECO, Inc., Salt Lake City, UT, Vancouver, Canada, 1998, pp. 167.

[148.] Li, X. Z., Liu, B., Cai, N., Wei, Q., Tian, J., and Cao, D. X., Progress in gas-loading D/Pd system: The feasibility of a self-sustaining heat generator, in *Tenth International Conference on Cold Fusion*, Hagelstein, P. L. andChubb, S. R. World Scientific Publishing Co., Cambridge, MA, 2003, pp. 113.

[149.] Wei, Q., Liu, B., Mo, Y. X., Li, X. Z., Zheng, S., Cao, D. X., Wang, X. M., and Tian, J., Deuterium (hydrogen) flux permeating through palladium and condensed matter nuclear science, in *11th International Converence on Cold Fusion*, Biberian, J.-P. World Scientific Co, Marseilles, France, 2004, pp. 351.

[150.] Li, X. Z., Liu, B., Tian, J., Wei, Q. M., Zhou, R., and Yu, Z. W., Correlation between abnormal deuterium flux and heat flow in a D/Pd system, *J. Phys. D: Appl. Phys.* 36, 3095, 2003.

[151.] McKubre, M. C. H., Crouch-Baker, S., Tanzella, F. L., Smedley, S. I., Williams, M., Wing, S., Maly-Schreiber, M., Rocha-Fiho, R. C., Searson, P. C., Pronko, J. G., and Kohler, D. A., Final Report No. EPRI TR-104195, 1994.

[152.] Loebich, J., D. and Raub, C. J., Das Zustandsdiagramm Lithium-Palladium und die Magnetischen Eigenschaften der Li-Pd Legierungen, *J. Less-Common Met.* 55, 67, 1977.

[153.] McKubre, M. C. H., Rocha-Filho, R. C., Smedley, S. I., Tanzella, F. L., Crouch-Baker, S., Passell, T. O., and Santucci, J., Isothermal flow calorimetric investigations of the D/Pd system, in *Second Annual Conference on Cold Fusion, "The Science of Cold Fusion"*, Bressani, T., Del Giudice, E., andPreparata, G. Societa Italiana di Fisica, Bologna, Italy, Como, Italy, 1991, pp. 419.

[154.] Donohue, D. L. and Petek, M., Isotopic measurements of palladium metal containing protium and deuterium by glow discharge mass spectrometry, *Anal. Chem.* 63, 740, 1991.

[155.] Guilinger, T. R., Kelly, J., Knapp, J. A., Walsh, D., and Doyle, B. L., Ion beam measurements of deuterium in palladium and calculation of hydrogen isotope separation, *J. Electrochem. Soc.* 138 (8), L26, 1991.

[156.] Dandapani, B. and Fleischmann, M., Electrolytic separation factors on palladium, *J. Electroanal. Chem.* 39, 323, 1972.

[157.] Cheek, G. T. and O'Grady, W. E., Measurement of H/D uptake characteristics at palladium using a quartz crystal microbalance, *J. Electroanal. Chem.* 368, 133, 1994.

[158.] Lewis, F. A., Leitch, W. F. N., and Murray, A., Electrolytic hydrogen isotope separation factors and efficiency of exchange between D_2O and hydrogen (protium) at palladium electrodes, *Sur. Technol.* 7, 385, 1978.

[159.] Redey, L., Myles, K. M., Dees, D., Krumpelt, M., and Vissers, D. R., Calorimetric measurements on electrochemical cells with Pd-D cathodes, *J. Fusion Energy* 9 (3), 249, 1990.

[160.] Storms, E. and Talcott-Storms, C., The effect of hydriding on the physical structure of palladium and on the release of contained tritium, *Fusion Technol.* 20, 246, 1991.

[161.] Storms, E., Cold fusion: An objective assessment, *www.LENR-CANR.org*, 2001.

[162.] Mills, R., *The grand unified theory of classical quantum mechanics* Cadmus Professional Communications, Ephrata, PA, 2006.

[163.] Mills, R. L. and Ray, P., The grand unified theory of classical quantum mechanics, *J. Hydrogen Eng.* 27, 565, 2002.

[164.] Mills, R. L. and Ray, P., Vibrational spectral emission of fractional-principal-quantum-energy-level hydrogen molecule ion, *J. Hydrogen Eng.* 27, 533, 2002.

[165.] Mills, R. L., Good, W. R., and Shaubach, R. M., Dihydrino molecule identification, *Fusion Technol.* 25, 103, 1994.

[166.] Mills, R. L. and Kneizys, P., Excess heat production by the electrolysis of an aqueous potassium carbonate electrolyte and the implications for cold fusion, *Fusion Technol.* 20, 65, 1991.

[167.] Dufour, J., Foos, J., Millot, J. P., and Dufour, X., From "cold fusion" to "Hydrex" and "Deutex" states of hydrogen, in *Sixth International Conference on Cold Fusion, Progress in New Hydrogen*

Energy, Okamoto, M. New Energy and Industrial Technology Development Organization, Tokyo Institute of Technology, Tokyo, Japan, Lake Toya, Hokkaido, Japan, 1996, pp. 482.

[168.] Dufour, J., Response to 'Comments on 'Interaction of palladium/hydrogen and palladium/deuterium to measure the excess energy per atom for each isotope', *Fusion Technol.* 33, 385, 1998.

[169.] Dufour, J., Foos, J. H., and Dufour, X. J. C., Formation and properties of Hydrex and Deutex, in *The Seventh International Conference on Cold Fusion*, Jaeger, F. ENECO, Inc., Salt Lake City, UT, Vancouver, Canada, 1998, pp. 113.

[170.] Dufour, J. J., Foos, J. H., and Dufour, X. J. C., Formation and properties of hydrex and deutex, *Infinite Energy* 4 (20), 53, 1998.

[171.] Dufour, J., Murat, D., Dufour, X., and Foos, J., Hydrex catallyzed transmutation of uranium and palladium: experimental part, in *8th International Conference on Cold Fusion*, Scaramuzzi, F. Italian Physical Society, Bologna, Italy, Lerici (La Spezia), Italy, 2000, pp. 153.

[172.] Dufour, J., Murat, D., Dufour, X., and Foos, J., The Hydrex concept-effect on heavy nuclei, in *8th International Conference on Cold Fusion*, Scaramuzzi, F. Italian Physical Society, Bologna, Italy, Lerici (La Spezia), Italy, 2000, pp. 431.

[173.] Dufour, J., Murat, D., Dufour, X., and Foos, J., Proton-electron reactions as precursors of anomalous nuclear events, *Fusion Technol.* 40, 91, 2001.

[174.] Mayer, F. J. and Reitz, J. R., Nuclear energy release in metals, *Fusion Technol.* 19, 552, 1991.

[175.] Moon, D., Excess heat vs transmutation (revised), *Elemental Energy* 26, 12, 1998.

[176.] Vigier, J. P., On cathodically polarized Pd/D systems, *Phys. Lett. A* 221, 138, 1996.

[177.] Dragic, A., Maric, Z., and Vigier, J. P., New quantum mechanical tight bound states and 'cold fusion', *Phys. Lett. A* 265, 163, 2000.

[178.] Antanasijevic, R. D., Konjevic, D. J., Maric, Z., Sevic, D. M., Vigier, J. P., and Zaric, A. J., "Cold fusion" in terms of new quantum chemistry: The role of magnetic interactions in dense physica media, in *5th International Conference on Cold Fusion*, Pons, S. IMRA Europe, Sophia Antipolis Cedex, France, Monte-Carlo, Monaco, 1995, pp. 505.

[179.] Dufour, J., Foos, J., Millot, J. P., and Dufour, X., Interaction of palladium/hydrogen and palladium/deuterium to measure the excess energy per atom for each isotope, *Fusion Technol.* 31, 198, 1997.

[180.] Kozima, H., Ohta, M., Fujii, M., Arai, K., and Kudoh, H., Possible explanation of ^{4}He production in a Pd/D_2 system by the TNCF model, *Fusion Sci. Technol.* 40 (July), 86, 2001.

[181.] Kozima, H., Warner, J., Cano, C. S., and Dash, J., Consistent explanation of topography changes and nuclear transmutation in surface layers of cathodes in electrolytic cold fusion experiments., in *The 9th International Conference on Cold Fusion, Condensed Matter Nuclear Science*, Li, X. Z. Tsinghua Univ. Press, Tsinghua Univ., Beijing, China, 2002, pp. 178.

[182.] Kozima, H., *Discovery of the cold fusion phenomenon* Ohotake Shuppan, Inc, Tokyo, Japan, 1998.

[183.] Kozima, H., TNCF model- A phenomenological approach, in *8th International Conference on Cold Fusion*, Scaramuzzi, F. Italian Physical Society, Bologna, Italy, Lerici (La Spezia), Italy, 2000, pp. 461.

[184.] Fisher, J. C., Theory of low-temperature particle showers, in *Tenth International Conference on Cold Fusion*, Hagelstein, P. L. andChubb, S. R. World Scientific Publishing Co., Cambridge, MA, 2003, pp. 915.

[185.] Fisher, J. C., Polyneutrons as agents for cold nuclear reactions, *Fusion Technol.* 22, 511, 1992.

[186.] Fisher, J. C., Outline of polyneutron physics (basics), in *6th International Workship on Anomalies in Hydrogen/Deuterium Loaded Metals*, Siena, Italy, 2005.

[187.] Oriani, R. A., Anomalous heavy atomic masses produced by electrolysis, *Fusion Technol.* 34, 76, 1998.

[188.] Oriani, R. A. and Fisher, J. C., Generation of nuclear tracks during electrolysis, *Jpn. J. Appl. Phys. A* 41, 6180-6183, 2002.

[189.] Oriani, R. A. and Fisher, J. C., Energetic charged particles produced during electrolysis, in *Tenth International Conference on Cold Fusion*, Hagelstein, P. L. andChubb, S. R. World Scientific Publishing Co., Cambridge, MA, 2003, pp. 567.

[190.] Daddi, L., Proton-electron reactions as precursors of anomalous nuclear events, *Fusion Technol.* 39, 249, 2001.

[191.] Schultz, R. and Kenny, J. P., Electronuclear catalysts and initiators: The di-neutron model for cold fusion, *Infinite Energy* 5 (29), 58, 1999.

[192.] Phipps Jr., T. E., Neutron formation by electron penetration of the nucleus, *Infinite Energy* 5 (26), 58, 1999.

[193.] Moon, D., Review of a cold fusion theory: Mechanisms of a disobedient science, *Infinite Energy* 5 (28), 33, 1999.

[194.] Yang, J., Tang, L., and Chen, X., Dineutron model research of cold fusion, *Acta Sci. Nat. Univ. Norm. Hunanensis* 19 (2), 25, 1996.

[195.] Pokropivnii, V. V., Bineutron theory of cold nuclear fusion, *Dokl. Akad. Nauk. Ukr.* (4), 86 (in Russian), 1993.

[196.] Yang, J., (2)(1)H-e touched capturing and (2)(1)H-(2)(0)N fusion, *Acta Sci. Nat. Univ. Norm. Hunanensis* 15 (1), 18, 1992.

[197.] Russell Jr., J. L., Virtual electron capture in deuterium, *Ann. Nucl. Energy* 18, 75, 1991.

[198.] Pokropivnii, V. V. and Ogorodnikov, V. V., The bineutron model of cold nuclear fusion in metals, *Pis'ma Zh. Teor. Fiz.* 16 (21), 31 (in Russian), 1990.

[199.] Cerofolini, G. F., Re, N., and Para, A. F., (D+D+)2e- binuclear atoms as activated precursors in cold fusion and warm fusion, in *Anomalous Nuclear Effects in Deuterium/Solid Systems, "AIP Conference Proceedings 228"*, Jones, S., Scaramuzzi, F., andWorledge, D. American Institute of Physics, New York, Brigham Young Univ., Provo, UT, 1990, pp. 668.

[200.] Conte, E. and Pieralice, M., An experiment indicates the nuclear fusion of a proton and electron into a neutron, *Infinite Energy* 4 (23), 67, 1999.

[201.] Baranov, D., Bazhutov, Y., Khokhov, N., Koretsky, V. P., Kuznetsov, A. B., Skuratnik, Y., and Sukovatkin, N., Experimental testing of the Erzion model by reacting of electron flux on the target, in *Fourth International Conference on Cold Fusion* Electric Power Research Institute 3412 Hillview Ave., Palo Alto, CA 94304, Lahaina, Maui, 1993, pp. 8.

[202.] Widom, A. and Larsen, L., Ultra low momentum neutron catalyzed nuclear reactions on metallic hydride surfaces, *Eur. Phys. J.* C46, 107, 2006.

[203.] Widom, A. and Larsen, L., Nuclear abundances in metallic hydride electrodes of electrolytic chemical cells, *arXiv:cond-mat/062472 v1*, 2006.

[204.] Widom, A. and Larsen, L., Absorption of nuclear gamma radiation by heavy electrons on metallic hydride surfaces, *arXiv:cond-mat*, 2006.

[205.] Koonin, S. E. and Nauenberg, M., Calculated fusion rates in isotopic hydrogen molecules, *Nature (London)* 339, 690, 1989.

[206.] Claytor, T. N., Seeger, P., Rohwer, R. K., Tuggle, D. G., and Doty, W. R., Tritium and neutron measurements of a solid state cell, in *NSF/EPRI Workshop on Anomalous Effects in Deuterated Materials* LA-UR-89-39-46, Washington, DC, 1989.

[207.] Bhattacharjee, J. K., Satpathy, L., and Waghmare, Y. R., A possible mechanism of cold fusion, *Pramana* 32, L841, 1989.

[208.] Antanasijevic, R. D., Lakicevic, I., Maric, Z., Zevic, D., Zaric, A., and Vigier, J. P., Preliminary observations on possible implications of new Bohr orbits (resulting from electromagnetic spin-spin and spin-orbit coupling) in 'cold' quantum mechanical fusion processes appearing in strong 'plasma focus' and 'capillary fusion' experiments, *Phys. Lett. A* 180, 25, 1993.

[209.] Gabovich, A. M., Possibility of cold fusion in palladium deuterides: screening effects and connection to superconducting properties, *Philos. Mag. B* 76, 107, 1997.

[210.] Das, D. and Ray, M. K. S., Fusion in condensed matter - a likely scenario, *Fusion Technol.* 24, 115, 1993.

[211.] Balin, D. V., Maev, E. M., Medvedev, V. I., Semenchuk, G. G., Smirenin, Y. V., Vorobyov, A. A., Vorobyov, A. A., and Zalite, Y. K., Experimental investigation of the muon catalyzed d-d fusion, *Phys. Lett. B* 141 (3/4), 173, 1984.

[212.] Jones, S. E., Muon-catalysed fusion revisited, *Nature (London)* 321, 327, 1986.

[213.] Szalewicz, K., Morgan III, J. D., and Monkhurst, H. J., Fusion rates for hydrogen isotopic molecules of relevance for 'cold fusion', *Phys. Rev. A* 40 (5), 2824, 1989.

[214.] Rabinowitz, M., High temperature superconductivity and cold fusion, *Mod. Phys. Lett. B* 4 (4), 233, 1990.

[215.] Vaidya, S. N., Deuteron screening, nuclear reactions in solids, and superconductivity, *Fusion Technol.* 29, 405, 1996.

[216.] Takahashi, A., Progress in condensed matter nuclear science, in *12 th International Conference on Cold Fusion*, Takahashi, A. World Scientific Co., Yokohama, Japan, 2006.

[217.] Schirber, J. E. and Northrup Jr., C. J. M., Concentration dependence of the superconducting transition temperature In Pd-H and Pd-D, *Phys. Rev. B* 10, 3818, 1974.

[218.] Miller, R. J. and Satterthwaite, C. B., Electronic model for the reverse isotope effect in superconducting Pd-H(D), *Phys. Rev. Lett.* 34 (3), 1975.

[219.] Harper, J. M. E., Effect of hydrogen concentration on superconductivity and clustering in palladium hydride, *Phys. Lett.* 47A (1), 69, 1974.

[220.] Lipson, A. G., Castano, C. H., Miley, G., Lyakhov, B. F., and Mitin, A. V., Emergence of a high-temperature superconductivity in hydrogen cycled Pd compounds as an evidence for superstoichiometric H/D sites, in *11th International Conference on Cold Fusion*, Biberian, J.-P. World Scientific Co., Marseilles, France, 2004, pp. 128.

[221.] Lipson, A. G., Castano, C. H., Miley, G., Lyakhov, B. F., Tsivadze, A. Y., and Mitin, A. V., Evidence of superstoichiometric H/D LENR active sites and high temperature superconductivity in a hydrogen-cycled Pd-PdO, in *12th International Conference on Cold Fusion*, Yokohama, Japan, 2005.

[222.] Tripodi, P., Di Gioacchion, D., Borelli, R., and Vinko, J. D., Possibility of high temperature superconducting phases in PdH, *Phys. C* 388-289, 571, 2003.

[223.] Flanagan, T. B. and Oates, W. A., The palladium-hydrogen system, *Annu. Rev. Mater. Sci.* 21, 269, 1991.

[224.] Anderson, I. S., Ross, D. K., and Carlile, C. J., The Structure of the g Phase of Palladium Deuteride, *Phys. Lett. A* 68, 249, 1978.

[225.] Mitacek, P. and Aston, J. G., The thermodynamic properties of pure palladium and its alloys with hydrogen between 30 and 300° K, *J. Am. Chem. Soc.* 85, 137, 1963.

[226.] Shoulders, K. and Shoulders, S., Charge clusters in action, in *Conference on Future Energy* Integrity Research Institute, Bethesda, MD, 1999, pp. 1.

[227.] Shoulders, K. and Shoulders, S., Observations on the role of charge clusters in nuclear cluster reactions, *J. New Energy* 1 (3), 111, 1996.

[228.] Shoulders, K., Permittivity transitions, *J. New Energy* 5 (2), 121, 2001.

[229.] Shoulders, K., Projectiles from the dark side, *Infinite Energy* 12 (70), 39, 2006.

[230.] Mesyats, G. A., Ecton processes at the cathode in a vacuum Ddscharge, in *XVIIth International Symp. of Discharges and Electrical Insulatiion in Vacuum*, Berkeley, CA, 1996, pp. 720.

[231.] Fox, H. and Jin, S. X., Low-energy nuclear reactions and high-density charge clusters, *J. New Energy* 3 (2/3), 56, 1998.

[232.] Fox, H. and X., J. S., Low-energy nuclear reactions and high-density charge clusters, *Infinite Energy* 4 (20), 26, 1998.

[233.] Fox, H., Bass, R. W., and Jin, S.-X., Plasma-injected transmutation, *J. New Energy* 1 (3), 222, 1996.

[234.] Jin, S.-X. and Fox, H., Characteristics of high-density charge clusters: A theoretical model, *J. New Energy* 1 (4), 5, 1996.

[235.] Bhadkamkar, A. and Fox, H., Electron charge cluster sparking in aqueous solutions, *J. New Energy* 1 (4), 62, 1996.

[236.] Lochak, G. and Urutskoev, L., Low-energy nuclear reactions and the leptonic monopole, in *11th International Conference on Cold Fusion*, Biberian, J.-P. World Scientific Co., Marseilles, France, 2004, pp. 421.

[237.] Lochak, G., Light monopoles theory: An overview of their effects in physics, chemistry, biology, and nuclear science (weak interactions), in *11th International Conference on Cold Fusion*, Biberian, J.-P. World Scientific Co., Marseilles, France, 2004, pp. 787.

[238.] Pryakhin, E. A., Tryapitsina, G. A., Urutskoyev, L. L., and Akleyev, A. V., Assessment of the biological effects of "strange" radiation, in *11th International Conference on Cold Fusion*, Biberian, J.-P. World Scientific Co., Marseilles, France, 2004, pp. 537.

[239.] Lewis, E., Comments on 'Transmutation in a gold-light water electrolysis system', *Fusion Technol.* 36, 242, 1999.

[240.] Lewis, E., Evidence of micrometre-sized plasmoid emission during electrolysis cold fusion, *Fusion Sci. Technol.* 40 (July), 107, 2001.

[241.] Lewis, E., The ball lightning state In cold fusion, in *Tenth International Conference on Cold Fusion*, Hagelstein, P. L. andChubb, S. R. World Scientific Publishing Co., Cambridge, MA, 2003, pp. 973.

[242.] Lewis, E. H., Evidence of microscopic ball lightning in cold fusion experiments, in *11th International Conference on Cold Fusion*, Biberian, J.-P. World Scientific Co., Marseilles, France, 2004, pp. 304.

[243.] Bostick, W., Plasmoids, *Sci. Am.* 197, 87, 1957.

[244.] Matsumoto, T., Observation of meshlike traces on nuclear emulsions during cold fusion, *Fusion Technol.* 23, 103, 1993.

[245.] Matsumoto, T., Prediction of new particle emission on cold fusion, *Fusion Technol.* 18, 647, 1990.

[246.] Rafelski, J., Sawicki, M., Gajda, M., and Harley, D., Nuclear reactions catalyzed by a massive negatively charged particle. How Cold Fusion Can Be Catalyzed, *Fusion Technol.* 18, 136, 1990.

[247.] McKibben, J. L., Can cold fusion be catalyzed by fractionally-charged ions that have evaded FC particle searches, *Infinite Energy* 1 (4), 14, 1995.

[248.] McKibben, J. L., Catalytic behavior of one (or two) subquarks bound to their nuclear hosts, *Infinite Energy* 3 (13/14), 103, 1997.

[249.] Filimonov, V. A., Neutrino-driven nuclear reactions of cold fusion and transmutation, in *11th International Conference on Cold Fusion*, Biberian, J.-P. World Scientific Co, Marseilles, France, 2004, pp. 776.

[250.] Zhang, Z.-L. and Zhang, W.-S., Possibility of electron capture by deuteron, in *The 9th International Conference on Cold Fusion, Condensed Matter Nuclear Science*, Li, X. Z. Tsinghua Univ. Press, Tsinghua Univ., Beijing, China, 2002, pp. 439.

[251.] Bazhutov, Y. N., Vereshkov, G. M., Kuz'min, R. N., and Frolov, A. M., Interpretation of cold nuclear fusion by means of Erzion catalysis, in *Fiz. Plazmy Nekotor. Vopr. Obshch. Fiz. M.*, 1990, pp. 67 (in Russian).

[252.] Matsumoto, T., Experiments of underwater spark discharge with pinched electrodes, *J. New Energy* 1 (4), 79, 1996.

[253.] Matsumoto, T., Cold fusion experiments using sparking discharges in water, in *5th International Conference on Cold Fusion*, Pons, S. IMRA Europe, Sophia Antipolis Cedex, France, Monte-Carlo, Monaco, 1995, pp. 583.

[254.] Sundaresan, R. and Bockris, J. O. M., Anomalous reactions during arcing between carbon rods in water, *Fusion Technol.* 26, 261, 1994.

[255.] Hanawa, T., X-ray spectroscropic analysis of carbon arc products in water, in *8th International Conference on Cold Fusion*, Scaramuzzi, F. Italian Physical Society, Bologna, Italy, Lerici (La Spezia), Italy, 2000, pp. 147.

[256.] Singh, M., Saksena, M. D., Dixit, V. S., and Kartha, V. B., Verification of the George Oshawa experiment for anomalous production of iron from carbon arc in water, *Fusion Technol.* 26, 266, 1994.

[257.] Jiang, X. L., Han, L. J., and Kang, W., Anomalous element production induced by carbon arcing under water, in *The Seventh International Conference on Cold Fusion*, Jaeger, F. ENECO, Inc., Salt Lake City, UT., Vancouver, Canada, 1998, pp. 172.

[258.] Stringham, R., Chandler, J., George, R., Passell, T. O., and Raymond, R., Predictable and reproducible heat, in *The Seventh International Conference on Cold Fusion*, Jaeger, F. ENECO, Inc., Salt Lake City, UT, Vancouver, Canada, 1998, pp. 361.

[259.] Stringham, R., The cavitation micro accelerator, in *8th International Conference on Cold Fusion*, Scaramuzzi, F. Italian Physical Society, Bologna, Italy, Lerici (La Spezia), Italy, 2000, pp. 299.

[260.] Stringham, R., The ^{4}He and excess heat production via cavitation, in *Am. Phys. Soc.*, Seattle, WA, 2001.

[261.] Stringham, R., Pinched cavitation jets and fusion events, in *The 9th International Conference on Cold Fusion, Condensed Matter Nuclear Science*, Li, X. Z. Tsinghua Univ. Press, Tsinghua Univ., Beijing, China, 2002, pp. 323.

[262.] Stringham, R., Cavitation and fusion, in *Tenth International Conference on Cold Fusion*, Hagelstein, P. L. andChubb, S. R. World Scientific Publishing Co., Cambridge, MA, 2003, pp. 233.

[263.] Stringham, R., Low mass 1.6 MHz sonofusion reactor, in *11th International Conference on Cold Fusion*, Biberian, J.-P. World Scientific Co., Marseilles, France, 2004, pp. 238.

[264.] Chubb, T. A. and Chubb, S. R., Cold fusion as an interaction between ion band states, *Fusion Technol.* 20, 93, 1991.

[265.] Chubb, T. A. and Chubb, S. R., The ion band state theory, in *5th International Conference on Cold Fusion*, Pons, S. IMRA Europe, Sophia Antipolis Cedex, France, Monte-Carlo, Monaco, 1995, pp. 315.

[266.] Chubb, S. R. and Chubb, T. A., Periodic order, symmetry, and coherence in cold fusion, in *The Seventh International Conference on Cold Fusion*, Jaeger, F. ENECO, Inc., Salt Lake City, UT., Vancouver, Canada, 1998, pp. 73.

[267.] Chubb, T., A., Is a quantum-of-mass always a particle?, *Infinite Energy* 12 (70), 24, 2006.

[268.] Hagelstein, P. L., Phonon-exchange models for anomalies in condensed matter systems with molecular deuterium, in *12th International Conference on Condensed Matter Nuclear Science* World Scientifiic Publishing Co., Yokohama, Japan, 2005.

[269.] Swartz, M. R., Phusons in nuclear reactions in solids, *Fusion Technol.* 31, 228, 1997.

[270.] Liu, F. S., The phonon mechanism of the cold fusion, *Mod. Phys. Lett. B* 10, 1129, 1996.

[271.] Komaki, H., production de proteins par 29 souches de microorganismes et augmentation du potassium en milieu de culture sodique sans potassium, *Revue de Pathologie Comparee* 67, 213, 1967.

[272.] Komaki, H., Formation de protines et variations minerales par des microorganismes en milieu de culture, sort avec or sans potassium, sort avec ou sans phosphore, *Revue de Pathologie Comparee* 69, 83, 1969.

[273.] Kervran, C. L., *Biological transmutations* Swan House Publishing Co., 1972.

[274.] Kervran, C. L., *Biological transmutation* Beekman Publishers, Inc, 1980.

[275.] Komaki, H., Observations on the biological cold fusion or the biological transformation of elements, in *Third International Conference on Cold Fusion, "Frontiers of Cold Fusion"*, Ikegami, H. Universal Academy Press, Inc., Tokyo, Japan, Nagoya Japan, 1992, pp. 555.

[276.] Komaki, H., An Approach to the Probable Mechanism of the Non-Radioactive Biological Cold Fusion or So-Called Kervran Effect (Part 2), in *Fourth International Conference on Cold Fusion* Electric Power Research Institute 3412 Hillview Ave., Palo Alto, CA 94304, Lahaina, Maui, 1993, pp. 44.

[277.] Vysotskii, V. I., Kornilova, A. A., and Samoylenko, I. I., Experimental discovery of phenomenon of low-energy nuclear transformation of isotopes (Mn^{55}=Fe^{57}) in growing biological cultures, in *Sixth International Conference on Cold Fusion, Progress in New Hydrogen Energy*, Okamoto, M. New Energy and Industrial Technology Development Organization, Tokyo Institute of Technology, Tokyo, Japan, Lake Toya, Hokkaido, Japan, 1996, pp. 687.

[278.] Vysotskii, V., Kornilova, A. A., and Samoylenko, I. I., Experimental discovery and investigation of the phenomenon of nuclear transmutation of isotopes in growing biological cultures, *Infinite Energy* 2 (10), 63, 1996.

[279.] Kornilova, A. A., Vysotskii, V. I., and Zykov, G. A., Investigation of combined influence of Sr, Cl and S on the effectiveness of nuclear transmutation of Fe-54 isotope in biological cultures., in *The 9th International Conference on Cold Fusion, Condensed Matter Nuclear Science*, Li, X. Z. Tsinghua Univ. Press, Tsinghua Univ., Beijing, China, 2002, pp. 174.

[280.] Vysotskii, V., Kornilova, A. A., Samoylenko, I. I., and Zykov, G. A., Experimental observation and study of controlled transmutation of intermediate mass isotopes in growing biological cultures, in *8th International Conference on Cold Fusion*, Scaramuzzi, F. Italian Physical Society, Bologna, Italy, Lerici (La Spezia), Italy, 2000, pp. 135.

[281.] Vysotskii, V., Kornilova, A. A., Samoylenko, I. I., and A., Z. G., Observation and mass-spectrometry. Study of controlled transmutation of intermediate mass isotopes in growing biological cultures, *Infinite Energy* 6 (36), 64, 2001.

[282.] Vysotskii, V. and Kornilova, A. A., *Nuclear fusion and transmutation of isotopes in biological systems* Mockba, Ukraine, 2003.

[283.] Pappas, P. T., Electrically induced nuclear fusion in the living cell, *J. New Energy* 3 (1), 4, 1998.

[284.] Pappas, P. T., The electrically induced nuclear fusion in a living cell, in *The Seventh International Conference on Cold Fusion*, Jaeger, F. ENECO, Inc., Salt Lake City, UT, Vancouver, Canada, 1998, pp. 460.

[285.] Oriani, R. A. and Fisher, J. C., Energetic particle showers in the vapor from electrolysis, in *11th International Conference on Cold Fusion*, Biberian, J.-P. World Scientific Co., Marseilles, France, 2004, pp. 281.

[286.] Vysotskii, V., Odintsov, A., Pavlovich, V. N., Tashirev, A., and Kornilova, A. A., Experiments on controlled decontamination of water mixture of long-lived active isotopes in biological cells, in *11th International Conference on Cold Fusion*, Biberian, J.-P. World Scientific Co., Marseilles, France, 2004, pp. 530.

[287.] Benford, M. S., Biological Nuclear Reactions: Empirical Data Describes Unexplained SHC Phenomenon, *J. New Energy* 3 (4), 19, 1999.

[288.] Ota, K., Kuratsuka, M., Ando, K., Iida, Y., Yoshitake, H., and Kamiya, N., Heat production at the heavy water electrolysis using mechanically treated cathode, in *Third International Conference on Cold Fusion, "Frontiers of Cold Fusion"*, Ikegami, H. Universal Academy Press, Inc., Tokyo, Japan, Nagoya Japan, 1992, pp. 71.

[289.] Miles, M. and Imam, M. A., United States 6764561, 2004, Palladium-boron alloys and methods making and using such alloys. July 20, 2004.

[290.] Miles, M., Imam, M. A., and Fleischmann, M., Excess heat and helium production in the palladium-boron system, *Trans. Am. Nucl. Soc.* 83, 371, 2000.

[291.] Miles, M. H., Fleischmann, M., and Imam, M. A., Report No. NRL/MR/6320--01-8526, 2001.

[292.] Miles, M. H., Imam, M. A., and Fleischmann, M., Calorimetric analysis of a heavy water electrolysis experiment using a Pd-B alloy cathode, *Proc. Electrochem. Soc.* 2001 (23), 194, 2001.

[293.] Miles, M. H. and Johnson, K. B., Electrochemical insertion of hydrogen into metals and alloys, *J. New Energy* 1 (2), 32, 1996.

[294.] Storms, E., Anomalous heat generated by electrolysis using a palladium cathode and heavy water, in *American Physical Society*, Atlanta, GA, 1999.

[295.] Warner, J. and Dash, J., Heat produced during the electrolysis of D_2O with titanium cathodes, in *8th International Conference on Cold Fusion*, Scaramuzzi, F. Italian Physical Society, Bologna, Italy, Lerici (La Spezia), Italy, 2000, pp. 161.

[296.] Storms, E., Ways to initiate a nuclear reaction in solid environments, *Infinite Energy* 8 (45), 45, 2002.

Chapter 6

[1.] Vysotskii, V. and Kornilova, A. A., *Nuclear fusion and transmutation of isotopes in biological systems* Mockba, Ukraine, 2003.

[2.] Vysotskii, V. I., Kornilova, A. A., and Samoylenko, I. I., Experimental discovery of phenomenon of low-energy nuclear transformation of isotopes ($Mn^{55}=Fe^{57}$) in growing biological cultures, in *Sixth International Conference on Cold Fusion, Progress in New Hydrogen Energy*, Okamoto, M. New Energy and Industrial Technology Development Organization, Tokyo Institute of Technology, Tokyo, Japan, Lake Toya, Hokkaido, Japan, 1996, pp. 687.

[3.] Vysotskii, V., Kornilova, A. A., and Samoylenko, I. I., Experimental discovery and investigation of the phenomenon of nuclear transmutation of isotopes in growing biological cultures, *Infinite Energy* 2 (10), 63, 1996.

[4.] Komaki, H., Observations on the biological cold fusion or the biological transformation of elements, in *Third International Conference on Cold Fusion, "Frontiers of Cold Fusion"*, Ikegami, H. Universal Academy Press, Inc., Tokyo, Japan, Nagoya Japan, 1992, pp. 555.

[5.] Komaki, H., An Approach to the Probable Mechanism of the Non-Radioactive Biological Cold Fusion or So-Called Kervran Effect (Part 2), in *Fourth International Conference on Cold Fusion* Electric Power Research Institute 3412 Hillview Ave., Palo Alto, CA 94304, Lahaina, Maui, 1993, pp. 44.

[6.] Kornilova, A. A., Vysotskii, V. I., and Zykov, G. A., Investigation of combined influence of Sr, Cl and S on the effectiveness of nuclear transmutation of Fe-54 isotope in biological cultures., in *The 9th International Conference on Cold Fusion, Condensed Matter Nuclear Science*, Li, X. Z. Tsinghua Univ. Press, Tsinghua Univ., Beijing, China, 2002, pp. 174.

[7.] Vysotskii, V., Kornilova, A. A., Samoylenko, I. I., and Zykov, G. A., Experimental observation and study of controlled transmutation of intermediate mass isotopes in growing biological cultures, in *8th International Conference on Cold Fusion*, Scaramuzzi, F. Italian Physical Society, Bologna, Italy, Lerici (La Spezia), Italy, 2000, pp. 135.

[8.] Arata, Y. and Zhang, Y. C., Formation of condensed metallic deuterium lattice and nuclear fusion, *Proc. Jpn. Acad., Ser. B* 78 (Ser. B), 57, 2002.

[9.] Arata, Y. and Zhang, Y. C., Observation of anomalous heat release and helium-4 production from highly deuterated fine particles, *Jpn. J. Appl. Phys. Part 2* 38 (7A), L774, 1999.

[10.] Arata, Y. and Zhang, Y. C., Picnonuclear fusion generated in "lattice-reactor" of metallic deuterium lattice within metal atom-clusters. II Nuclear fusion reacted inside a metal by intense sonoimplantion effect, in *The 9th International Conference on Cold Fusion, Condensed Matter Nuclear Science*, Li, X. Z. Tsinghua Univ. Press, Tsinghua Univ., Beijing, China, 2002, pp. 5.

[11.] Arata, Y., Zhang, Y.-C., Fujita, H., and Inoue, A., Discovery of solid deuterium nuclear fusion of pycnodeuterium-lumps solidified locally within nano-Pd particles, *Koon Gakkaishi* 29, 68, 2003.

[12.] Case, L. C., Catalytic fusion of deuterium into helium-4, in *The Seventh International Conference on Cold Fusion*, Jaeger, F. ENECO, Inc., Salt Lake City, UT, Vancouver, Canada, 1998, pp. 48.

[13.] Kowalski, L., Jones, S., Letts, D., and Cravens, D., Charged particles from Ti and Pd foils, in *11th International Conference on Cold Fusion*, Biberian, J.-P. World Scientific Co., Marseille, France, 2004, pp. 269.

[14.] Keeney, F., Jones, S. E., Johnson, A., Buehler, D. B., Cecil, F. E., Hubler, G. K., Hagelstein, P. L., Scott, M., and Ellsworth, J., Charged-particle emissions from metal deuterides, in *Tenth International Conference on Cold Fusion*, Hagelstein, P. L. andChubb, S. R. World Scientific Publishing Co., Cambridge, MA, 2003, pp. 509.

[15.] Mo, W., Liu, Y. S., Zhou, L. Y., Dong, S. Y., Wang, K. L., Wang, S. C., and Li, X. Z., Search for precursor and charged particles in "cold fusion", in *Second Annual Conference on Cold Fusion*,

"The Science of Cold Fusion", Bressani, T., Del Giudice, E., andPreparata, G. Societa Italiana di Fisica, Bologna, Italy, Como, Italy, 1991, pp. 123.

[16.] Cecil, F. E., Liu, H., Beddingfield, D. H., and Galovich, C. S., Observation of charged-particle bursts from deuterium-loaded thin-titanium foils, in *Anomalous Nuclear Effects in Deuterium/Solid Systems, "AIP Conference Proceedings 228"*, Jones, S., Scaramuzzi, F., andWorledge, D. American Institute of Physics, New York, Brigham Young Univ., Provo, UT, 1990, pp. 375.

[17.] Li, X. Z., Yue, W. Z., Huang, G. S., Shi, H., Gao, L., Liu, M. L., and Bu, F. S., "Excess heat" measurement in gas-loading D/Pd system, *J. New Energy* 1 (4), 34, 1996.

[18.] Li, X. Z., Liu, B., Cai, N., Wei, Q., Tian, J., and Cao, D. X., Progress in gas-loading D/Pd system: The feasibility of a self-sustaining heat generator, in *Tenth International Conference on Cold Fusion*, Hagelstein, P. L. andChubb, S. R. World Scientific Publishing Co., Cambridge, MA, 2003, pp. 113.

[19.] Campari, E. G., Fasano, G., Focardi, S., Lorusso, G., Gabbani, V., Montalbano, V., Piantelli, F., Stanghini, C., and Veronesi, S., Photon and particle emission, heat production and surface transformation in Ni-H system, in *11th International Conference on Cold Fusion*, Biberian, J.-P. World Scientific Co., Marseilles, France, 2004, pp. 405.

[20.] Focardi, S., Gabbani, V., Montalbano, V., Piantelli, F., and Veronesi, F., Evidence of electromagnetic radiation from Ni-H systems, in *11th International Conference on Cold Fusion*, Biberian, J.-P. World Scientific Co., Marseilles, France, 2004, pp. 70.

[21.] Focardi, S., Gabbani, V., Montalbano, V., Piantelli, F., and Veronesi, S., Large excess heat production in Ni-H systems, *Nuovo Cimento* 111A (11), 1233, 1998.

[22.] Cammarota, G., Collis, W. J. M. F., Rizzo, A., and Stremmenos, C., A flow calorimeter study of the Ni/H system, in *Asti Workshop on Anomalies in Hydrogen/Deuterium Loaded Metals*, Collis, W. J. M. F. Italian Phys. Soc., Villa Riccardi, Rocca d'Arazzo, Italy, 1997, pp. 1.

[23.] Arata, Y. and Zhang, Y.-C., Development of "DS-reactor" as a practical reactor of "cold fusion" based on the "DS-cell" with "DS-cathode", in *12th International Conference on Cold Fusion*, Yokohama, Japan, 2005.

[24.] Bartolomeo, C., Feischmann, M., Larramona, G., Pons, S., Roulett, J., Sugiura, H., and Preparata, G., Alfred Coehn and after: The alpha, beta and gamma of the palladium-hydrogen system, in *Fourth International Conference on Cold Fusion* Electric Power Research Institute 3412 Hillview Ave., Palo Alto, CA 94304, Lahaina, Maui, 1993, pp. 19.

[25.] Celani, F., Spallone, A., Tripodi, P., Petrocchi, A., di Gioacchino, D., Marini, P., di Stefano, V., Pace, S., and Mancini, A., Deuterium overloading of palladium wires by means of high power microsecond pulsed electrolysis and electromigration: suggestions of a "phase transition" and related excess heat, *Phys. Lett. A* 214, 1, 1996.

[26.] Del Giudice, E., De Ninno, A., Franttolillo, A., Preparata, G., Scaramuzzi, F., Bulfone, A., Cola, M., and Giannetti, C., The Fleischmann-Pons effect in a novel electrolytic configuration, in *8th International Conference on Cold Fusion*, Scaramuzzi, F. Italian Physical Society, Bologna, Italy, Lerici (La Spezia), Italy, 2000, pp. 47.

[27.] Mizuno, T., Enyo, M., Akimoto, T., and Azumi, K., Anomalous heat evolution from $SrCeO_3$-type proton conductors during absorption/desorption in alternate electric field, in *Fourth International Conference on Cold Fusion* Electric Power Research Institute 3412 Hillview Ave., Palo Alto, CA 94304, Lahaina, Maui, 1993, pp. 14.

[28.] Biberian, J.-P., Excess heat measurements in $AlLaO_3$ doped with deuterium, in *5th International Conference on Cold Fusion*, Pons, S. IMRA Europe, Sophia Antipolis Cedex, France, Monte-Carlo, Monaco, 1995, pp. 49.

[29.] Mizuno, T., Akimoto, T., Azumi, K., Kitaichi, M., and Kurokawa, K., Anomalous heat evolution from a solid-state electrolyte under alternating current in high-temperature D_2 gas, *Fusion Technol.* 29, 385, 1996.

[30.] Mizuno, T., Akimoto, T., Azumi, K., Kitaichi, M., Kuroiwa, K., and Enyo, M., Excess heat evolution and analysis of elements for solid state electrolyte in deuterium atmosphere during applied electric field, *J. New Energy* 1 (1), 79, 1996.

[31.] Oriani, R. A., An investigation of anomalous thermal power generation from a proton-conducting oxide, *Fusion Technol.* 30, 281, 1996.

[32.] Samgin, A. L., Finodeyev, O., Tsvetkov, S. A., Andreev, V. S., Khokhlov, V. A., Filatov, E. S., Murygin, I. V., Gorelov, V. P., and Vakarin, S. V., Cold fusion and anomalous effects in deuteron conductors during non-stationary high-temperature electrolysis, in *5th International Conference on Cold Fusion*, Pons, S. IMRA Europe, Sophia Antipolis Cedex, France, Monte-Carlo, Monaco, 1995, pp. 201.

[33.] Biberian, J.-P., Lonchampt, G., Bonnetain, L., and Delepine, J., Electrolysis of LaAlO₃ single crystals and ceramics in a deuteriated atmosphere, in *The Seventh International Conference on Cold Fusion*, Jaeger, F. ENECO, Inc., Salt Lake City, UT, Vancouver, Canada, 1998, pp. 27.

[34.] Dillon, C. T. and Kennedy, B. J., The electrochemically formed palladium-deuterium system. I. Surface composition and morphology, *Aust. J. Chem.* 46, 663, 1993.

[35.] Celani, F., Spallone, A., Marini, P., Stefano, V., Nakamura, M., Mancini, A., Pace, S., Tripodi, P., Di Gioacchino, D., Catena, C., D'Agostaro, G., Petraroli, R., Quercia, P., Righi, E., and Trenta, G., High hydrogen loading into thin palladium wires through precipitate of alkaline-earth carbonate on the surface of cathode: Evidence of new phases in the Pd-H system and unexpected problems due to bacteria contamination in the heavy-water, in *8th International Conference on Cold Fusion*, Scaramuzzi, F. Italian Physical Society, Bologna, Italy, Lerici (La Spezia), Italy, 2000, pp. 181.

[36.] Celani, F., Spallone, A., Marini, P., di Stefano, V., Nakamura, M., Mancini, A., D'Agostaro, G., Righi, E., Trenta, G., Quercia, P., and Catena, C., Electrochemical D loading of palladium wires by heavy ethyl-alcohol and water electrolyte, related to Ralstonia bacteria problematics., in *The 9th International Conference on Cold Fusion, Condensed Matter Nuclear Science*, Li, X. Z. Tsinghua Univ. Press, Tsinghua Univ., Beijing, China, 2002, pp. 29.

[37.] Swartz, M. R. and Verner, G., Excess heat from low-electrical conducting heavy water spiral-wound Pd/D₂O/Pt and Pd/D₂O-PdCl₂/Pt devices, in *Tenth International Conference on Cold Fusion*, Chubb, P. I. H. a. S. R. World Scientific Publishing Co., Cambridge, MA, 2003, pp. 29.

[38.] Dardik, I., Zilov, T., Branover, H., El-Boher, A., Greenspan, E., Khachatorov, B., Krakov, V., Lesin, S., and Tsirlin, M., Excess heat in electrolysis experiments at Energetics Technologies, in *11th International Conference on Cold Fusion*, Biberian, J.-P. World Scientific Co., Marseilles, France, 2004, pp. 84.

[39.] Mengoli, G., Bernardini, M., Manduchi, C., and Zannoni, G., Calorimetry close to the boiling temperature of the D₂O/Pd electrolytic system, *J. Electroanal. Chem.* 444, 155, 1998.

[40.] Fleischmann, M. and Pons, S., Calorimetry of the Pd-D₂O system: From simplicity via complications to simplicity., in *Third International Conference on Cold Fusion, "Frontiers of Cold Fusion"*, Ikegami, H. Universal Academy Press, Inc., Tokyo, Japan, Nagoya Japan, 1992, pp. 47.

[41.] Lonchampt, G., Biberian, J.-P., Bonnetain, L., and Delepine, J., Excess heat measurement with Pons and Fleischmann Type cells, in *The Seventh International Conference on Cold Fusion*, Jaeger, F. ENECO, Inc., Salt Lake City, UT, Vancouver, Canada, 1998, pp. 202.

[42.] Yuan, L. J., Wan, C. M., Liang, C. Y., and Chen, S. K., Neutron monitoring on cold-fusion experiments, in *Third International Conference on Cold Fusion, "Frontiers of Cold Fusion"*, Ikegami, H. Universal Academy Press, Inc., Tokyo, Japan, Nagoya Japan, 1992, pp. 461.

[43.] Liaw, B. Y., Tao, P.-L., Turner, P., and Liebert, B. E., Elevated temperature excess heat production using molten-salt electrochemical techniques, in *8th World Hydrogen Energy Conf* Hawaii Natural Energy Institute, 2540 Dole St., Holmes Hall 246, Honolulu, HI 96822, Honolulu, HI, 1990, pp. 49.

[44.] Liaw, B. Y., Tao, P.-L., Turner, P., and Liebert, B. E., Elevated-temperature excess heat production in a Pd + D system, *J. Electroanal. Chem.* 319, 161, 1991.

[45.] Okamoto, H. and Nezu, S., Measurements of hydrogen loading ratio of Pd anodes polarized in LiH-LiCl-KCl molten salt systems, in *Fourth International Conference on Cold Fusion* Electric Power Research Institute 3412 Hillview Ave., Palo Alto, CA 94304, Lahaina, Maui, 1993, pp. 27.

46. Chubb, T. A., Production of excited surface states by reactant starved electrolysis, in *The 9th International Conference on Cold Fusion, Condensed Matter Nuclear Science*, Li, X. Z. Tsinghua Univ. Press, Tsinghua Univ., Beijing, China, 2002, pp. 64.

47. Miley, G. H. and Shrestha, P. J., Review of transmutation reactions in solids, in *Tenth International Conference on Cold Fusion*, Hagelstein, P. L. andChubb, S. R. World Scientific Publishing Co., Cambridge, MA, 2003, pp. 361.

48. Mills, R. L. and Kneizys, P., Excess heat production by the electrolysis of an aqueous potassium carbonate electrolyte and the implications for cold fusion, *Fusion Technol.* 20, 65, 1991.

49. Mizuno, T., Ohmori, T., Akimoto, T., and Takahashi, A., Production of heat during plasma electrolysis, *Jpn. J. Appl. Phys. A* 39, 6055, 2000.

50. Celani, F., Achilli, M., Battaglia, A., Cattaneo, C., Buzzanca, C., Sona, P. G., and Mancini, A., Preliminary results with "Cincinnati Group Cell" on thorium "transmutation" under 50 Hz AC excitation, in *The Seventh International Conference on Cold Fusion*, Jaeger, F. ENECO, Inc., Salt Lake City, UT., Vancouver, Canada, 1998, pp. 56.

51. Matsumoto, T., Cold fusion experiments by using electrical discharge in water, in *Fourth International Conference on Cold Fusion* Electric Power Research Institute 3412 Hillview Ave., Palo Alto, CA 94304, Lahaina, Maui, 1993, pp. 10.

52. Ohmori, T., Yamada, H., Narita, S., Mizuno, T., and Aoki, Y., Enrichment of ^{41}K isotope in potassium formed on and in a rhenium electrode during plasma electrolysis in K_2CO_3/H_2O and K_2CO_3/D_2O solutions., *J. Appl. Electrochem.* 33, 643, 2003.

53. Sundaresan, R. and Bockris, J. O. M., Anomalous reactions during arcing between carbon rods in water, *Fusion Technol.* 26, 261, 1994.

54. Cirillo, D. and Iorio, V., Transmutation of metal at low energy in a confined plasma in water, in *11th International Conference on Cold Fusion*, Biberian, J.-P. World Scientific Co., Marseilles, France, 2004, pp. 492.

55. Naudin, J. L., Experimental study of glow discharge in light water with W electrodes, hhtp://jlnlabs.imars.com/cfr/, 2006.

56. Mizuno, T., Akimoto, T., Azumi, K., Ohmori, T., Aoki, Y., and Takahashi, A., Hydrogen evolution by plasma electrolysis in aqueous solution, *Jpn. J. Appl. Phys. A* 44 (1A), 396, 2005.

57. Passell, T. O., Glow discharge calorimetry, in *6th International Workship on Anomalies in Hydrogen/Deuterium Loaded Metals*, Siena, Italy, 2005.

58. Savvatimova, I. and Gavritenkov, D. V., Results of analysis of Ti foil after glow discharge with deuterium, in *11th International Conference on Cold Fusion*, Biberian, J.-P. World Scientific Co., Marseilles, France, 2004, pp. 438.

59. Karabut, A., Research into characteristics of X-ray emission laser beams from solid-state cathode medium of high-current glow discharge, in *11th International Conference on Cold Fusion*, Biberian, J.-P. World Scientific Co., Marseilles, France, 2004, pp. 253.

60. Karabut, A., Excess heat production in Pd/D during periodic pulse discharge current in various conditions, in *11th International Conference on Cold Fusion*, Biberian, J.-P. World Scientific Co., Marseilles, France, 2004, pp. 178.

61. Benson, T. B. and Passell, T. O., Calorimetry of energy-efficient glow discharge apparatus design and calibration, in *11th International Conference on Cold Fusion*, Biberian, J.-P. World Scientific Co., Marseilles, France, 2004, pp. 147.

62. Narita, S., Yamada, H., Arapi, A., Sato, N., Kato, D., Yamamura, M., and Itagaki, M., Gamma ray detection and surface analysis on palladium electrode in DC glow-like discharge experiment, in *Tenth International Conference on Cold Fusion*, Hagelstein, P. L. andChubb, S. R. World Scientific Publishing Co., Cambridge, MA, 2003, pp. 603.

63. Lipson, A. G., Roussetski, A. S., Karabut, A. B., and Miley, G. H., Strong enhancement of DD-reaction accompanied by X-ray generation in a pulsed low voltage high-current deuterium glow discharge with a titanuim cathode, in *Tenth International Conference on Cold Fusion*, Hagelstein, P. L. andChubb, S. R. World Scientific Publishing Co., Cambridge, MA, 2003, pp. 635.

[64.] Arapi, A., Ito, R., Sato, N., Itagaki, M., Narita, S., and Yamada, H., Experimental observation of the new elements production in the deuterated and/or hydride palladium electrodes, exposed to low energy DC glow discharge, in *The 9th International Conference on Cold Fusion, Condensed Matter Nuclear Science*, Li, X. Z. Tsinghua Univ. Press, Tsinghua Univ., Beijing, China, 2002, pp. 1.

[65.] Karabut, A. B., Excess heat power, nuclear products and X-ray emission in relation to the high current glow discharge experimental parameters, in *The 9th International Conference on Cold Fusion, Condensed Matter Nuclear Science*, Li, X. Z. Tsinghua Univ. Press, Tsinghua Univ., Beijing, China, 2002, pp. 151.

[66.] Dash, J., Savvatimova, I., Frantz, S., Weis, E., and Kozima, H., Effects of glow discharge with hydrogen isotope plasmas on radioactivity of uranium, in *The Ninth International Conference on Cold Fusion*, Li, X. Z. Tsinghua University, Beijing, China, 2002, pp. 77.

[67.] Yamada, H., Uchiyama, K., Kawata, N., Kurisawa, Y., and Nakamura, M., Producing a radioactive source in a deuterated palladium electrode under direct-current glow discharge, *Fusion Technol.* 39, 253, 2001.

[68.] Savvatimova, I. B. and Karabut, A. B., Nuclear reaction products detected at the cathode after a glow discharge in deuterium, *Poverkhnost(Surface)* (1), 63 (in Russian), 1996.

[69.] Kennel, E. and Kalandarachvili, A. G., Investigation of deuterium glow discharge of the Kucherov type, in *Fourth International Conference on Cold Fusion* Electric Power Research Institute 3412 Hillview Ave., Palo Alto, CA 94304, Lahaina, Maui, 1993, pp. 41.

[70.] Claytor, T. N., Schwab, M. J., Thoma, D. J., Teter, D. F., and Tuggle, D. G., Tritium production from palladium alloys, in *The Seventh International Conference on Cold Fusion*, Jaeger, F. ENECO, Inc., Salt Lake City, UT., Vancouver, Canada, 1998, pp. 88.

[71.] Romodanov, V. A., Tritium generation during the interaction of plasma glow discharge with metals and a magnetic field, in *Tenth International Conference on Cold Fusion*, Hagelstein, P. L. andChubb, S. R. World Scientific Publishing Co., Cambridge, MA, 2003, pp. 325.

[72.] Dufour, J., Murat, D., Dufour, X., and Foos, J., Experimental observation of nuclear reactions in palladium and uranium - possible explanation by Hydrex mode, *Fusion Sci. & Technol.* 40, 91, 2001.

[73.] Dufour, J., Foos, J., and Millot, J. P., Cold fusion by sparking in hydrogen isotopes. Energy balances and search for fusion by-products, *Trans. Fusion Technol.* 26 (#4T), 375, 1994.

[74.] Violante, V., Castagna, E., Sibilia, C., Paoloni, S., and Sarto, F., Analysis of Ni-hydride thin film after surface plasmons generation by laser technique, in *Tenth International Conference on Cold Fusion*, Hagelstein, P. L. andChubb, S. R. World Scientific Publishing Co., Cambridge, MA, 2003, pp. 421.

[75.] Bazhutov, Y., Bazhutova, S. Y., Nekrasov, V. V., Dyad'kin, A. P., and Sharkov, V. F., Calorimetric and neutron diagnostics of liquids during laser irradiation, in *11th International Conference on Cold Fusion*, Biberian, J.-P. World Scientific Co., Marseilles, France, 2004, pp. 374.

[76.] Letts, D. and Cravens, D., Cathode frabrication methods to reproduce the Letts-Cravens effect, in *ASTI-5* www.iscmns.org/, Asti, Italy, 2004.

[77.] Storms, E., Use of a very sensitive Seebeck calorimeter to study the Pons-Fleischmann and Letts effects, in *Tenth International Conference on Cold Fusion*, Hagelstein, P. L. andChubb, S. R. World Scientific Publishing Co., Cambridge, MA, 2003, pp. 183.

[78.] McKubre, M. C., The conditions for excess heat production in the D/Pd system, in *ASTI-5* www.iscmns.org, Asti, Italy, 2004.

[79.] Swartz, M. R., Photo-induced excess heat from laser-irradiated electrically polarized palladium cathodes in D_2O, in *Tenth International Conference on Cold Fusion*, Hagelstein, P. L. andChubb, S. R. World Scientific Publishing Co., Cambridge, MA, 2003, pp. 213.

[80.] Castellano, Di Giulio, M., Dinescu, M., Nassisi, V., Conte, A., and Pompa, P. P., Nuclear transmutation in deutered Pd films irradiated by an UV laser, in *8th International Conference on Cold Fusion*, Scaramuzzi, F. Italian Physical Society, Bologna, Italy, Lerici (La Spezia), Italy, 2000, pp. 287.

[81.] Nassisi, V., Transmutation of elements in saturated palladium hydrides by an XeCl excimer laser, *Fusion Technol.* 33, 468, 1998.

[82.] Nassisi, V., Incandescent Pd and Anomalous Distribution of Elements in Deuterated Samples Processed by an Excimer Laser, *J. New Energy* 2 (3/4), 14, 1997.

[83.] Nassisi, V. and Longo, M. L., Experimental results of transmutation of elements observed in etched palladium samples by an excimer laser, *Fusion Technol.* 37 (May), 247, 2000.

[84.] Roussetski, A. S., Lipson, A. G., and Andreanov, V. P., Nuclear emissions from titanium hydride/deuteride, induced by powerful picosecond laser beam, in *Tenth International Conference on Cold Fusion*, Chubb, P. I. H. a. S. R. World Scientific Publishing Co., Cambridge, MA, 2003, pp. 559.

[85.] Clark, E. L., Krushelnick, M., Zepf, M., Beg, F. N., Tatarakis, M., Machacek, A., Santala, M. I. K., Watts, I., Norreys, P. A., and Dangor, A. E., Energetic heavy-ion and proton generation from ultraintense laser-plasma interactions with solids, *Phys. Rev. Lett.* 85 (8), 1654, 2000.

[86.] Bass, R. W., Optimal laser wavelength for resonant transmission through the coulomb barrier, in *Tenth International Conference on Cold Fusion*, Hagelstein, P. L. andChubb, S. R. World Scientific Publishing Co., Cambridge, MA, 2003, pp. 1009.

[87.] Stringham, R., Cavitation and fusion, in *Tenth International Conference on Cold Fusion*, Hagelstein, P. L. andChubb, S. R. World Scientific Publishing Co., Cambridge, MA, 2003, pp. 233.

[88.] Griggs, J. L., A brief introduction to the hydrosonic pump and the associated "excess energy" phenomenon, in *Fourth International Conference on Cold Fusion* Electric Power Research Institute 3412 Hillview Ave., Palo Alto, CA 94304, Lahaina, Maui, 1993, pp. 43.

[89.] Jorné, J., Ultrasonic irradiation of deuterium-loaded palladium particles suspended in heavy water, *Fusion Technol.* 29, 83, 1996.

[90.] Wark, A. W., Crouch-Baker, S., McKubre, M. C. H., and Tanzella, F. L., The effect of ultra sound on the electrochemical loading of hydrogen in palladium, *J. Electroanal. Chem.* 418, 199, 1996.

[91.] Taleyarkhan, R. P., West, C. D., Cho, J. S., Lahey Jr., R. T., Nigmatulin, R. I., and Block, R. C., Evidence for nuclear emissions during acoustic cavation, *Science* 295, 1868, 2002.

[92.] Azumi, K., Ishiguro, S., Mizuno, T., and Seo, M., Acoustic emission from a palladium electrode during hydrogen charging and its release in a LiOH electrolyte, *J. Electroanal. Chem.* 347, 111, 1993.

[93.] Menlove, H. O., Fowler, M. M., Garcia, E., Miller, M. C., Paciotti, M. A., Ryan, R. R., and Jones, S. E., Measurement of neutron emission from Ti and Pd in pressurized D_2 gas and D_2O electrolysis cells, *J. Fusion Energy* 9 (4), 495, 1990.

[94.] Karabut, A. B., Kolomeychenko, S. A., and Savvatimova, I. B., High energy phenomena in glow discharge experiments, in *5th International Conference on Cold Fusion*, Pons, S. IMRA Europe, Sophia Antipolis Cedex, France, Monte-Carlo, Monaco, 1995, pp. 241.

Chapter 7

[1.] Ilic, R., Rant, J., Sutej, T., Kristof, E., Skvarc, J., Kozelj, M., Najzer, M., Humar, M., Cercek, M., Glumac, B., Cvikl, B., Fajgelj, A., Gyergyek, T., Trkov, A., Loose, A., Peternelj, J., Remec, I., and Ravnik, M., A search for neutrons, protons, tritons, (3)He ions, gamma- and x-rays from deuterium-deuterium nuclear reaction in electrochemically charged palladium, *Nucl. Tracks Radiat. Meas.* 17, 483, 1990.

[2.] Jordan, K. C., Blanke, B. C., and Dudley, W. A., Half-life of tritium, *J. Inorg. Nucl. Chem.* 29, 2129, 1967.

[3.] Cedzynska, K. and Will, F. G., Closed-system analysis of tritium in palladium, *Fusion Technol.* 22, 156, 1992.

[4.] Claytor, T. N., Jackson, D. D., and Tuggle, D. G., Tritium production from low voltage deuterium discharge on palladium and other metals, *Infinite Energy* 2 (7), 39, 1996.

[5.] Yamada, H., Narita, S., Inamura, I., Nakai, M., Iwasaki, K., and Baba, M., Tritium production in palladium deuteride/hydride in evacuated chamber, in *8th International Conference on Cold Fusion*, Scaramuzzi, F. Italian Physical Society, Bologna, Italy, Lerici (La Spezia), Italy, 2000, pp. 241.

[6.] Sankaranarayanan, T. K., Sprinivasan, M., Bajpai, M. B., and Gupta, D. S., Evidence for tritium generation in self-heated nickel wires subjected to hydrogen gas absorption/desorption cycles., in *5th International Conference on Cold Fusion*, Pons, S. IMRA Europe, Sophia Antipolis Cedex, France, Monte-Carlo, Monaco, 1995, pp. 173.

[7.] Itoh, T., Iwamura, Y., Gotoh, N., and Toyoda, I., Observation of nuclear products in gas release experiments with electrochemically deuterated palladium, in *Sixth International Conference on Cold Fusion, Progress in New Hydrogen Energy*, Okamoto, M. New Energy and Industrial Technology Development Organization, Tokyo Institute of Technology, Tokyo, Japan, Lake Toya, Hokkaido, Japan, 1996, pp. 410.

[8.] Clarke, W. B., Jenkins, W. J., and Top, Z., Determination of tritium by mass spectrometric measurement of ^{3}He, *Int. J. Appl. Radia. Isot.* 27, 515, 1976.

[9.] Clarke, B. W., Oliver, B. M., McKubre, M. C. H., Tanzella, F. L., and Tripodi, P., Search for ^{3}He and ^{4}He in Arata-style palladium cathodes II: Evidence for tritium production, *Fusion Sci. & Technol.* 40, 152, 2001.

[10.] Corrigan, D. A. and Schneider, E. W., Tritium separation effects during heavy water electrolysis: implications for reported observations of cold fusion, *J. Electroanal. Chem.* 281, 305, 1990.

[11.] Ritley, K. A., Lynn, K. G., Dull, P., Weber, M. H., Carroll, M., and Hurst, J. J., A search for tritium production in electrolytically deuterided palladium, *Fusion Technol.* 19 (1), 192., 1991.

[12.] Sevilla, J., Escarpizo, B., Fernandez, F., Cuevas, F., and Sanchez, C., Time-evolution of tritium concentration in the electrolyte of prolonged cold fusion experiments and its relation to Ti cathode surface treatment, in *Third International Conference on Cold Fusion, "Frontiers of Cold Fusion"*, Ikegami, H. Universal Academy Press, Inc., Tokyo, Japan, Nagoya Japan, 1992, pp. 507.

[13.] Boucher, G. R., Collins, F. E., and Matlock, R. L., Separation factors for hydrogen isotopes on palladium, *Fusion Technol.* 24, 200, 1993.

[14.] Szpak, S., Mosier-Boss, P. A., and Boss, R. D., Comments on the analysis of tritium content in electrochemical cells, *J. Electroanal. Chem.* 373, 1, 1994.

[15.] Will, F. G., Cedzynska, K., and Linton, D. C., Reproducible tritium generation in electrochemical cells employing palladium cathodes with high deuterium loading, *J. Electroanal. Chem.* 360, 161, 1993.

[16.] Storms, E. and Talcott, C. L., Electrolytic tritium production, *Fusion Technol.* 17, 680, 1990.

[17.] Szpak, S., Mosier-Boss, P. A., Boss, R. D., and Smith, J. J., On the behavior of the Pd/D system: evidence for tritium production, *Fusion Technol.* 33, 38, 1998.

[18.] Packham, N. J. C., Wolf, K. L., Wass, J. C., Kainthla, R. C., and Bockris, J. O. M., Production of tritium from D$_2$O electrolysis at a palladium cathode, *J. Electroanal. Chem.* 270, 451, 1989.

[19.] Chien, C.-C., Hodko, D., Minevski, Z., and Bockris, J. O. M., On an electrode producing massive quantities of tritium and helium, *J. Electroanal. Chem.* 338, 189, 1992.

[20.] Iyengar, P. K. and Srinivasan, M., Report No. B.A.R.C.1500, 1989.

[21.] Chêne, J. and Brass, A. M., Tritium production during the cathodic discharge of deuterium on palladium, *J. Electroanal. Chem.* 280, 199, 1990.

[22.] Wolf, K. L., Whitesell, L., Jabs, H., and Shoemaker, J., Tritium and tritons in cold fusion, in *Anomalous Nuclear Effects in Deuterium/Solid Systems, "AIP Conference Proceedings 228"*, Jones, S., Scaramuzzi, F., andWorledge, D. American Institute of Physics, New York, Brigham Young Univ., Provo, UT, 1990, pp. 552.

[23.] Cedzynska, K., Barrowes, S. C., Bergeson, H. E., Knight, L. C., and Will, F. G., Tritium analysis in palladium with an open system analytical procedure, *Fusion Technol.* 20, 108, 1991.

[24.] Friedlander, G., Kennedy, J. W., and Miller, J. M., *Nuclear and radiochemistry* John Wiley & Sons, Inc., New York, 1955.

[25.] Swartz, M. R. and Verner, G., Bremsstrahlung in hot and cold fusion, *J. New Energy* 3 (4), 90-101, 1999.

[26.] Jones, S. E., Bartlett, T. K., Buehler, D. B., Czirr, J. B., Jensen, G. L., and Wang, J. C., Preliminary results from the BYU charged-particle spectrometer, in *Anomalous Nuclear Effects in Deuterium/Solid Systems, "AIP Conference Proceedings 228"*, Jones, S., Scaramuzzi, F., andWorledge, D. American Institute of Physics, New York, Brigham Young Univ., Provo, UT, 1990, pp. 397.

[27.] Iwamura, Y., Itoh, T., Sakano, M., Yamazaki, N., Kuribayashi, S., Terada, Y., Ishikawa, T., and Kasagi, J., Observation of nuclear transmutation reactions induced by D_2 gas permeation through Pd complexes, in *ICCF-11, International Conference on Condensed Matter Nuclear Science*, Biberian, J. P. World Scientific, Marseilles, France, 2004, pp. 339.

[28.] Afonichev, D., Ascending diffusion or transmutation, in *Tenth International Conference on Cold Fusion*, Hagelstein, P. L. andChubb, S. R. World Scientific Publishing Co., Cambridge, MA, 2003, pp. 483.

[29.] Abell, G. C., Matson, L. K., Steinmeyer, R. H., Bowman Jr., R. C., and Oliver, B. M., Helium release from aged palladium tritide, *Phys. Rev. B: Mater. Phys.* 41 (2), 1220, 1990.

[30.] Jung, P., Diffusion and Clustering of Helium in Noble Metals, in *Fundamental Aspects of Inert Gasses in Solids*, Donnelly, S. E. andEvans, J. H. Plenum Press, NY, 1991, pp. 59.

[31.] Morrey, J. R., Caffee, M. W., Farrar IV, H., Hoffman, N. J., Hudson, G. B., Jones, R. H., Kurz, M. D., Lupton, J., Oliver, B. M., Ruiz, B. V., Wacker, J. F., and van Veen, A., Measurements of helium in electrolyzed palladium, *Fusion Technol.* 18, 659, 1990.

[32.] Camp, W. J., Helium detrapping and release from metal tritides, *J. Vac. Sci. Technol.* 14, 514, 1977.

[33.] Case, L. C., Catalytic fusion of deuterium into helium-4, in *The Seventh International Conference on Cold Fusion*, Jaeger, F. ENECO, Inc., Salt Lake City, UT, Vancouver, Canada, 1998, pp. 48.

[34.] McKubre, M. C. H., Tanzella, F. L., Tripodi, P., and Hagelstein, P. L., The emergence of a coherent explanation for anomalies observed in D/Pd and H/Pd system: evidence for ^{4}He and ^{3}He production, in *8th International Conference on Cold Fusion*, Scaramuzzi, F. Italian Physical Society, Bologna, Italy, Lerici (La Spezia), Italy, 2000, pp. 3.

[35.] Miles, M., Bush, B. F., and Lagowski, J. J., Anomalous effects involving excess power, radiation, and helium production during D_2O electrolysis using palladium cathodes, *Fusion Technol.* 25, 478, 1994.

[36.] Miles, M., Correlation of excess enthalpy and helium-4 production: A review, in *Tenth International Conference on Cold Fusion*, Hagelstein, P. L. andChubb, S. R. World Scientific Publishing Co., Cambridge, MA, 2003, pp. 123.

[37.] Takahashi, A., ^{3}He/^{4}He production ratios by tetrahedral symmetric condensation, in *11th International Conference on Cold Fusion*, Biberian, J.-P. World Scientific Co., Marseilles, France, 2004, pp. 730.

[38.] Schwinger, J., Cold fusion: a hypothesis, *Z. Naturforsch.* 45A, 756, 1990.

[39.] Hemminger, W. and Höhne, G., *Calorimetry, fundamentals and practice* Verlag Chemie GmbH, Weinheim, Germany, 1984.

[40.] Storms, E., Calorimetry 101 for cold fusion, www.LENR-CANR.org, 2004.

[41.] Storms, E., How to make a cheap and effective Seebeck calorimeter, in *Tenth International Conference on Cold Fusion*, Hagelstein, P. L. andChubb, S. R. World Scientific Publishing Co., Cambridge, MA, 2003, pp. 269.

[42.] Storms, E., Description of a sensitive Seebeck calorimeter used for cold fusion studies, in *12th International Conference on Condensed Matter Nuclear Science*, Yokohama, Japan, 2005.

[43.] Gür, T. M., Schreiber, M., Lucier, G., Ferrante, J. A., Chao, J., and Huggins, R. A., An isoperibolic calorimeter to study electrochemical insertion of deuterium into palladium, *Fusion Technol.* 25, 487, 1994.

[44.] Benson, T. B. and Passell, T. O., Calorimetry of energy-efficient glow discharge apparatus design and calibration, in *11th International Conference on Cold Fusion*, Biberian, J.-P. World Scientific Co., Marseilles, France, 2004, pp. 147.

[45.] Barrowes, S. C. and Bergeson, H. E., Linear, high precision, redundant calorimeter, in *Fourth International Conference on Cold Fusion* Electric Power Research Institute 3412 Hillview Ave., Palo Alto, CA 94304, Lahaina, Maui, 1994, pp. 21.

[46.] Närger, U., Hayden, M. E., Booth, J. L., Hardy, W. N., Whitehead, L. A., Carolan, J. F., Balzarini, D. A., Wishnow, E. H., and Blake, C. C., High precision calorimetric apparatus for studying electrolysis reactions, *Rev. Sci. Instr.* 61 (5), 1504, 1990.

[47.] Ferrari, C., Papucci, F., Salvetti, G., Tognoni, E., and Tombari, E., A calorimeter for the electrolytic cell and other open systems, *Nuovo Cimento Soc. Ital. Fis. D* 18 (11), 1333, 1996.

[48.] McKubre, M. C. H., Rocha-Filho, R. C., Smedley, S. I., Tanzella, F. L., Crouch-Baker, S., Passell, T. O., and Santucci, J., Isothermal flow calorimetric investigations of the D/Pd system, in *Second Annual Conference on Cold Fusion, "The Science of Cold Fusion"*, Bressani, T., Del Giudice, E., andPreparata, G. Societa Italiana di Fisica, Bologna, Italy, Como, Italy, 1991, pp. 419.

[49.] McKubre, M. C. H., Crouch-Baker, S., Rocha-Filho, R. C., Smedley, S. I., Tanzella, F. L., Passell, T. O., and Santucci, J., Isothermal flow calorimetric investigations of the D/Pd and H/Pd systems, *J. Electroanal. Chem.* 368, 55, 1994.

[50.] Swartz, M. R., Thermal conduction and non-differential temperature corrections to the enthalpic flow equation, *J. New Energy* 3 (1), 10-13, 1998.

[51.] Holst-Hansen, P. and Britz, D., Can current fluctuations account for the excess heat claims of Fleischmann and Pons?, *J. Electroanal. Chem.* 388, 11, 1995.

[52.] Swartz, M. R., Generality of optimal operating point behavior in low energy nuclear systems, *J. New Energy* 4 (2), 218-228, 1999.

[53.] Pons, S. and Fleischmann, M., The calorimetry of electrode reactions and measurements of excess enthalpy generation in the electrolysis of D_2O using Pd-based cathodes, in *Second Annual Conference on Cold Fusion, "The Science of Cold Fusion"*, Bressani, T., Del Giudice, E., andPreparata, G. Societa Italiana di Fisica, Bologna, Italy, Como, Italy, 1991, pp. 349.

[54.] Fleischmann, M., Pons, S., Le Roux, M., and Roulette, J., Calorimetry of the Pd-D_2O system: The search for simplicity and accuracy, *Trans. Fusion Technol.* 26 (4T), 323, 1994.

[55.] Miles, M. H., Imam, M. A., and Fleischmann, M., "Case studies" of two experiments carried out with the ICARUS systems, in *8th International Conference on Cold Fusion*, Scaramuzzi, F. Italian Physical Society, Bologna, Italy, Lerici (La Spezia), Italy, 2000, pp. 105.

[56.] Bush, R. T. and Eagleton, R. D., Evidence for electrolytically induced transmutation and radioactivity correlated with excess heat in electrolytic cells with light water rubidium salt electrolytes, *Trans. Fusion Technol.* 26 (4T), 344, 1994.

[57.] Hansen, W. N., Report to the Utah State Fusion/Energy Council on the analysis of selected Pons Fleischmann calorimetric data, in *Second Annual Conference on Cold Fusion, "The Science of Cold Fusion"*, Bressani, T., Del Giudice, E., andPreparata, G. Societa Italiana di Fisica, Bologna, Italy, Como, Italy, 1991, pp. 491.

[58.] Swartz, M. R., Some lessons from optical examination of the PFC phase-II calorimetric curves, in *Fourth International Conference on Cold Fusion* Electric Power Research Institute 3412 Hillview Ave., Palo Alto, CA 94304, Lahaina, Maui, 1993, pp. 19.

[59.] Hansen, W. N. and Melich, M. E., Pd/D calorimetry- The key to the F/P effect and a challenge to science, *Trans. Fusion Technol.* 26 (4T), 355, 1994.

[60.] Miles, M. H., Park, K. H., and Stilwell, D. E., Electrochemical calorimetric studies of the cold fusion effect, in *The First Annual Conference on Cold Fusion*, Will, F. National Cold Fusion Institute, University of Utah Research Park, Salt Lake City, Utah, 1990, pp. 328.

[61.] Schreiber, M., Gür, T. M., Lucier, G., Ferrante, J. A., Chao, J., and Huggins, R. A., Recent measurements of excess energy production in electrochemical cells containing heavy water and palladium, in *The First Annual Conference on Cold Fusion*, Will, F. National Cold Fusion Institute, University of Utah Research Park, Salt Lake City, Utah, 1990, pp. 44.

[62.] Miles, M. H. and Johnson, K. B., Improved, open cell, heat conduction, isoperibolic calorimetry, in *Sixth International Conference on Cold Fusion,Progress in New Hydrogen Energy*, Okamoto, M. New Energy and Industrial Technology Development Organization, Tokyo Institute of Technology, Tokyo, Japan, Lake Toya, Hokkaido, Japan, 1996, pp. 496.

[63.] Storms, E., Description of a dual calorimeter, *Infinite Energy* 6 (34), 22, 2000.

[64.] McKubre, M. C. H., Crouch-Baker, S., Riley, A. M., Smedley, S. I., and Tanzella, F. L., Excess power observations in electrochemical studies of the D/Pd system; the influence of loading, in *Third International Conference on Cold Fusion, "Frontiers of Cold Fusion"*, Ikegami, H. Universal Academy Press, Inc., Tokyo, Japan, Nagoya Japan, 1992, pp. 5.

[65.] Takahashi, A., Inokuchi, T., Chimi, Y., Ikegawa, T., Kaji, N., Nitta, Y., Kobayashi, K., and Taniguchi, M., Experimental correlation between excess heat and nuclear products, in *5th International Conference on Cold Fusion*, Pons, S. IMRA Europe, Sophia Antipolis Cedex, France, Monte-Carlo, Monaco, 1995, pp. 69.

[66.] Swartz, M. R., Can a Pd/D_2O/Pt device be made portable to demnonstrate the optimal operating point?, in *Tenth International Conference on Cold Fusio*, Chubb, P. I. H. a. S. R. World Scientific Publishing Co., Cambridge, MA, 2003, pp. 45.

[67.] Dash, J. and Ambadkar, A., Co-deposition of palladium with hydrogen isotopes, in *11th International Conference on Cold Fusion*, Biberian, J.-P. World Scientific Co., Marseilles, France, 2004, pp. 477.

[68.] Oriani, R. A., Nelson, J. C., Lee, S.-K., and Broadhurst, J. H., Calorimetric measurements of excess power output during the cathodic charging of deuterium into palladium, *Fusion Technol.* 18, 652, 1990.

[69.] Bertalot, L., Bettinali, L., De Marco, F., Violante, V., De Logu, P., Dikonimos, T., and La Barbera, A., Analysis of tritium and heat excess in electrochemical cells with Pd cathodes, in *Second Annual Conference on Cold Fusion, "The Science of Cold Fusion"*, Bressani, T., Del Giudice, E., andPreparata, G. Societa Italiana di Fisica, Bologna, Italy, Como, Italy, 1991, pp. 3.

[70.] Olofsson, G., Wadsoe, I., and Eberson, L., Design and testing of a calorimeter for measurements on electrochemical reactions with gas evolution, *J. Chem. Thermodyn.* 23, 95, 1991.

[71.] Bush, B. F. and Lagowski, J. J., Methods of generating excess heat with the Pons and Fleischmann effect: rigorous and cost effective calorimetry, nuclear products analysis of the cathode and helium analysis, in *The Seventh International Conference on Cold Fusion*, Jaeger, F. ENECO, Inc., Salt Lake City, UT., Vancouver, Canada, 1998, pp. 38.

[72.] Zhang, W.-S., Zhang, Z.-F., and Zhang, Z.-L., Primary calorimetric results on closed Pd/D_2O electrolysis systems by calvet calorimetry, in *The 9th International Conference on Cold Fusion, Condensed Matter Nuclear Science*, Li, X. Z. Tsinghua Univ. Press, Tsinghua Univ., Beijing, China, 2002, pp. 431.

[73.] Tian, J., Li, X. Z., Yu, W. Z., Cao, D. X., Zhou, R., Yu, Z. W., Ji, Z. F., Liu, Y., T., H. J., and Zhou, R. X., Anomanous heat flow and its correlation with deuterium flux in a gas-loading deuterium-palladium system, in *The 9th International Conference on Cold Fusion, Condensed Matter Nuclear Science*, Li, X. Z. Tsinghua Univ. Press, Tsinghua Univ., Beijing, China, 2002, pp. 353.

[74.] Warner, J., Dash, J., and Frantz, S., Electrolysis of D_2O with titanium cathodes: enhancement of excess heat and further evidence of possible transmutation, in *The Ninth International Conference on Cold Fusion*, Li, X. Z. Tsinghua University, Beijing, China: Tsinghua University, 2002, pp. 404.

[75.] Storms, E., *Description of a Seebeck calorimeter* www.LENR.org, 2005.

[76.] Storms, E., A critical evaluation of the Pons-Fleischmann effect: Part 1, *Infinite Energy* 6 (31), 10, 2000.

[77.] Storms, E., A critical evaluation of the Pons-Fleischmann effect: Part 2, *Infinite Energy* 6 (32), 52, 2000.

[78.] Shanahan, K., A systematic error in mass flow calorimetery demonstrated, *Thermochim. Acta* 382 (2), 95-101, 2002.

[79.] Storms, E., Comment on papers by K. Shanahan that propose to explain anomalous heat generated by cold fusion, *Thermochim. Acta* 441 (2), 207, 2006.

[80.] Swartz, M. R., Potential for positional variation in flow calorimetric systems, *J. New Energy* 1 (1), 126-130, 1996.

[81.] Jones, J. E., Hansen, L. D., Jones, S. E., Shelton, D. S., and Thorne, J. M., Faradaic efficiencies less than 100% during electrolysis of water can account for reports of excess heat in 'cold fusion' cells, *J. Phys. Chem.* 99, 6973, 1995.

[82.] Shkedi, Z., McDonald, R. C., Breen, J. J., Maguire, S. J., and Veranth, J., Calorimetry, excess heat, and Faraday efficiency in Ni-H$_2$O electrolytic cells, *Fusion Technol.* 28, 1720, 1995.

[83.] Shkedi, Z., Response to "Comments on 'Calorimetry, excess heat, and Faraday efficiency in Ni-H$_2$O electrolytic cells'". *Fusion Technol.* 30, 133, 1996.

[84.] Jones, S. E. and Hansen, L. D., Examination of claims of Miles et al in Pons-Fleischmann-Type cold fusion experiments, *J. Phys. Chem.* 99, 6966, 1995.

[85.] Jones, S. E., Hansen, L. D., and Shelton, D. S., An assessment of claims of excess heat in cold fusion calorimetry, *J. Phys. Chem. B* 102, 3647, 1998.

[86.] Miles, M. H., Bush, B. F., and Stilwell, D. E., Calorimetric principles and problems in measurements of excess power during Pd-D$_2$O electrolysis, *J. Phys. Chem.* 98, 1948, 1994.

[87.] Miles, M. and Jones, C. P., Cold fusion experimenter Miles responds to critic, *21st Century Sci. & Technol.* Spring, 75, 1992.

[88.] Miles, M. H., Reply to 'An assessment of claims of excess heat in cold fusion calorimetry', *J. Phys. Chem. B* 102, 3648, 1998.

[89.] Miles, M. H., Reply to 'Examination of claims of Miles *et al.* in Pons-Fleischmann-type cold fusion experiments', *J. Phys. Chem. B* 102, 3642, 1998.

[90.] Storms, E., Cold fusion: An objective assessment, *www.LENR-CANR.org*, 2001.

[91.] Shelton, D. S., Hansen, L. D., Thorne, J. M., and Jones, S. E., An assessment of claims of 'excess heat' in 'cold fusion' calorimetry, *Thermochim. Acta* 297, 7, 1997.

[92.] Keesing, R. G., Greenhow, R. C., Cohler, M. D., and McQuillan, A. J., Thermal, thermoelectric, and cathode poisoning effects in cold fusion experiments, *Fusion Technol.* 19, 375, 1991.

Chapter 8

[1.] Fleischmann, M., An overview of cold fuson phenomena, in *The First Annual Conference on Cold Fusion*, Will, F. National Cold Fusion Institute, University of Utah Research Park, Salt Lake City, Utah, 1990, pp. 344.

[2.] Fleischmann, M., Pons, S., and Preparata, G., Possible theories of cold fusion, *IL Nuovo Cimento* 107A (1), 143, 1994.

[3.] Fleischmann, M., Background to cold fusion: The genesis of a concept, in *Tenth International Conference on Cold Fusion*, Hagelstein, P. L. andChubb, S. R. World Scientific Publishing Co., Cambridge, MA, 2003, pp. 1.

[4.] Takahashi, A., Progress in condensed matter nuclear science, in *12 th International Conference on Cold Fusion*, Takahashi, A. World Scientific Publishing., Yokohama, Japan, 2006, pp. 1.

[5.] Chechin, V. A., Tsarev, V. A., Rabinowitz, M., and Kim, Y. E., Critical review of theoretical models for anomalous effects in deuterated metals, *Int. J. Theo. Phys.* 33, 617, 1994.

[6.] Preparata, G., Some theories of 'cold' nuclear fusion: a review, *Fusion Technol.* 20, 82, 1991.

[7.] Kirkinskii, V. A. and Novikov, Y. A., A new approach to theoretical modelling of nuclear fusion in palladium deuteride, *Europhys. Lett.* 46, 448, 1999.

[8.] Louis, E., Moscardo, F., San-Fabian, E., and Perez-Jorda, J. M., Calculation of hydrogen-hydrogen potential energies and fusion rates in palladium hydride (Pd$_x$H$_2$) clusters (x=2,4), *Phys. Rev. B: Mater. Phys.* 42, 4996, 1990.

[9.] Raiola, F., Gang, L., Bonomo, C., Gyürky, G., Aliotta, M., Becker, H. W., Bonetti, R., Broggini, C., Corvisiero, P., D'Onofrio, A., Fülöp, Z., Gervino, G., Gialanella, L., Junker, M., Prati, P., Roca,

V., Rolfs, C., Romano, M., Somorjai, E., Terrasi, F., Fiorentini, G., Langanke, K., and Winter, J., Enhanced electron screening in d(d,p)t for deuterated metals, *Eur. Phys. J. A* 19, 283, 2004.

[10.] Huke, A., Czerski, K., and Heide, P., Accelerator experiments and theoretical models for the electron screening effect in metallic environments, in *11th International Converence on Cold Fusion*, Biberian, J.-P. World Scientific Co., Marseilles, France, 2004, pp. 194.

[11.] Czerski, K., Heide, P., and Huke, A., Electron screening constraints for the cold fusion, in *11th International Conference on Cold Fusion*, Biberian, J.-P. World Scientific Co., Marseilles, France, 2004, pp. 228.

[12.] Kasagi, J., Yuki, H., Baba, T., and Noda, T., Low energy nuclear fusion reactions in solids, in *8th International Conference on Cold Fusion*, Scaramuzzi, F. Italian Physical Society, Bologna, Italy, Lerici (La Spezia), Italy, 2000, pp. 305.

[13.] Schwinger, J., Cold fusion: a hypothesis, *Z. Naturforsch.* 45A, 756, 1990.

[14.] Kim, Y. E., New cold nuclear fusion theory and experimental tests, *J. Fusion Energy* 9 (4), 423, 1990.

[15.] Kim, Y. E. and Passell, T. O., Alternative interpretations of low-energy nuclear reaction processes with deuterated metals based on the Bose-Einstein condensation mechanism, in *11th International Conference on Cold Fusion*, Biberian, J.-P. World Scientific Co., Marseilles, France, 2004, pp. 718.

[16.] Frodl, P., Roessler, O. E., Hoffmann, M., and Wahl, F., Possible participation of lithium in Fleischmann-Pons reaction is testable, *Z. Naturforsch.* 45A, 757, 1990.

[17.] Kozima, H., *Discovery of the cold fusion phenomenon* Ohotake Shuppan, Inc, Tokyo, Japan, 1998.

[18.] Taplin, H., "Light element fission", The lithium-fast proton nuclear reaction, in *The Seventh International Conference on Cold Fusion*, Jaeger, F. ENECO, Inc., Salt Lake City, UT, Vancouver, Canada, 1998, pp. 478.

[19.] Passell, T. O., Search for nuclear reaction products in heat-producing Pd, in *The Seventh International Conference on Cold Fusion*, Jaeger, F. ENECO, Inc., Salt Lake City, UT, Vancouver, Canada, 1998, pp. 309.

[20.] Takahashi, A., Deuteron cluster fusion and ash, in *ASTI-5* www.iscmns.org/, Asti, Italy, 2004.

[21.] Takahashi, A., Production of stable isotopes by selective channel photofission of Pd, *Jpn. J. Appl. Phys. A* 40 (12), 7031-7046, 2001.

[22.] Mizuno, T., Ohmori, T., Kurokawa, K., Akimoto, T., Kitaichi, M., Inoda, K., Azumi, K., Shimokawa, S., and Enyo, M., Anomalous isotopic distribution of elements deposited on palladium induced by cathodic electrolysis, *Denki Kagaku oyobi Kogyo Butsuri Kagaku* 64, 1160 (in Japanese), 1996.

[23.] Hanawa, T., X-ray spectroscropic analysis of carbon arc products in water, in *8th International Conference on Cold Fusion*, Scaramuzzi, F. Italian Physical Society, Bologna, Italy, Lerici (La Spezia), Italy, 2000, pp. 147.

[24.] Jiang, X. L., Han, L. J., and Kang, W., Anomalous element production induced by carbon arcing under water, in *The Seventh International Conference on Cold Fusion*, Jaeger, F. ENECO, Inc., Salt Lake City, UT., Vancouver, Canada, 1998, pp. 172.

[25.] Sundaresan, R. and Bockris, J. O. M., Anomalous reactions during arcing between carbon rods in water, *Fusion Technol.* 26, 261, 1994.

[26.] Vysotskii, V. and Kornilova, A. A., *Nuclear fusion and transmutation of isotopes in biological systems* Mockba, Ukraine, 2003.

[27.] Widom, A. and Larsen, L., Ultra low momentum neutron catalyzed nuclear reactions on metallic hydride surfaces, *Eur. Phys. J.* C46, 107, 2006.

[28.] Mills, R., *The grand unified theory of classical quantum mechanics* Cadmus Professional Communications, Ephrata, PA, 2006.

[29.] Conte, E. and Pieralice, M., An experiment indicates the nuclear fusion of a proton and electron into a neutron, *Infinite Energy* 4 (23), 67, 1999.

Chapter 9

[1] O'Brian, E. D., The U. S. Patent & Trademark Office: A bureaucracy out of control, *Infinite Energy* 12 (70), 31, 2006.

Appendix A

[1] Jones, J. E., Hansen, L. D., Jones, S. E., Shelton, D. S., and Thorne, J. M., Faradaic efficiencies less than 100% during electrolysis of water can account for reports of excess heat in 'cold fusion' cells, *J. Phys. Chem.* 99, 6973, 1995.
[2] Storms, E., Comment on papers by K. Shanahan that propose to explain anomalous heat generated by cold fusion, *Thermochim. Acta* 441 (2), 207, 2006.

Appendix B

[1] Sakamoto, Y., Imoto, M., Takai, K., Yanaru, T., and Ohshima, K., Calorimetric enthalpies for palladium-hydrogen (deuterium) systems at H(D) contents up to about [H]([D])/[Pd] = 0.86, *J. Phys.: Condens. Mater.* 8, 3229, 1996.
[2] Sakamoto, Y., Imoto, M., Takai, K., and Yanaru, T., Calorimetric enthalpies in the b-phase regions of Pd black-H(D) systems, in *Sixth International Conference on Cold Fusion,Progress in New Hydrogen Energy*, Okamoto, M. New Energy and Industrial Technology Development Organization, Tokyo Institute of Technology, Tokyo, Japan, Lake Toya, Hokkaido, Japan, 1996, pp. 162.
[3] Shelton, D. S., Hansen, L. D., Thorne, J. M., and Jones, S. E., An assessment of claims of 'excess heat' in 'cold fusion' calorimetry, *Thermochim. Acta* 297, 7, 1997.
[4] Flanagan, T. B., Luo, W., and Clewley, J. D., Calorimetric enthalpies of absorption and desorption of protium and deuterium by palladium, *J. Less-Common Met.* 172-174, 42, 1991.

Appendix C

[1] Berrondo, M., Computer simulation of D atoms in a Pd lattice, in *Anomalous Nuclear Effects in Deuterium/Solid Systems, "AIP Conference Proceedings 228"*, Jones, S., Scaramuzzi, F., andWorledge, D. American Institute of Physics, New York, Brigham Young Univ., Provo, UT, 1990, pp. 653.
[2] Switendick, A. C., Electronic structure and stability of palladium hydrogen (deuterium) systems, PdH(D)n, 1≤n≤3, *J. less-Common Met.* 172-174, 1363, 1991.
[3] Bertalot, L., DeMarco, F., DeNinno, A., Felici, R., LaBarbera, A., Scaramuzzi, F., and Violante, V., Deuterium charging in palladium by electrolysis of heavy water: Measurement of lattice parameter, in *Fourth International Conference on Cold Fusion* Electric Power Research Institute 3412 Hillview Ave., Palo Alto, CA 94304, Lahaina, Maui, 1993, pp. 29.
[4] Warner, J., Dash, J., and Frantz, S., Electrolysis of D_2O with titanium cathodes: enhancement of excess heat and further evidence of possible transmutation, in *The Ninth International Conference on Cold Fusion*, Li, X. Z. Tsinghua University, Beijing, China: Tsinghua University, 2002, pp. 404.
[5] Warner, J. and Dash, J., Heat produced during the electrolysis of D_2O with titanium cathodes, in *8th International Conference on Cold Fusion*, Scaramuzzi, F. Italian Physical Society, Bologna, Italy, Lerici (La Spezia), Italy, 2000, pp. 161.
[6] Zhang, Q., Gou, Q., Zhu, Z., Luo, J., Liu, F., Sun, J., Miao, B., Ye, A., and Cheng, X., The excess heat experiments on cold fusion in titanium lattice, *Chin. J. At. Mol. Phys.* 12 (2), 165, 1995.
[7] Menlove, H. O., Fowler, M. M., Garcia, E., Mayer, A., Miller, M. C., Ryan, R. R., and Jones, S. E., The measurement of neutron emission from Ti plus D2 gas, *J. Fusion Energy* 9, 215, 1990.

[8.] Focardi, S., Habel, R., and Piantelli, F., Anomalous heat production in Ni-H systems, *Nuovo Cimento* 107A, 163, 1994.

[9.] Focardi, S., Gabbani, V., Montalbano, V., Piantelli, F., and Veronesi, S., Large excess heat production in Ni-H systems, *Nuovo Cimento* 111A (11), 1233, 1998.

[10.] Campari, E. G., Focardi, S., Gabbani, V., Montalbano, V., Piantelli, F., and Veronesi, S., Surface analysis of hydrogen-loaded nickel alloys, in *11th International Conference on Cold Fusion*, Biberian, J.-P. World Scientific Co., Marseilles, France, 2004, pp. 414.

[11.] Bush, R. T., A light water excess heat reaction suggests that 'cold fusion' may be 'alkali-hydrogen fusion', *Fusion Technol.* 22, 301, 1992.

[12.] Ohmori, T. and Enyo, M., Excess heat evolution during electrolysis of H_2O with nickel, gold, silver, and tin cathodes, *Fusion Technol.* 24, 293, 1993.

[13.] Sankaranarayanan, T. K., Sprinivasan, M., Bajpai, M. B., and Gupta, D. S., Evidence for tritium generation in self-heated nickel wires subjected to hydrogen gas absorption/desorption cycles., in *5th International Conference on Cold Fusion*, Pons, S. IMRA Europe, Sophia Antipolis Cedex, France, Monte-Carlo, Monaco, 1995, pp. 173.

[14.] Srinivasan, M., Shyam, A., Sankaranarayanan, T. K., Bajpai, M. B., Ramamurthy, H., Mukherjee, U. K., Krishnan, M. S., Nayar, M. G., and Naik, Y. P., Tritium and excess heat generation during electrolysis of aqueous solutions of alkali salts with nickel cathode, in *Third International Conference on Cold Fusion, "Frontiers of Cold Fusion"*, Ikegami, H. Universal Academy Press, Inc., Tokyo, Japan, Nagoya Japan, 1992, pp. 123.

[15.] Tian, J., Jin, L. H., Weng, Z. K., Song, B., Zhao, X. L., Xiao , Z. J., Chen, G., and Du, B. Q., "Excess heat" during electrolysis in platinum/K_2CO_3/nickel light water system, in *11th International Conference on Cold Fusion*, Biberian, J.-P. World Scientific Co., Marseilles, France, 2004, pp. 102.

[16.] Swartz, M. R., Verner, G. M., and Frank, A. H., The impact of heavy water (D_2O) on nickel-light water cold fusion systems, in *The 9th International Conference on Cold Fusion, Condensed Matter Nuclear Science*, Li, X. Z. Tsinghua Univ. Press, Tsinghua Univ., Beijing, China, 2002, pp. 335.

[17.] Chiba, M., Shirakawa, T., Fujii, M., Ikebe, T., Yamaoka, S., Sueki, K., Nakahara, H., and Hirose, T., Measurement of neutron emission from LiNbO3 fracture process in D2 and H2 atmosphere, *Nuovo Cimento Soc. Ital. Fis. A* 108, 1277, 1995.

[18.] Takahashi, R., Synthesis of substance and generation of heat in charcoal cathode in electrolysis of H_2O and D_2O using various alkalihydrooxides, in *5th International Conference on Cold Fusion*, Pons, S. IMRA Europe, Sophia Antipolis Cedex, France, Monte-Carlo, Monaco, 1995, pp. 619.

[19.] Dufour, J., Cold fusion by sparking in hydrogen isotopes, *Fusion Technol.* 24, 205, 1993.

[20.] Dalard, F., Ulman, M., Augustynski, J., and Selvam, P., Electrochemical incorporation of lithium into palladium from aprotic electrolytes, *J. Electroanal. Chem.* 270, 445, 1989.

[21.] Brillas, E., Esteve, J., Sardin, G., Casado, J., Domenech, X., and Sanchez-Cabeza, J. A., Product analysis from D_2O electrolysis with Pd and Ti cathodes, *Electrochim. Acta* 37 (2), 215, 1992.

[22.] Czerwinski, A., Influence of lithium cations on hydrogen and deuterium electrosorption in palladium, *Electrochim. Acta* 39, 431, 1994.

[23.] Astakhov, I. I., Davydov, A. D., Katargin, N. V., Kazarinov, V. E., Kiseleva, I. G., Kriksunov, L. B., Kudryavtsev, D. Y., Lebedev, I. A., Myasoedov, B. F., Shcheglov, O. P., Teplitskaya, G. L., and Tsionskii, V. M., An attempt to detect neutron and gamma radiations in heavy water electrolysis with a palladium cathode, *Electrochim. Acta* 36 (7), 1127, 1991.

[24.] Augustynski, J., Commentaire: Pourquoi les experiences de 'fusion froide' de deuterium sont-elles si difficiles a reproduire, *Chimia* 43, 99 (in French). 1989.

[25.] Kazarinov, V. E., Astakhov, I. I., Teplitskaya, G. L., Kiseleva, I. G., Davydov, A. D., Nekrasova, N. V., Kudryavtsev, D. Y., and Zhukova, T. B., Cathodic behaviour of palladium in electrolytic solutions containing alkali metal ions, *Elektrokhimiya* 27, 9 (in Russian), 1991.

[26.] Miyamoto, S., Sueki, K., Kobayashi, K., Fujii, M., Chiba, M., Nakahara, H., Shirakawa, T., Kobayashi, T., Yanokura, M., and Aratani, M., Movement of Li during electrolysis of 0.1M-

LiOD/D$_2$O solution, in *Fourth International Conference on Cold Fusion* Electric Power Research Institute 3412 Hillview Ave., Palo Alto, CA 94304, Lahaina, Maui, 1993, pp. 28.

[27.] Nakada, M., Kusunoki, T., Okamoto, M., and Odawara, O., A role of lithium for the neutron emission in heavy water electrolysis, in *Third International Conference on Cold Fusion, "Frontiers of Cold Fusion"*, Ikegami, H. Universal Academy Press, Inc., Tokyo, Japan, Nagoya Japan, 1992, pp. 581.

[28.] Ota, K., Yoshitake, H., Yamazaki, O., Kuratsuka, M., Yamaki, K., Ando, K., Iida, Y., and Kamiya, N., Heat measurement of water electrolysis using Pd cathode and the electrochemistry, in *Fourth International Conference on Cold Fusion* Electric Power Research Institute 3412 Hillview Ave., Palo Alto, CA 94304, Lahaina, Maui, 1993, pp. 5.

[29.] Oya, Y., Aida, M., Iinuma, K., and Okamoto, M., A role of alkaline ions for dynamic movement of hydrogen isotopes in Pd, in *The Seventh International Conference on Cold Fusion*, Jaeger, F. ENECO, Inc., Salt Lake City, UT, Vancouver, Canada, 1998, pp. 302.

[30.] Yamazaki, O., Yoshitake, H., Kamiya, N., and Ota, K., Hydrogen absorption and Li inclusion in a Pd cathode in LiOH solution, *J. Electroanal. Chem.* 390, 127, 1995.

[31.] Blaser, J. P., Haas, O., Ptitjean, C., Barbero, C., Bertl, W., Lou, K., Mathias, M., Baumann, P., Daniel, H., Hartmann, J., Hechtl, E., Ackerbauer, P., Kammel, P., Scrinzi, A., Zmeskal, H., Kozlowski, T., Kipfer, R., Baur, H., Signr, P., and Wieler, R., Experimental investigation of cold fusion phenomena in palladium, *Chimia* 43, 262, 1989.

[32.] Oates, W. A. and Flanagan, T. B., Thermodynamic properties of hydrogen in palladium and its alloys under conditions of constant volume, *J. Chem. Soc. Faraday Trans.* 1 (73), 993, 1977.

[33.] Flanagan, T. B., Absorption of deuterium by palladium, *J. Phys. Chem.* 65, 280, 1961.

[34.] Oates, W. A. and Flanagan, T. B., Formation of nearly stoichiometric palladium-hydrogen systems, *Nature Phys. Sci.* 231, 19, 1971.

[35.] Flanagan, T. B. and Oates, W. A., The palladium-hydrogen system, *Annu. Rev. Mater. Sci.* 21, 269, 1991.

[36.] Flanagan, T. B., Luo, W., and Clewley, J. D., Calorimetric enthalpies of absorption and desorption of protium and deuterium by palladium, *J. Less-Common Met.* 172-174, 42, 1991.

[37.] Sakamoto, Y., Ohira, K., Kokubu, M., and Flanagan, T. B., Thermodynamic properties for solutions of hydrogen in Pd-Pt-Rh alloys, *J. Alloys and Compounds* 253-254, 212, 1997.

[38.] Lewis, F. A., The hydrides of palladium and palladium alloys, *Pt. Met. Rev.* 4-5, 132, 1960.

[39.] Lewis, F. A., *The palladium hydrogen system* Academic Press, New York, 1967.

[40.] Musket, G., Effects of contamination on the interaction of H gas with Pd : A review, *J. Less-Common Met.* 45, 173, 1976.

[41.] Fukai, Y. and Okuma, N., Formation of superabundant vacancies in Pd hydride under high hydrogen pressures., *Phys. Rev. Lett.* 73, 1640, 1994.

[42.] Nace, D. M. and Aston, J. G., Palladium hydride. III. The thermodynamic study of Pd$_2$H between 15 and 303° K. evidence for the Tetragonal PdH$_4$ structure in palladium hydride, *J. Am. Chem. Soc.* 79, 3627, 1957.

[43.] Ferguson, G. A., Schindler, A. I., Tanaka, T., and Morita, T., Neutron diffraction study of temperature-dependent properties of palladium containing absorbed hydrogen, *Phys. Rev.* 137 (2A), 483, 1965.

[44.] Lipson, A. G., Castano, C. H., Miley, G., Lyakhov, B. F., and Mitin, A. V., Emergence of a high-temperature superconductivity in hydrogen cycled Pd compounds as an evidence for superstoichiometric H/D sites, in *11th International Conference on Cold Fusion*, Biberian, J.-P. World Scientific Co., Marseilles, France, 2004, pp. 128.

[45.] Gillespie, L. J. and Downs, W. R., The palladium-deuterium equilibrium, *J. Am. Chem. Soc.* 61, 2494, 1939.

[46.] McKubre, M. C. and Tanzella, F., Using resistivity to measure H/Pd and D/Pd loading: method and significance, in *12th International Conference on Condensed Matter Nuclear Science*, Takahashi, A. World Scientific Publiishing, Yokohama, Japan, 2005.

[47] Spallone, A., Measurements of resistance temperature coefficient at H/Pd overloadings, in *12th International Conference on Condensed Matter Nuclear Science*, Takahashi, A., Yokohama, Japan, 2005.

[48] Storms, E., The nature of the energy-active state in Pd-D, *Infinite Energy* (#5 and #6), 77, 1995.

[49] Wei, S. H. and Zunger, A., Stability of atomic and diatomic hydrogen in fcc palladium, *Solid State Commun.* 73, 327, 1990.

[50] Moizhes, B. Y., Formation of a compact D_2 molecule in interstitial sites - a possible explanation for cold nuclear fusion, *Sov. Tech. Phys. Lett.* 17, 540, 1991.

[51] Uhm, H. S. and Lee, W. M., High concentration of deuterium in palladium, *Fusion Technol.* 21, 75., 1992.

[52] Takahashi, A., Mass-8-and-charge-4 increased transmutation by octahedral resonance fusion model, in *JCF-4*, Morioka, Japan, 2002.

[53] Muguet, F. F. and Bassez-Muguet, M.-P., Ab initio computations of one and two hydrogen or deuterium atoms in the palladium tetrahedral site, *J. Fusion Energy* 9 (4), 383, 1990.

[54] McKubre, M. C. H., Crouch-Baker, S., Hauser, A. K., Smedley, S. I., Tanzella, F. L., Williams, M. S., and Wing, S. S., Concerning reproducibility of excess power production, in *5th International Conference on Cold Fusion*, Pons, S. IMRA Europe, Sophia Antipolis Cedex, France, Monte-Carlo, Monaco, 1995, pp. 17.

[55] Minato, J., Nakata, T., Denzumi, S., Yamamoto, Y., Takahashi, A., Aida, H., Tsuchida, Y., Akita, H., and Kunimatsu, K., Materials/surface aspects of hydrogen/deuterium loading into Pd cathode, in *5th International Conference on Cold Fusion*, Pons, S. IMRA Europe, Sophia Antipolis Cedex, France, Monte-Carlo, Monac, 1995, pp. 383.

[56] Celani, F., Spallone, A., Righi, E., Trenta, G., Catena, C., D'Agostaro, G., Quercia, P., Andreassi, V., Marini, P., Di Stefano, V., Nakamura, M., Mancini, A., Sona, P. G., Fontana, F., Gamberale, L., Garbelli, D., Celia, E., Falcioni, F., Marchesini, M., Novaro, E., and Mastromatteo, U., Innovative procedure for the, in situ, measurement of the resistive thermal coefficient of H(D)/Pd during electrolysis; cross-comparison of new elements detected in the Th-Hg-Pd-D(H) electroytic cells, in *11th International Conference on Cold Fusion*, Biberian, J.-P. World Scientific Co., Marseilles, France, 2004, pp. 108.

[57] Kamimura, H., Senjuh, T., Miyashita, S., and Asami, N., Excess heat in fuel cell type cells from pure Pd cathodes annealed at high temperatures, in *Sixth International Conference on Cold Fusion,Progress in New Hydrogen Energy*, Okamoto, M. New Energy and Industrial Technology Development Organization, Tokyo Institute of Technology, Tokyo, Japan, Lake Toya, Hokkaido, Japan, 1996, pp. 45.

[58] Asami, N., Senjuh, T., Uehara, T., Sumi, M., Kamimura, H., Miyashita, S., and Matsui, K., Material behavior of highly deuterated palladium, in *The Seventh International Conference on Cold Fusion*, Jaeger, F. ENECO, Inc., Salt Lake City, UT., Vancouver, Canada, 1998, pp. 15.

[59] Yasuda, K., Nitta, Y., and Takahashi, A., Study of excess heat and nuclear products with closed D_2O electrolysis systems, in *Sixth International Conference on Cold Fusion, Progress in New Hydrogen Energy*, Okamoto, M. New Energy and Industrial Technology Development Organization, Tokyo Institute of Technology, Tokyo, Japan, Lake Toya, Hokkaido, Japan, 1996, pp. 36.

[60] Nakata, T., Kobayashi, M., Nagahama, M., Akita, H., Hasegawa, N., and Kunimatsu, K., Excess heat measurement at high cathode loading by deuterium during electrolysis of heavy water using Pd cathode, in *Sixth International Conference on Cold Fusion, Progress in New Hydrogen Energy*, Okamoto, M. New Energy and Industrial Technology Development Organization, Tokyo Institute of Technology, Tokyo, Japan, Lake Toya, Hokkaido, Japan, 1996, pp. 121.

[61] De Marco, F., De Ninno, A., Frattolillo, A., La Barbera, A., Scaramuzzi, F., and Violante, V., Progress report on the research activities on cold fusion at ENEA Frascati, in *Sixth International Conference on Cold Fusion, Progress in New Hydrogen Energy*, Okamoto, M. New Energy and Industrial Technology Development Organization, Tokyo Institute of Technology, Tokyo, Japan, Lake Toya, Hokkaido, Japan, 1996, pp. 145.

[62.] Celani, F., Spallone, A., Tripodi, P., Nuvoli, A., Petrocchi, A., Di Gioacchino, D., Boutet, M., Marini, P., and Di Stefano, V., High power µs pulsed electrolysis for large deuterium loading in Pd plates, *Trans. Fusion Technol.* 26 (#4T), 127, 1994.

[63.] Dardik, I., Zilov, T., Branover, H., El-Boher, A., Greenspan, E., Khachatorov, B., Krakov, V., Lesin, S., and Tsirlin, M., Excess heat in electrolysis experiments at Energetics Technologies, in *11th International Conference on Cold Fusion*, Biberian, J.-P. World Scientific Co., Marseilles, France, 2004, pp. 84.

[64.] Dominguez, D. D., Hagans, P. L., and Imam, M. A., The effect of microstructure on deuterium loading in palladium cathodes, in *Sixth International Conference on Cold Fusion,Progress in New Hydrogen Energy*, Okamoto, M. New Energy and Industrial Technology Development Organization, Tokyo Institute of Technology, Tokyo, Japan, Lake Toya, Hokkaido, Japan, 1996, pp. 239.

[65.] Dominguez, D. D., Hagans, P. L., and Imam, M. A., Report No. NRL/MR/6170-96-7803, 1996.

[66.] De Ninno, A., La Barbera, A., and Violante, V., Selection of palladium metallurgical parameters to achieve very high loading rations, in *Sixth International Conference on Cold Fusion, Progress in New Hydrogen Energy*, Okamoto, M. New Energy and Industrial Technology Development Organization, Tokyo Institute of Technology, Tokyo, Japan, Lake Toya, Hokkaido, Japan, 1996, pp. 192.

[67.] De Ninno, A., La Barbera, A., and Violante, V., Deformations induced by high loading ratios in palladium-deuterium compounds, *J. Alloys and Compounds* 253-254, 181, 1997.

[68.] Miles, M. H. and Johnson, K. B., Electrochemical insertion of hydrogen into metals and alloys, *J. New Energy* 1 (2), 32, 1996.

[69.] Miles, M. H., Johnson, K. B., and Imam, M. A., Electrochemical loading of hydrogen and deuterium into palladium and palladium-boron alloys, in *Sixth International Conference on Cold Fusion, Progress in New Hydrogen Energy*, Okamoto, M. New Energy and Industrial Technology Development Organization, Tokyo Institute of Technology, Tokyo, Japan, Lake Toya, Hokkaido, Japan, 1996, pp. 208.

[70.] Santandrea, R. P. and Behrens, R. G., A review of the thermodynamics and phase relationships in the palladium- hydrogen, palladium-deuterium and palladium-tritium systems., *High Temperature Materials and Processes* 7, 149, 1986.

[71.] De Ninno, A., Violante, V., and La Barbera, A., Consequences of lattice expansive strain gradients on hydrogen loading in palladium, *Phys. Rev. B* 56 (5), 2417, 1997.

[72.] Enyo, M. and Biswas, P. C., Hydrogen pressure equivalent to overpotential on Pd + Ag alloy electrodes in acidic solutions in the presence of thiourea, *J. Electroanal. Chem.* 357, 67, 1993.

[73.] Letts, D. and Cravens, D., Laser stimulation of deuterated palladium, *Infinite Energy* 9 (50), 10, 2003.

[74.] Ulman, M., Liu, J., Augustynski, J., Meli, F., and Schlapbach, L., Surface and electrochemical characterization of Pd cathodes after prolonged charging in LiOD + D_2O solutions, *J. Electroanal. Chem.* 286, 257, 1990.

[75.] Szpak, S., Mosier-Boss, P. A., and Smith, J. J., Deuterium uptake during Pd-D codeposition, *J. Electroanal. Chem.* 379, 121, 1994.

[76.] Swartz, M. R., Codeposition of palladium and deuterium, *Fusion Technol.* 32, 126, 1997.

[77.] Storms, E., Excess power production from platinum cathodes using the Pons-Fleischmann effect, in *8th International Conference on Cold Fusion*, Scaramuzzi, F. Italian Physical Society, Bologna, Italy, Lerici (La Spezia), Italy, 2000, pp. 55.

[78.] Szpak, S., Mosier-Boss, P. A., Miles, M., and Fleischmann, M., Thermal behavior of polarized Pd/D electrodes prepared by co-deposition, *Thermochim. Acta* 410, 101, 2004.

[79.] Dash, J. and Ambadkar, A., Co-deposition of palladium with hydrogen isotopes, in *11th International Conference on Cold Fusion*, Biberian, J.-P. World Scientific Co., Marseilles, France, 2004, pp. 477.

[80.] Rolison, D. R. and Trzaskoma, P. P., Morphological differences between hydrogen-loaded and deuterium-loaded palladium as observed by scanning electron microscopy, *J. Electroanal. Chem.* 287, 375, 1990.

[81.] Storms, E., A study of those properties of palladium that influence excess energy production by the Pons-Fleischmann effect, *Infinite Energy* 2 (8), 50, 1996.

[82.] Storms, E., The nature of the energy-active state in Pd-D, in *II Workshop on the Loading of Hydrogen/Deuterium in Metals, Characterization of Materials and Related Phenomena*, Asti, Italy, 1995.

[83.] Julin, P. and Bursill, L. A., Dendritic surface morphology of palladium hydride produced by electrolytic deposition, *J. Solid State Chem.* 93, 403, 1991.

[84.] Bockris, J. and Reddy, A. K. N., *Modern electrochemistry* Plenum Press, New York, 1977.

[85.] Lee, I., Chan, K.-Y., and Phillips, D. L., Two dimensional dendrites and 3 dimensional growth of electrodeposited platinum nanoparticles, *Japanese J. Appl. Phys.* 43 (2), 767, 2004.

[86.] Silver, D. S. and Dash, J., Surface studies of palladium after interaction with hydrogen isotopes, in *The Seventh International Conference on Cold Fusion*, Jaeger, F. ENECO, Inc., Salt Lake City, UT, Vancouver, Canada, 1998, pp. 351.

[87.] Dash, J., Noble, G., and Diman, D., Surface morphology and microcomposition of palladium cathodes after electrolysis in acified light and heavy water: correlation with excess heat, *Trans. Fusion Technol.* 26 (4T), 299, 1994.

[88.] Nassisi, V. and Longo, M. L., Experimental results of transmutation of elements observed in etched palladium samples by an excimer laser, *Fusion Technol.* 37 (May), 247, 2000.

[89.] Savvatimova, I. and Gavritenkov, D. V., Results of analysis of Ti foil after glow discharge with deuterium, in *11th International Conference on Cold Fusion*, Biberian, J.-P. World Scientific Co., Marseilles, France, 2004, pp. 438.

[90.] Jiang, X.-L., Chen, C.-Y., Fu, D.-F., Han, L.-J., and Kang, W., Tip effect and nuclear-active sites, in *The Seventh International Conference on Cold Fusion*, Jaeger, F. ENECO, Inc., Salt Lake City, UT., Vancouver, Canada, 1998, pp. 175.

[91.] Iwamura, Y., Itoh, T., Sakano, M., Yamazaki, N., Kuribayashi, S., Terada, Y., and Ishikawa, T., Observation of surface distribution of products by X-ray fluorescence spectrometry during D_2 gas permeation through Pd complexes, in *12th International Conference on Condensed Matter Nuclear Science*, Takahashi, A., Yokohama, Japan, 2005.

[92.] Storms, E., Measurements of excess heat from a Pons-Fleischmann-type electrolytic cell using palladium sheet, *Fusion Technol.* 23, 230, 1993.

[93.] Storms, E. and Talcott-Storms, C., The effect of hydriding on the physical structure of palladium and on the release of contained tritium, *Fusion Technol.* 20, 246, 1991.

[94.] Storms, E., Formation of b-PdD containing high deuterium concentration using electrolysis of heavy-water, *J. Alloys Comp.* 268, 89, 1998.

Appendix E

[1.] Williams, D. E. G., Findlay, D. J. S., Craston, D. H., Sene, M. R., Bailey, M., Croft, S., Hooton, B. W., Jones, C. P., Kucernak, A. R. J., Mason, J. A., and Taylor, R. I., Upper bounds on 'cold fusion' in electrolytic cells, *Nature (London)* 342, 375, 1989.

[2.] Sona, P. G., Parmigiani, F., Barberis, F., Battaglia, A., Berti, R., Buzzanca, G., Capelli, A., Capra, D., and Ferrari, M., Preliminary tests on tritium and neutrons in cold nuclear fusion within palladium cathodes, *Fusion Technol.* 17, 713, 1990.

[3.] Lin, G. H., Kainthla, R. C., Packham, N. J. C., Velev, O. A., and Bockris, J., On electrochemical tritium production, *Int. J. Hydrogen Energy* 15 (8), 537, 1990.

[4.] Adzic, R. G., Bae, I., Cahan, B., and Yeager, E., Tritium measurements and deuterium loading in D_2O electrolysis with a palladium cathode, in *The First Annual Conference on Cold Fusion*, Will,

F. National Cold Fusion Institute, University of Utah Research Park, Salt Lake City, Utah, 1990, pp. 261.

[5.] Brillas, E., Sardin, G., Casado, J., Doménech, X., and Sánchez, J., Product analysis from D_2O electrolysis with palladium and titanium cathodes, in *Second Annual Conference on Cold Fusion, "The Science of Cold Fusion"*, Bressani, T., Del Giudice, E., andPreparata, G. Societa Italiana di Fisica, Bologna, Italy, Como, Italy, 1991, pp. 9.

[6.] Szpak, S., Mosier-Boss, P. A., and Smith, J. J., Comments on methodology of excess tritium determination, in *Third International Conference on Cold Fusion, "Frontiers of Cold Fusion"*, Ikegami, H. Universal Academy Press, Inc., Tokyo, Japan, Nagoya Japan, 1992, pp. 515.

[7.] Sevilla, J., Escarpizo, B., Fernandez, F., Cuevas, F., and Sanchez, C., Time-evolution of tritium concentration in the electrolyte of prolonged cold fusion experiments and its relation to Ti cathode surface treatment, in *Third International Conference on Cold Fusion, "Frontiers of Cold Fusion"*, Ikegami, H. Universal Academy Press, Inc., Tokyo, Japan, Nagoya Japan, 1992, pp. 507.

[8.] Hodko, D. and Bockris, J., Possible excess tritium production on Pd codeposited with deuterium, *J. Electroanal. Chem.* 353, 33, 1993.

[9.] Boucher, G. R., Collins, F. E., and Matlock, R. L., Separation factors for hydrogen isotopes on palladium, *Fusion Technol.* 24, 200, 1993.

[10.] Corrigan, D. A. and Schneider, E. W., Tritium separation effects during heavy water electrolysis: implications for reported observations of cold fusion, *J. Electroanal. Chem.* 281, 305, 1990.

[11.] Szpak, S. and Mosier-Boss, P. A., Nuclear and thermal events associated with Pd + D co-deposition, *J. New Energy* 1 (3), 54, 1996.

[12.] Roy, L. P., Influence of temperature on the electrolytic separation factor of hydrogen isotopes, *Can. J. Chem.* 40, 1452, 1962.

Appendix F

[1.] Tanzella, F. L., Crouch-Baker, S., McKeown, A., McKubre, M. C. H., Williams, M., and Wing, S., Parameters affecting the loading of hydrogen isotopes into palladium cathodes, in *Sixth International Conference on Cold Fusion, Progress in New Hydrogen Energy*, Okamoto, M. New Energy and Industrial Technology Development Organization, Tokyo Institute of Technology, Tokyo, Japan, Lake Toya, Hokkaido, Japan, 1996, pp. 171.

[2.] Algueró, M., Fernández, J. F., Cuevas, F., and Sánchez, C., An experimental method to measure the rate of H(D)-absorption by a Pd cathode during the electrolysis of an aqueous solution: Advantages and disadvantages, in *5th International Conference on Cold Fusion*, Pons, S. IMRA Europe, Sophia Antipolis Cedex, France, Monte-Carlo, Monaco, 1995, pp. 441.

[3.] McKubre, M. C. and Tanzella, F., Using resistivity to measure H/Pd and D/Pd loading: method and significance, in *12th International Conference on Condensed Matter Nuclear Science*, Takahashi, A. World Scientific Publiishing, Yokohama, Japan, 2005.

[4.] Zhang, W.-S., Zhang, Z.-F., and Zhang, Z.-L., Some problems on the resistance method in the in situ measurement of hydrogen content in palladium electrode, *J. Electroanal. Chem.* 528, 1, 2002.

[5.] Fazlekibria, A. K. M., Tanaka, T., and Sakamoto, Y., Pressure-composition and electrical resistance-composition isotherms of palladium-deuterium system, *Int. J. Hydrogen Energy* 23 (10), 891, 1998.

[6.] Bennington, S. M., Benham, M. J., Stonadge, P. R., Fairclough, J. P. A., and Ross, D. K., In-situ measurements of deuterium uptake into a palladium electrode using time-of-flight neutron diffractometry, *J. Electroanal. Chem.* 281, 323, 1990.

[7.] Yamamoto, T., Oka, T., and Taniguchi, R., In-situ observation of deuteride formation in palladium electrochemical cathode by x-ray diffraction method, *Annu. Rep. Osaka Prefect. Radiat. Res. Inst.* 30, 79, 1990.

[8.] Batalla, E., Zwartz, E. G., and Judd, B. A., In-situ X-ray diffraction of palladium cathodes in electrolytic cells, *Solid State Commun.* 71, 805, 1989.

[9.] Felici, R., Bertalot, L., De Ninno, A., La Barbera, A., and Violante, V., In situ measurement of the deuterium (hydrogen) charging of a palladium electrode during electrolysis by energy dispersive x-ray diffraction, *Rev. Sci. Instr.* 66 (5), 3344, 1995.

[10.] Bertalot, L., DeMarco, F., DeNinno, A., Felici, R., LaBarbera, A., Scaramuzzi, F., and Violante, V., Deuterium charging in palladium by electrolysis of heavy water: Measurement of lattice parameter, in *Fourth International Conference on Cold Fusion* Electric Power Research Institute 3412 Hillview Ave., Palo Alto, CA 94304, Lahaina, Maui, 1993, pp. 29.

[11.] Lawson, A. C., Conant, J. W., Robertson, R., Rohwer, R. K., Young, V. A., and Talcott, C. L., Debye-Waller factors of PdDx materials by neutron powder diffraction, *J. Alloys and Compounds* 183, 174, 1992.

[12.] Chang, C. P., Wu, J. K., Yao, Y. D., Wang, C. W., and Lin, E. K., Hydrogen and deuterium in palladium, *Int. J. Hydrogen Energy* 16, 491, 1991.

[13.] Mukhopadhyay, R., Dasannacharya, B. A., Nandan, D., Singh, A. J., and Iyer, R. M., Real time deuterium loading investigation in palladium using neutron diffraction, *Solid State Commun.* 75, 359, 1990.

[14.] Schirber, J. E. and Morosin, B., Lattice constants of beta-Pd-Hx and beta-PdDx with x near 1.0, *Phys. Rev. B* 12, 117, 1975.

INDEX